THE POWER OF CHANGE

Innovation for Development and Deployment of Increasingly Clean Electric Power Technologies

Committee on Determinants of Market Adoption of Advanced Energy Efficiency and Clean Energy Technologies

Board on Science, Technology, and Economic Policy
Policy and Global Affairs

Board on Energy and Environmental Systems
Division on Engineering and Physical Sciences

A Report of
The National Academies of
SCIENCES · ENGINEERING · MEDICINE

THE NATIONAL ACADEMIES PRESS
Washington, DC
www.nap.edu

THE NATIONAL ACADEMIES PRESS 500 Fifth Street, NW Washington, DC 20001

This activity was supported by Contract Number DE-PI0000010, Order Number DE-DT0004510, with the U.S. Department of Energy; the National Academy of Sciences, Engineering, and Medicine's Presidents' Circle Fund, and the National Academy of Sciences Thomas Lincoln Casey Fund. Any opinions, findings, conclusions, or recommendations expressed in this publication do not necessarily reflect the views of any organization or agency that provided support for the project.

International Standard Book Number-13: 978-0-309-37142-1
International Standard Book Number-10: 0-309-37142-2
Digital Object Identifier: 10.17226/21712

Additional copies of this publication are available for sale from the National Academies Press, 500 Fifth Street, NW, Keck 360, Washington, DC 20001; (800) 624-6242 or (202) 334-3313; http://www.nap.edu.

Printed in the United States of America

Suggested citation: National Academies of Sciences, Engineering, and Medicine. 2016. *The Power of Change: Innovation for Development and Deployment of Increasingly Clean Electric Power Technologies*. Washington, DC: The National Academies Press. doi: 10.17226/21712.

The National Academies of

SCIENCES · ENGINEERING · MEDICINE

The **National Academy of Sciences** was established in 1863 by an Act of Congress, signed by President Lincoln, as a private, nongovernmental institution to advise the nation on issues related to science and technology. Members are elected by their peers for outstanding contributions to research. Dr. Marcia McNutt is president.

The **National Academy of Engineering** was established in 1964 under the charter of the National Academy of Sciences to bring the practices of engineering to advising the nation. Members are elected by their peers for extraordinary contributions to engineering. Dr. C. D. Mote, Jr., is president.

The **National Academy of Medicine** (formerly the Institute of Medicine) was established in 1970 under the charter of the National Academy of Sciences to advise the nation on medical and health issues. Members are elected by their peers for distinguished contributions to medicine and health. Dr. Victor J. Dzau is president.

The three Academies work together as the **National Academies of Sciences, Engineering, and Medicine** to provide independent, objective analysis and advice to the nation and conduct other activities to solve complex problems and inform public policy decisions. The Academies also encourage education and research, recognize outstanding contributions to knowledge, and increase public understanding in matters of science, engineering, and medicine.

Learn more about the National Academies of Sciences, Engineering, and Medicine at **www.national-academies.org**.

The National Academies of
SCIENCES · ENGINEERING · MEDICINE

Reports document the evidence-based consensus of an authoring committee of experts. Reports typically include findings, conclusions, and recommendations based on information gathered by the committee and committee deliberations. Reports are peer reviewed and are approved by the National Academies of Sciences, Engineering, and Medicine.

Proceedings chronicle the presentations and discussions at a workshop, symposium, or other convening event. The statements and opinions contained in proceedings are those of the participants and are not necessarily endorsed by other participants, the planning committee, or the National Academies of Sciences, Engineering, and Medicine.

For information about other products and activities of the Academies, please visit nationalacademies.org/whatwedo.

COMMITTEE ON DETERMINANTS OF MARKET ADOPTION OF ADVANCED ENERGY EFFICIENCY AND CLEAN ENERGY TECHNOLOGIES

Charles "Chad" Holliday (NAE), *Chair*
Chairman
Royal Dutch Shell, PLC

Jerome "Jay" Apt
Professor, Tepper School of Business
Co-director, Carnegie Mellon Electricity Industry Center, Carnegie Mellon University

Frances Beinecke
President (ret.), Natural Resources Defense Council

Nora Brownell
Co-founder, ESPY Energy Solutions

Paul Centolella
President, Paul Centolella & Associates
Senior Consultant, Tabors Caramanis Rudkevich

David Garman
Principal and Managing Partner (ret.), Decker Garman Sullivan and Associates, LLC

Clark Gellings (NAE)
Independent Energy Consultant
Fellow (ret.), Electric Power Research Institute

Bart Gordon
Partner, K&L Gates LLP
Former U.S. Representative, Tennessee, U.S. House of Representatives

William "Bill" Hogan
Raymond Plank Professor of Global Energy Policy and Research Director, Harvard Electricity Policy Group, Harvard Kennedy School of Government

Richard K. Lester
Japan Steel Industry Professor, Department of Nuclear Science and Engineering, and Associate Provost, Massachusetts Institute of Technology

August "Bill" Ritter
Co-Founder, Center for the New Energy Economy, Colorado State University
Former Governor, State of Colorado

James Rogers
President and CEO (ret.), Duke Energy

continued

Theodore "Ted" Roosevelt
Managing Director and Chairman, Clean Tech Initiative Co-chair, Military Services Network, Barclays Bank

Peter Rothstein
President, NECEC

Adm. Gary Roughead (Ret.)
Annenberg Distinguished Fellow, Hoover Institution, Stanford University

Maxine Savitz (NAE)
General Manager for Technology Partnerships (ret.), Honeywell, Inc.

Mark Williams (deceased)

PROJECT STAFF

Paul Beaton
Study Director

Gail Cohen
Director

Stephen Merrill
Director Emeritus

David Ammerman
Financial Officer

Aqila Coulthurst
Associate Program Officer

Alan Crane
Senior Scientist

Christopher J. Jones
Christine Mirzayan Science and Technology Policy Graduate Fellow

Karolina Konarzewska
Program Coordinator

Frederic Lestina
Senior Program Assistant

Kavitha Ramane
Christine Mirzayan Science and Technology Policy Graduate Fellow

Erik Saari
Senior Program Assistant

David Visi
Christine Mirzayan Science and Technology Policy Graduate Fellow

BOARD ON SCIENCE, TECHNOLOGY, AND ECONOMIC POLICY

For the National Academies of Sciences, Engineering, and Medicine, this project was overseen by the Board on Science, Technology, and Economic Policy (STEP), a standing board established in 1991, with the collaboration of the Board on Energy and Environmental Systems (BEES). The mandate of the STEP Board is to advise federal, state, and local governments and inform the public about economic and related public policies to promote the creation, diffusion, and application of new scientific and technical knowledge to enhance the productivity and competitiveness of the U.S. economy and foster economic prosperity for all Americans. The STEP Board and its committees marshal research and the expertise of scholars, industrial managers, investors, and former public officials in a wide range of policy areas that affect the speed and direction of scientific and technological changes and their contributions to the growth of the U.S. and global economies. Results are communicated through reports, conferences, workshops, briefings, and electronic media subject to the procedures of the National Academies to ensure their authoritativeness, independence, and objectivity. The members of the STEP Board and Academies staff involved with this project are listed below:

RICHARD K. LESTER, *Chair*, Massachusetts Institute of Technology
JEFF BINGAMAN, Former U.S. Senator, New Mexico
ELLEN DULBERGER, Ellen Dulberger Enterprises, LLC, Mahopac, New York
ALAN M. GARBER (NAM), Harvard University
RALPH E. GOMORY (NAS/NAE), New York University
MICHAEL GREEENSTONE, The University of Chicago
JOHN L. HENNESSY (NAS/NAE), Stanford University
LUIS M. PROENZA, University of Akron
KATHRYN L. SHAW, Stanford University
JAY WALKER, Walker Innovation, Inc., Stamford, Connecticut

STEP STAFF

GAIL COHEN, Director
DAVID AMMERMAN, Financial Officer
PAUL BEATON, Senior Program Officer
DAVID DIERKSHEIDE, Program Officer
FREDERIC LESTINA, Senior Program Assistant
ERIK SAARI, Senior Program Assistant
SUJAI SHIVAKUMAR, Senior Program Officer

BOARD ON ENERGY AND ENVIRONMENTAL SYSTEMS

The Board on Energy and Environmental Systems (BEES) is a unit of the Division on Engineering and Physical Sciences (DEPS) of the National Academies of Sciences, Engineering, and Medicine. Since 1975, the BEES Board (formerly the Energy Engineering Board [EEB]) has conducted a diverse program of studies and related activities (workshops, symposia, etc.) to produce authoritative, independent recommendations about the science and technology aspects of public policy questions in energy, the environment, national security, and defense.

ANDREW BROWN, JR. (NAE), *Chair,* Delphi Corporation (retired), Troy, Michigan
DAVID ALLEN, The University of Texas at Austin
W. TERRY BOSTON (NAE), PJM Interconnection, LLC, Audubon, Pennsylvania
WILLIAM BRINKMAN (NAS), Princeton University, Princeton, New Jersey
EMILY A. CARTER (NAS/NAE), Princeton University, Princeton, New Jersey
JARED COHON (NAE), Carnegie Mellon University, Pittsburgh, Pennsylvania
BARBARA KATES-GARNICK, Tufts University, Boston, Massachusetts
DEBBIE NIEMEIER, University of California, Davis
MARGO OGE, McLean, Virginia
JACKALYNE PFANNENSTIEL, Independent Consultant, Piedmont, California
MICHAEL RAMAGE (NAE), ExxonMobil Research and Engineering Company (retired), New York City
DOROTHY ROBYN, Consultant, Washington, DC
GARY ROGERS, Roush Industries, Livonia, Michigan
ALISON SILVERSTEIN, Consultant, Pflugerville, Texas
MARK THIEMENS (NAS), University of California, San Diego
JOHN WALL (NAE), Cummins Engine Company (retired), Belvedere, California
ROBERT WEISENMILLER, California Energy Commission, Sacramento, California
MARY LOU ZOBACK (NAS), Stanford University, Stanford, California

BEES STAFF

JAMES ZUCCHETTO, Director
DANA CAINES, Financial Associate
LINDA CASOLA, Senior Program Assistant

continued

Preface

Reliable access to affordable energy is vital to any economy. Growing economic activity in America and around the globe has led to ever greater demands for energy. Energy intensity (energy consumption per unit of national output) has decreased substantially over the past 40 years in the United States, and energy-efficiency measures have played an important role in reducing the growth in demand for electricity. Nonetheless, the rise in demand and growing recognition of the need to control the pollutants emitted as a result of energy consumption due to increased economic activity have generated a growing need for increasingly clean electric power. One approach to meeting this need has been to install pollution control technologies that capture pollutants after fuel is burned, effectively making the electricity production cleaner; investments in such pollution control technologies have increased significantly since 1990. An additional approach is to use energy sources such as wind, solar, or geothermal that innately produce little to no pollution. Investments in technologies that enable the use of such fuels also have increased recently, more than doubling from 1999 to 2005 and then rising more than six-fold from 2006 to 2012.

The tremendous growth in investment in and use of these various technologies has resulted in dramatic decreases in emissions of pollutants that cause smog, ground-level ozone, and acid rain, and these decreases have resulted in significantly cleaner air across the United States. Despite these gains, however, greenhouse gas emissions have remained relatively constant. A primary challenge is that, absent a price on carbon dioxide, fossil fuels remain the cheapest abundant source of energy, while technologies that make it possible to capture and utilize or store carbon emissions remain costly and nascent. Advanced technologies for capturing or reducing carbon pollution hold great promise for changing the equation, yet many of these technologies can be developed only to the early prototype stage because private-sector financing cannot accommodate the enormous capital requirements and multidecade lag before return on investment can be realized. Technologies for the use of renewable fuel sources such as wind and solar remain costlier still. Nuclear power accounts for two-thirds of the zero- or low-carbon U.S. electricity supply, but the nuclear fleet is beginning to face age-related attrition issues.

It is within this context that the Department of Energy, with the support of the U.S. Senate, requested that the National Academies convene a committee of experts to analyze the determinants that can enable market adoption of advanced energy efficiency and increasingly clean energy. Specifically, the committee's task was to "determine whether and how federal policies can accelerate the market adoption of advanced energy efficiency and low- or non-polluting energy technologies." The committee was asked to focus on the post-research and development (R&D) stages of the electric power supply chain, including scaled-up deployment and widespread adoption, and to consider a range of policy instruments, such as subsidies, tax incentives, demonstration projects, loan guarantees and other financial instruments, procurement, and regulation.

Since 1991, the National Research Council (NRC), under the auspices of the Board on Science, Technology, and Economic Policy, has undertaken a program of activities designed to improve policy makers' understanding of the interconnections among science, technology, and economic policy and their importance for the American economy and its international competitive position. The board's activities have corresponded with increased policy recognition of the importance of knowledge and technology to economic growth. New economic growth theory emphasizes the role of technology creation, which is believed to be characterized by significant growth externalities.

Under the auspices of the Board on Energy and Environmental Systems, the NRC has undertaken a program of studies and other activities to provide independent advice to the executive and legislative branches of government and the private sector on issues in energy and environmental technology and related public policy. The board directs expert attention to issues surrounding energy supply and demand technologies and systems, including resource extraction through mining and drilling; energy conversion, distribution and delivery, and efficiency of use; environmental consequences of energy-related activities; environmental systems and controls in areas related to the production, energy conversion, transmission, and use of fuels; and related issues in national security and defense.

A central focus of NRC analysis has been the importance of energy innovation to the growth of the U.S. economy and to the reduction of negative environmental, public health, and other consequences of energy-related activities. Many performance gains remain to be achieved in energy technologies, such as the capture of carbon from the use of fossil fuels, advanced nuclear power, renewable fuels for electricity generation and for vehicles, and increasingly efficient use of energy. Yet undertaking the efforts required to produce the innovations needed to transform the performance of the energy sector so as to mitigate the risks from greenhouse gases and other pollutants may be the greatest challenge humanity has ever faced. It is a worldwide challenge that will require tremendous effort and leadership.

Throughout history, the United States has consistently demonstrated that its greatest resource is its people and their talent for innovation and leadership. There has never been a greater need or opportunity for American leadership than that posed by the challenge of achieving increasingly clean electric power, a challenge that is the subject of this report.

ACKNOWLEDGMENTS

On behalf of the National Academies of Sciences, Engineering, and Medicine, the committee expresses its appreciation for and recognition of the insights, information, experiences, and perspectives made available by the many participants in workshops and roundtables held during the course of this study. We would particularly like to recognize Nidhi Santen, Scot Holliday, Vignesh Gowrishankar, Xin "Charlotte" Wang, David Taylor, and Nathaniel Green for their invaluable research and technical assistance in the preparation of this report. We also thank Frederic Lestina, Erik Saari, Alisa Decatur, and Rona Briere for their assistance in preparing this report for publication.

We would also like to recognize the contributions of committee member Mark Williams who passed away on March 6, 2016. Mark made numerous contributions to the committee, including the fundamental approach to this report's organization. The quality of this report reflects his invaluable contributions.

Reviewers

This report has been reviewed in draft form by individuals chosen for their diverse perspectives and technical expertise, in accordance with procedures approved by the National Academies of Sciences, Engineering, and Medicine's Report Review Committee. The purpose of this independent review is to provide candid and critical comments that will assist the institution in making its published report as sound as possible and to ensure that the report meets institutional standards for objectivity, evidence, and responsiveness to the study charge. The review comments and draft manuscript remain confidential to protect the integrity of the process.

We wish to thank the following individuals for their review of this report: Joseph Aldy, Harvard University; George Apostolakis, Massachusetts Institute of Technology; William Brinkman, Princeton University; David Cash (Retired), Massachusetts Executive Office of Environmental Affairs; Ahmad Chatila, SunEdison, Inc.; Linda Cohen, University of California, Irvine; Michael Corradini, University of Wisconsin-Madison; Lewis Davis, GE Power Generation Products; Michael Ettenberg, Dolce Technologies; Peter Fox-Penner, The Brattle Group; Kenneth Gillingham, Yale University;

Robert Repetto, International Institute for Sustainable Development; and Catherine Wolfram, University of California, Berkeley.

Although the reviewers listed above have provided many constructive comments and suggestions, they were not asked to endorse the conclusions or recommendations, nor did they see the final draft of the report before its release. The review of this report was overseen by Elisabeth Drake (Retired), Massachusetts Institute of Technology, and Christopher Whipple (Retired), ENVIRON. Appointed by the Academies, they were responsible for making certain that an independent examination of this report was carried out in accordance with institutional procedures and that all review comments were carefully considered. Responsibility for the final content of this report rests entirely with the authoring committee and the institution.

Chad Holliday Paul Beaton

Contents

Box, Tables, and Figures

BOX

TABLES

FIGURES

Summary

Electricity, supplied reliably and affordably, is foundational to the U.S. economy and is utterly indispensable to modern society. The National Academy of Engineering has called electrification the greatest engineering achievement of the 20th century (Constable and Somerville, 2003). Generating electricity also creates pollution, however, especially emissions of air pollutants. While the most severe and life-threatening pollution from electric power plants is largely a thing of the past in America, power plant emissions of particulates as well as oxides of nitrogen and sulfur (NO_x and SO_x)[1] still cause harms and contribute to increases in morbidity and mortality (Bell et al., 2008; Laden et al., 2006; Pope et al., 2009). Those harms include premature deaths, contributions to illnesses such as asthma, and increased hospitalizations, and electricity prices do not fully incorporate the costs of those harms (NRC, 2010b). Harms from greenhouse gas (GHG) emissions—to which the power sector is an important contributor, accounting for nearly 40 percent of all domestic emissions (EPA, 2016)—remain almost completely unpriced and thus above the level they would be if market prices reflected their full costs.

While the precise impacts of climate change are uncertain, plausible extreme and costly economic and environmental harms create a growing urgency to reduce GHG emissions substantially. Uncertainty is not a reason for inaction in this as in many other areas of life, such as buying home insurance even though it may never be needed (NRC, 2011). Rather, the challenge for society is to acknowledge uncertainty and respond accordingly. As has been the case in prior Academies reports, this report focuses on the United States while recognizing that climate change is inherently an international concern. Effectively addressing climate concerns may require responses from all countries, as well as technologies that are globally scalable and affordable.

Intense interest in low- and nonpolluting electric power generation technologies started in earnest during the oil embargoes of the 1970s. The desire to mitigate climate change impacts has both revived and intensified that interest.

---

[1]These are often called "criteria pollutants" because of their regulated status under the Clean Air Act.

Yet wind produced less than 5 percent, solar produced less than 1 percent, and other renewables combined (mostly hydroelectric) produced about 8 percent of all U.S. electricity in 2015, while nuclear accounted for 20 percent, coal 33 percent, and natural gas 33 percent.

In this context, the Department of Energy (DOE) commissioned the National Academies of Sciences, Engineering, and Medicine to convene a committee to undertake a study examining the determinants of market adoption of advanced energy-efficiency and increasingly clean energy technologies, focusing primarily on the electric power sector. The principal goal was to understand what barriers exist to greater market penetration of such technologies and what actions governments—federal and state—can take to reduce or eliminate those barriers and accelerate market adoption. To carry out its task, the committee studied the widest possible range of technologies currently available for the production of electricity, as well as a robust suite of technologies for increasing the efficiency of use of electric power. Key considerations included whether a technology is sufficiently mature, as well as the expected price to consumers of the electricity produced. Also in accordance with its statement of task, the committee deliberated on what policies, legislation, or other actions—current and plausible—would best encourage adoption of increasingly clean power technologies, taking into account market conditions, likelihood of impact, and at what cost.

During the course of the study, the committee concluded that a binary categorization of technologies as "clean" or "dirty" may be counterproductive given that producers are compelled to use the most abundant and affordable primary energy resources they can readily access and use for power generation. All electricity generation technologies have some environmental effects. Thus for purposes of this report, the committee classifies an "increasingly clean" technology only on the basis of emissions of criteria pollutants and GHGs produced in the generation of electricity (rather than other environmental effects or those associated with the mining or extraction and transport of the primary energy source). By that token, solar, wind, nuclear, and fossil fuel-fired combustion with carbon capture and storage (CCS) are low-polluting technologies; conventional natural gas is a medium-polluting technology for criteria pollutants such as NO_x and particulate matter and emits less carbon dioxide (CO_2) than conventional coal-fired generation; and conventional coal-fired generation is a high-polluting technology.

The committee's findings and recommendations fall into three prioritized categories: overarching, key, and other. In the first category are 2 recommendations that the committee concludes are more important than all the others. Also included in this summary are 10 key recommendations and 8 key findings. The 12 other recommendations are presented in the appropriate report chapters.

OVERARCHING RECOMMENDATIONS

The committee concluded that there are two significant barriers to accelerating greater penetration of increasingly clean electricity technologies. First, as noted above, the market prices for electricity do not include "hidden" costs from pollution, stemming mainly from negative impacts on human health, agriculture, and the environment. Levels of criteria pollutants declined over the past three decades, but still cause harms. Harms from GHGs are difficult to estimate, but if accounted for in the market, could be considered by consumers.

In most locations within the United States, prices for increasingly clean power technologies are higher than those for less clean, incumbent technologies. While costs have declined over the past several years for some increasingly clean technologies—notably solar photovoltaics—natural gas supplies have opened up, causing dramatic decreases in natural gas prices. There are notable locations where unsubsidized wind- and solar-generated electricity is competitive with or cheaper than electricity from other sources. Yet for most of the country, most of the time, the prices of dirtier incumbent electric power generation technologies are lower than those of increasingly clean technologies, in part because their price does not include their full costs. Thus they are built and utilized more often and in turn produce more pollution than would be the case if their prices were correct.

Inaccurate price is an example of a "market failure" where government action is often justified. In this case, the solution to correct the market failure is intellectually simple but politically difficult: governments can require that market actors include the price of pollution in their decision making. This has been done in some form with SO_x and NO_x since the early 1990s and in limited ways for GHGs since the late 2000s.

The second barrier is that the scale of the climate change challenge is so large that it necessitates a significant switch to increasingly clean power sources. In most of the United States, however, even with a price on pollution, most increasingly clean technologies would lack cost and performance profiles that would result in the levels of adoption required. In most cases, their levelized costs are higher than those of dirtier technologies, and there are significant challenges and costs entailed in integrating them into the grid at high levels. This means that reducing the harmful effects of emissions due to electricity generation will require a broader range of low-cost, low- and zero-emission energy options than is currently available, as well as significant changes to the technologies and functionality of the electricity grid and the roles of utilities, regulators, and third parties.

Lastly, the committee notes that even if the technological and institutional barriers to greater adoption of increasingly clean power technologies were overcome but their prices were not competitive, an adequate scale of deployment would require tremendous public outlays, and in many parts of the world would be unlikely to occur. While learning by doing can lower some costs, deployment incentives are likely to be insufficient as the primary policy mechanism for achieving timely cost and performance improvements.

The committee formulated two overarching recommendations to address the above challenges.

Recommendation 2-1:[2] The U.S. federal government and state governments should significantly increase their emphasis on supporting innovation in increasingly clean electric power generation technologies.

Simply put, the best way to encourage market uptake is first to have technologies with competitive cost and performance profiles. The need for increased innovation and expanded technology options is especially important given the global picture. In many parts of the world, coal remains the cheapest fuel for electricity generation. China, India, and the nations of Southeast Asia are expected to continue rapidly adding new electricity generation facilities, most of them coal-fired and with minimal pollution controls. Thus there is a need for technological innovations that are affordable outside the United States as well. These improvements in performance and cost will be essential to achieve long-term GHG reductions, such as the reduction called for in the COP21 agreement,[3] without significantly increasing electricity prices. While the challenge may be great, it also creates an opportunity for the United States to continue to lead in the pursuit of increasingly clean, more efficient electricity generation through innovation in advanced technologies.

Recommendation 2-2: Congress should consider an appropriate price on pollution from power production to level the playing field; create consistent market pull; and expand research, development, and commercialization of increasingly clean energy resources and technologies.

Correcting market prices will encourage more deployment of increasingly clean technologies. Where such technologies are already the lowest-price choice, they will become even more so; in other locations, a pollution price will make these technologies the most affordable option or narrow the gap. In

---

[2]The committee's findings and recommendations are numbered according to the chapter of the full report in which they appear.

[3]COP21 refers to the 21st yearly session of the Conference of the Parties to the 1992 United Nations Framework Convention on Climate Change. Under that agreement, the "United States intends to achieve an economy-wide target of reducing its greenhouse gas emissions by 26%-28% below its 2005 level in 2025 and to make best efforts to reduce its emissions by 28%." Full text available at http://www4.unfccc.int/Submissions/INDC/Published%20Documents/United%20States%20of%20America/1/U.S.%20Cover%20Note%20INDC%20and%20Accompanying%20Information.pdf.

addition to providing this market pull for the deployment of mature increasingly clean technologies, pollution pricing can be expected to spur the development of new, even more effective and competitively priced technologies.

KEY FINDINGS AND RECOMMENDATIONS

In addition to the above overarching recommendations, the committee formulated key findings and recommendations related to a number of important, specific barriers to innovation in increasingly clean energy technologies.

Energy Technology Innovation Process

The first set of barriers relates to the energy technology innovation process (ETIP). Overcoming these barriers and empowering private-sector flows of capital and research, development, and demonstration (RD&D) activity are key because it is clear that reducing the cost and improving the performance of increasingly clean energy technologies in many cases will require more than incremental changes to current technology. Entirely new technologies, sufficiently compelling in cost and performance to be globally deployable, will likely be needed, along with changes to the way the electricity grid is engineered and operated.

The ETIP is a complex network of market and nonmarket institutions and incentives, and each stage of the innovation process presents a range of obstacles to the would-be innovator. The most important priorities for strengthening the system relate to identifying and creating new options, developing and demonstrating the efficacy of these options, and setting the stage for early adoption of those that are most promising.

> **Finding 3-1:** Market failures and nonmarket barriers for increasingly clean power technologies exist at all stages of the innovation process.

> **Finding 3-5:** Regional efforts that leverage regional energy markets and initiatives by states, universities, entrepreneurs, industry, and others can complement federal actions to help bridge funding and commercialization gaps.

> **Finding 3-6:** Funding and commercialization gaps for innovations in energy technologies tend to be most acute in, and most closely associated with, the early to intermediate innovation stages.

Proof-of-concept and pilot projects need to have clear missions and goals. A proven means to this end is sector-specific road mapping and challenge funding developed with specific technology development milestones. DOE could advance innovation in energy technologies by using these techniques for sponsored projects, recognizing that doing so might require redirecting DOE and national laboratory research and development (R&D) programs toward the achievement of more ambitious cost and performance objectives. DOE also could consider further use of inducement prizes featuring specific milestones and goals, possibly through a dedicated Office of Innovation Prizes within the Office of the Under Secretary, as a complement to patents, grants, procurement contracts, and other types of support for energy innovation. While not suited to all research and innovation objectives, prizes can spur innovation when the objective is clear even if the pathway to achieving that objective is unclear.

The intermediate stages of innovation are among the most critical and often overlooked, and are where promising technologies face their greatest challenges. Once a concept has been proven, it faces a range of scale-up, systems integration, manufacturing, regulatory, and market challenges to commercialization. Private investment often is restricted because capital requirements typically increase rapidly and significantly, while times to return often are longer than private investors can wait. The Small Business Administration's Small Business Investment Company (SBIC) program has a tremendous opportunity to help overcome these funding barriers to demonstration, early-adoption, and scale-up activities. For example, allocating up to 20 percent of current SBIC funding to create new venture capital funds focused on early-stage increasingly clean power technologies could stimulate significant levels of private investment.

Regional variation within the United States is important, and the federal government could leverage that variation by supporting a network of local, state, or regional public/private partnerships, called regional energy innovation and development institutes (REIDIs), that would help spur the development of innovations showing the most promise. Where capabilities already exist, this network would facilitate access; where capabilities do not already exist, it would help identify likely development needs for promising technologies and fund or plan and create the support capabilities, physical infrastructure (where applicable), and translational relationships that might be needed for simulation, testing, standards development, and certification.

Simulation and testing are key capabilities, and it would be important for DOE to take the lead in assessing the availability of public and private simulation and testing capabilities, identifying any gaps and weaknesses, and supporting or incentivizing the creation of capabilities needed to fill those gaps. Linking simulation and testing facilities into a network that worked closely with federal road mapping and challenge funding would help align these facilities with and achieve targeted objectives. This initiative would provide streamlined access to new and existing federal, state, regional, and private testing resources; simulation modeling and testing laboratories, and preconfigured test sites.

Recommendation 3-1: DOE should direct funds to a broader portfolio of projects than will ultimately prove viable and should tolerate the inevitable failure of some experiments, while at the same time winnowing at each stage of the innovation process.

In addition to being essential to limit costs, downselecting at each stage would provide opportunities to identify at earlier stages of the innovation process technologies that are unlikely to succeed commercially (in their current form). The most important objective would not be to avoid failure, but to ensure that failure is recognized, understood, and addressed without delay. This could be accomplished by ending funding for projects that failed to meet preset cost and performance improvement targets.

Energy Efficiency

Beyond technologies for generating or delivering electricity, the committee focused on the promise and opportunities of reducing use. Americans today spend almost $400 billion annually on electricity to power their homes, offices, and factories, with a large share of electricity being used in residential and commercial buildings. There is evidence that energy-efficiency measures have been effective at reducing energy consumption. At the same time, the committee considered evidence for an "energy-efficiency gap"—the difference between projected savings from avoided energy use due to energy-efficiency measures and the actual measures undertaken. The committee noted that more work is needed to improve measures of projected savings and to ensure that programs are cost-effective. The committee also identified potential barriers to fully utilizing opportunities for energy efficiency and formulated recommendations to remove those barriers.

Recommendation 4-5: The federal government, state and local governments, and the private sector should take steps to remove barriers to, provide targeted support for, and place a high priority on the development and deployment of all cost-effective energy-efficiency measures.

One barrier to higher utilization of energy-efficiency measures is the above-noted failure of electricity prices to incorporate the costs of pollution. Second, even if prices were corrected to include the costs of pollution, other market imperfections might limit consumers' purchases. Information about energy use and price is not always readily available to consumers, and when it is, they may be unable to translate it into actual costs or savings. Additionally, consumers may be reluctant to make new purchases because of inertia or limited attention. Moreover, the effectiveness of increases in the price of electricity in inducing conservation is limited by the very low measured price elasticity of demand for electricity, especially in the short term. The committee found evidence that appliance standards can help overcome these problems by improving the efficiency of all appliances available to consumers.

Recommendation 4-1: DOE should on an ongoing basis set new standards for home appliances and commercial equipment at the maximum levels that are technologically feasible and economically justified.

The committee also found great opportunity for innovation in the energy-efficiency sector. One such opportunity is to improve the accuracy of predictive models of energy savings. Seeking how to do so, DOE has issued a request for information (RFI), and it could do more in this regard. DOE also is ideally poised to support research on how to translate insights from behavioral science into interventions that reduce electricity usage. That knowledge would be valuable for designing effective and cost-effective policies where appropriate and could be made available to relevant stakeholders.

Recommendation 4-3: DOE should increase its investments in innovative energy-efficiency technologies; improve its ability to forecast energy savings from these technologies; and, in conjunction with other agencies, obtain data with which to develop behavioral interventions for improving energy efficiency.

Beyond DOE, the rest of the federal government is positioned to lead by example through direct efforts to promote energy efficiency. The federal government owns or operates more building space than any other entity in the world, and the administration has issued an executive order requiring the head of each federal agency to promote building energy conservation, efficiency, and management. The federal government could carry out this order by

- continuing to lead in the development of procurement practices for appliances and equipment that take life-cycle costs into account;
- evaluating the benefits of improving the energy efficiency of the Department of Housing and Urban Development's 1.2 million units of public housing; and
- taking the lead on contracting for services that provides incentives to third parties to invest in energy efficiency.

Nuclear Power, Fossil Fuels, and Renewable Energy

The committee also examined specific challenges for developing the next generation of power generation technologies utilizing nuclear, fossil, and renewable fuels. An expansion of nuclear power is almost certainly required to produce the reduction in GHGs likely needed to avoid the most costly climate change scenarios. Nonetheless, nuclear power faces three major obstacles to expansion and innovation.

First, absent a price on GHG pollution, current nuclear technologies are more expensive than technologies based on other fuels, especially natural gas and wind in some areas of the United States. These high costs highlight the need for significant innovation in next-generation reactor designs. Second, the business and regulatory risks of designing innovative nuclear technologies are currently quite high. Capital costs of R&D for any energy technology are typically much higher than those for other sectors, and nuclear power is the extreme example of this.

> **Finding 5-2:** Pilot- or full-scale nuclear reactor demonstration projects are likely to cost hundreds of millions of dollars or more.

In addition, the licensing process is currently an open-ended, all-or-nothing regulatory development process designed for existing light water technologies without certainty of outcomes or even clear milestones along the way. Developers face having to spend up to several hundred million dollars without knowing until the very end whether they will be granted a license.

> **Recommendation 5-1: The U.S. Nuclear Regulatory Commission, on an accelerated basis, should prepare for a rulemaking on the licensing of advanced nuclear reactors that would establish (1) a risk-informed regulatory pathway for considering advanced non-light water reactor technologies, and (2) a staged licensing process, with clear milestones and increasing levels of review at each stage, from conceptual design to full-scale commercial deployment.**

A third obstacle that uniquely deters nuclear innovation in the United States is the continued lack of progress in resolving the spent fuel management issue. The absence of a national policy and plan for interim storage and final disposal of spent fuel is a major impediment to private investment in the development of advanced nuclear power plant technologies.

Credible forecasts also suggest that fossil fuels, especially natural gas, will continue to be available in high quantities and at low prices for decades, and thus will make up a significant fraction of the fuels used to generate electric power for years to come. Coupled with the dramatic reductions in GHGs that can be realized through CCS technologies, the development, demonstration, and deployment of these technologies for both coal and natural gas generators remain critical. While some prototype carbon capture units have been built or are under construction or in development, continued efforts will be needed to bring down the costs of the current technologies and to develop, pilot, and demonstrate novel technologies. Continued efforts also will be needed to resolve institutional challenges, including liability and ownership issues for CO_2 stored in deep saline aquifers or other underground structures.

Current and past federal support for RD&D efforts has been either insufficiently funded or insufficiently robust given the scope of the challenge. One way to generate funding would involve an industry-led CCS technology development and demonstration program supported by funding from utility ratepayers. Given the size of the U.S. electricity market, even a tiny fee levied against every kilowatt hour (kWh) of electricity sold in retail markets could yield billions of dollars for RD&D of a range of increasingly clean energy technologies with minimal impact on the electricity bills of residential ratepayers.[4]

Finding 5-6: The risks involved in transporting and storing CO_2 and the lack of a regulatory regime are key barriers to developing and deploying technically viable and commercially competitive CCS technologies for the power sector at scale.

Recommendation 5-3: Congress should direct the Environmental Protection Agency to develop a set of long-term performance standards for the transport and storage of captured CO_2. This effort should include establishing management plans for long-term stewardship and liability for storage sites once they have been closed, as well as GHG accounting programs.

Expanding the deployment of renewable generation technologies to make them a major source of energy will also be critical to addressing the pollution challenge. Doing so will require new technologies for the generation of electricity, as well as new grid technologies for its transmission and delivery (NRC, 2010b).

---

[4]The United States saw approximately 3.7 billion megawatt hours of retail electricity sales in 2014. A one-tenth of a cent charge on each kWh sold would yield $3.7 billion. The impact of such a charge on a typical residential ratepayer consuming 911 kWh per month (the U.S. average in 2014 according to the Energy Information Administration) would be less than a dollar per month.

The diversity of U.S. renewables markets due to the range of renewable resources, regional electricity markets, state-specific policies, regulatory and market structures, and several thousand utility jurisdictions provides opportunities to learn from the most robust markets. Leveraging these opportunities through ongoing government support for innovation and encouraging private-sector investment can create opportunities for the United States to be a technology leader in rapidly growing global markets for renewable technologies. Domestically, prices continue to decline, but some prices, particularly for solar photovoltaics, remain high compared with those in other countries, including developed economies in Europe.

Many incentives are in place at the state level. While states have a range of pricing and procurement policies, incentives, standards, and models, many parts of the United States encourage competition for wind projects to win power purchase contracts and enable low-cost financing for their construction. Another common option is the renewable portfolio standard (RPS), which requires a minimum quantity of renewable energy supply or capacity. Many RPSs include a set-aside or carve-out that requires a minimum portion of the overall standard to be met using a specific technology, typically solar energy. In early assessments, RPSs have been found to reduce emissions while incurring only modest increases in electricity rates. Still, in regions with the most cost-effective renewable resources and market development efforts, competitive proposals for wind, solar, and other resources, including natural gas, may produce more efficient results. Pricing pollution, such as GHGs, would produce less costly reductions in GHG emissions and provide better incentives for innovation.

Across all technologies and scales, it is important to emphasize that deployment of renewables needs to take place in an increasingly competitive market, and to continue to reward learning and economies of scale, as well as projects with the best economics. Effective federal, state, and local policies need to be consistent with growing market signals that look forward at least 5 years to encourage innovation and development investment that will continue to bring down costs.

Finding 5-8: Consistent siting, streamlined permitting, clear and responsive interconnection processes and costs, training in installation best practices, and reductions in other soft costs can have a significant impact on lowering the cost of solar and other distributed generation renewable technologies.

Recommendation 5-5: As renewable technologies approach becoming economically competitive, states should seek to expand competitive solicitation processes for the most cost-effective renewable generation projects and consider the long-term power purchase agreements (PPAs) necessary to enable low-cost capital for project financing.

Recommendation 5-6: DOE and national laboratory programs should provide technical support to states, cities, regulators, and utilities for identifying and adopting best practices—such as common procurement methods, soft cost reduction approaches, PPA contracts, structures for subsidies and renewable energy certificates, and common renewables definitions (taking into account regional resources)—that could align regional policies to enable more consistent and efficient markets that would support the adoption of renewables.

Electric Power System

Developing and deploying cost-effective increasingly clean energy technologies will require an electric power sector with systems, regulation, and infrastructure that encourage and accommodate those technologies. Developing such a power sector will, in turn, require technological changes to the power system so that it is capable of integrating these new technologies and in greater quantities. To this end, utility regulators will need to incentivize utilities to become fully engaged in innovation and the demonstration of new technologies, with rules that permit reasonable and nondiscriminatory access to the transmission and delivery systems.

These shifts are under way, and as a result, the electric industry faces significant new expectations and requirements to replace aging infrastructure, possibly at costs of hundreds of billions of dollars. The industry also must work to mitigate the effects of storms and other disruptive events while securing the electric power system and critical infrastructure against cyber and physical attacks. Utilities and system operators must maintain system stability while retiring coal and some nuclear generation and integrating increasing amounts of variable and distributed resources. At the same time, current utility business models often rely on volumetric increases in sales to provide funds for new investments. Slowly growing or declining sales mean many utilities lack the revenue growth used historically to fund new investments. This trend could leave the United States with an outdated power system and prove costly to consumers.

While these challenges are substantial, there are also significant opportunities for improvement. Distributed resources, such as combined heat and power, photovoltaics, and efficient fuel cells, can improve reliability if integrated under appropriate regulatory and technological regimes. Technological innovation can reduce costs and improve load factors and asset utilization.

Finding 6-1: To expedite innovative solutions, it will be necessary to redesign business models and regulatory incentives currently designed for a centrally controlled system so they are built on a customer-driven model with multiple solutions.

Finding 6-3: Many state regulatory commissions require additional analytical tools, training, and other resources to develop and implement effectively regulatory models that support and encourage the development of increasingly clean energy and energy-efficiency technologies.

For example, DOE could provide additional resources and training, and perhaps serve as both a coordinator and repository for best practices and lessons learned, as states undertake regulatory reforms. Moreover, the electric power industry typically budgets very small amounts for innovation compared with other technological industries.

Recommendation 6-4: State regulators and policy makers should implement policies designed to support innovation. For example, they could evaluate approaches in which utility or energy customer funds are set aside to support state and regional innovation programs.

Two emerging parallel and potentially complementary business models for distribution utilities and/or other market participants are being considered—distribution system operators (DSOs) and customer energy service providers (CESPs). DSOs could efficiently integrate distributed energy technologies, distribution automation, volt/volt ampere reactive (VAR) optimization, and other characteristics of a smarter power grid with the robustness and flexibility necessary to maintain reliability and security. CESPs might be able to provide similar value, focused on customer-facing aspects of the industry. Full development and implementation of both of these models, however, would require overcoming a number of challenges.

Recommendation 6-5: DOE should undertake a multiyear R&D program to ensure the timely development of the capabilities needed for effective DSOs or CESPs through policy analysis; dialogue; and the sharing of experience and best practices among regulators, utilities, and other stakeholders to advance understanding of the emerging business models. DOE should strongly consider prioritizing the development of robust, well-designed

systems that incorporate appropriate security measures to guard against and respond to cyber attacks.

Utilities also face significant workforce challenges. Large numbers of skilled employees are eligible to retire soon. The anticipated industry changes discussed here imply that the future workforce likely will require a different set of skills and abilities, especially greater "niche" skills to support the implementation, maintenance, and operation of systems with many digital components. Power providers and system operators will need to provide new training programs, guidance documents, and training manuals. Industry and government could partner to develop programs that would help bridge the immediate gap in the skilled workforce and to attract talent in the future by creating and communicating a vision of the electric power industry as one that is attractive, stimulating, and worth celebrating for its vital role in people's lives and the nation's prosperity.

Financing Energy Technologies

Finally, with respect to government support for innovation in energy technologies and technological shifts, history suggests that such supports as direct subsidies and tax exemptions tend to continue well after technologies have matured and are market-competitive. While subsidies can serve important public policy functions in helping to establish industries, they work best when they are predictable and structured to be performance- or outcome-oriented without regard to specific technologies, and to include sunset provisions so they expire either after a specified length of time or once a certain performance has been achieved, as is the case with the recently renewed production tax credits for power from wind and solar. By contrast, the many subsidies for oil and natural gas have no sunset provisions despite the maturity of those industries.

1

Introduction

Stable access to energy is a key factor in economic stability and growth; electric power is particularly important for advanced economies. Accordingly, as the U.S. economy has grown, particularly following World War II, demand for energy has increased almost in lock step. In addition to the many benefits energy brings, its production, use, and consumption often entail negative consequences, usually in the form of pollution, which society increasingly has sought to reduce or eliminate. Doing so essentially requires bringing to market technologies and practices that can provide the energy needed but with fewer or no harmful impacts from pollution. That is, society needs to develop increasingly clean energy sources and practices.

Several barriers challenge full market deployment of increasingly clean energy technologies, however. First, many of the negative impacts of pollution are not reflected in market prices for energy supplies or services. Second, newer, increasingly clean technologies frequently have different performance characteristics from those of incumbent technologies. They may not perform as well, or may just perform differently. For example, solar photovoltaic panels can generate electricity with little or no pollution and with no fuel cost, but only when enough light strikes them. Third, performance challenges can create difficulties with integrating these newer technologies into existing energy systems and infrastructure. The net result is that cleaner technologies almost universally continue to have higher market prices, and market adoption of many increasingly clean energy sources and technologies has proceeded slowly.

One way to address such barriers is through government action. Accordingly, the Department of Energy's Offices of Energy Efficiency and Renewable Energy, Electricity Delivery and Energy Reliability, Fossil Energy, and Nuclear Energy tasked the Academies' Board on Science, Technology, and Economic Policy and Board on Energy and Environmental Systems with examining what policies and actions could accelerate wide market adoption of increasingly clean electric power generation and end-use efficiency technologies.

STATEMENT OF TASK

In response to this request, the National Research Council appointed the Committee on Determinants of Market Adoption of Advanced Energy Efficiency and Clean Energy Technologies. The statement of task shown in Box 1-1 was developed and used as a departure point for the committee's work.

STUDY SCOPE

The statement of task for this study included no guidance on how far upstream or downstream to account for the effects of pollution. Upstream pollution is certainly important. Estimates of its damages, however, entail much greater uncertainty than is the case with downstream pollution. In 2010 the National Research Council published a report currently considered the most authoritative reference regarding "unpriced consequences of energy production and use" from a life-cycle analysis perspective (NRC, 2010b). This study attempted to characterize all pollution associated with energy production and use; however, it was only able to monetize impacts due to emissions of particulate matter, oxides of sulfur, oxides of nitrogen, and greenhouse gases.

BOX 1-1
Statement of Task

An *ad hoc* committee of experts with industrial, financial, academic, and public policy backgrounds will undertake a consensus study to determine whether and how federal policies can accelerate the market adoption of advanced energy efficiency and low- or non-polluting energy technologies. As part of the study the committee will hold workshops, commission research, and prepare a report with recommendations. The committee will consider technologies for the generation, transmission, and storage of electric power and for energy efficiency such as renewable and advanced nuclear and fossil fuel sources, storage and transmission technologies, and building heating and lighting technologies. The study will consider market conditions that may advantage traditional technologies and disadvantage technologies with lower external costs to the environment, public health, and national security. It will focus on the post-R&D stages of the energy supply chain, including scaled-up deployment and widespread adoption. It may consider policy instruments such as subsidies, tax incentives, demonstration projects, loan guarantees and other financial instruments, procurement, and regulation. Although the focus will be on developing recommendations for consideration by Congress, the White House, Department of Energy, and other federal agencies, recommendations may also address actions by States and regional entities.

Given this lack of quantitative analysis of more upstream pollution and the large scope of its task, the committee focused its investigation on the downstream pollution caused by the emissions characterized in that earlier report. The committee still recognizes the importance of upstream pollution—for example, methane leaks during the production and delivery of natural gas that contribute to climate change—and hopes that future work will address the topic with the depth it deserves.

Electric power markets are not only fundamentally important but also enormously complex and complicated. A number of areas deserve further inquiry beyond what resources allowed during the course of this study. The magnitude and scope of climate change, for example, are global, and addressing the problem will require developing technologies that are affordable not only in the United States but also in the rest of the world, especially in rapidly growing economies. A full understanding of how technological developments in the United States will impact those countries and the climate would require analysis of intellectual property, trade, and other technology transfer matters beyond the scope of this study.

STUDY APPROACH

To gather evidence and augment its members' knowledge of the industry, technologies, regulation, financing, and economics of electric power, the committee conducted an extensive search of the relevant literature and convened three workshops to elicit the perspectives of industry leaders, academics, and senior government officials. In addition, the committee conducted several site visits and held numerous consultations with regulators, industry leaders, and investors in electric power and energy-efficiency technologies and companies.

ORGANIZATION OF THE REPORT

In the course of its work, the committee identified five key themes that underlie efforts to accelerate the market adoption of increasingly clean energy and energy-efficiency technologies:

1. **expanding the portfolio of increasingly clean energy technology options;**
2. **leveraging the advantages of energy efficiency;**
3. **facilitating the development of increasingly clean technologies, including nuclear power, cleaner fossil fuels, and renewables;**
4. **improving existing technologies, systems, and infrastructure; and**
5. **leveling the playing field for increasingly clean energy technologies.**

These themes informed the basic structure of this report. Chapter 2 reviews the capabilities of currently available technologies to produce increasingly clean electric power and of current policies to encourage their market adoption, as well as the impact of their deployment on technology innovation. Chapter 3 analyzes challenges and barriers within the energy innovation system to expanding the portfolio of cleaner energy technology options. Chapter 4 examines opportunities to leverage the advantages of energy efficiency. Chapter 5 analyzes unique barriers to market adoption for the most well-developed technologies for increasingly clean power generation from nuclear power, fossil fuels, and renewables. Chapter 6 considers improvements to existing technologies, systems, and infrastructure needed to accommodate the market adoption of increasingly clean power generation and energy-efficiency technologies. Finally, Chapter 7 describes how existing institutions, infrastructure, and policies favor incumbent over innovative and cleaner technologies, as well as the challenges investors and firms face in financing the innovation, development, and deployment of increasingly clean power generation and energy-efficiency technologies. The report also includes five appendixes: Appendix A contains biographical information on members of the Committee on Determinants of Market Adoption of Advanced Energy Efficiency and Clean Energy Technologies; Appendix B details the underlying principles used to calculate the levelized cost of electricity; Appendix C describes recent developments in economic models used to estimate the effects of deployment on costs and technological improvement; Appendix D provides assessments of the technology readiness levels of a comprehensive suite of technologies; and Appendix E is a glossary of acronyms and abbreviations used in the report.

2

Assessment of Current Technologies for and Policies Supporting Increasingly Clean Electric Power Generation

The United States has made significant progress in reducing air pollution and its harmful effects since pollution control laws such as the Clean Air Act (originally passed in 1963, with major amendments in 1970, 1977, and 1990) were introduced. "Killer fog" in America is, at present, a thing of the past. Tragedies such as the Donora smog of 1948 and the "Great Smog" of 1952 that killed thousands of people are essentially unheard of in developed nations. Acid rain and even the once-famous smog in Los Angeles have significantly dissipated. Notwithstanding the measured decreases since the 1960s, however, pollution from the production of electric power continues to cause tangible harm, nor does the price of electricity currently include all of the societal costs of electricity generation.

A 2010 National Research Council study, for example, found that air pollution from coal-fired electric power plants in the aggregate still caused significant harms to human health, including, among others, asthma and premature deaths (NRC, 2010b). These harms arise from sulfur dioxide (SO_2), oxides of nitrogen (NO_x), particulate matter ($PM_{2.5}$ and PM_{10}), ammonia (NH_3), and volatile organic compounds (VOCs), referred to collectively as *criteria pollutants* as they are regulated under the Clean Air Act. The 2010 National Research Council study estimates that in 2005, the emissions of criteria pollutants from coal-fired power plants caused damages costing, on average, $0.032/kilowatt hour (kWh) of electricity generated. The human health harms from all coal-generated electricity thus cost about 33 percent of the value of all electric power produced that year.[1] The 2005 emissions from gas-fired plants

[1]The National Research Council (2010b) study reports damages for the year 2005 but in 2007 dollars. The average retail price of electricity in the United States in 2005 was $0.0814/kWh (EIA, 2007). The Bureau of Labor Statistics' Consumer Price Index

caused human health damages costing approximately $0.0016/kWh of electricity generated, representing about 2 percent of the average retail price of all electric power sold that year.

Electric power plants also produce 39 percent of all U.S. emissions of greenhouse gases (GHGs) (which trap heat in the earth's atmosphere)—the largest share of any source (EPA, 2016). Translating GHG emissions into climate-related damages depends on estimates of damages per ton of carbon dioxide (CO_2) equivalents. The above NRC (2010b) study estimates the climate-related damages to be 1.0-10.0 cents per kWh of electricity produced by coal-fired plants and 0.5-5.0 cents per kWh for natural gas-fired plants, corresponding to damages of $10-100 per ton of CO_2 equivalents.

Reducing emissions further to ameliorate these harms will require a technological shift to increasingly clean—that is low- or nonpolluting—technologies for the generation of electric power. The magnitude of ongoing harms, including those likely due to climate change, makes it imperative to effect this shift as quickly as is efficient. Such increasingly clean technologies rely either on non-fossil fuel sources, such as wind, nuclear, or solar, or on "tailpipe" solutions—technologies that capture or otherwise prevent emission of the pollution from fossil fuels. Effecting this technological shift will in turn require ensuring that newly built generating assets (power plants) are increasingly clean (low- or nonpolluting) compared with those currently operating or recently retired. This means not only building increasingly clean power plants in response to new demand, but also encouraging the retirement of more polluting assets in favor of those running on increasingly clean technologies. The latter strategy is particularly important given that new asset builds in response to demand are likely to remain small. Although electricity demand in the United States continues to grow, the rate of increase has been in secular decline since the 1950s (see Figure 2-1). Consequently, it is reasonable to expect that most new power plants in the United States will be built to replace retiring plants rather than to increase total generating capacity in response to rising demand (EIA, 2015a).

TECHNOLOGIES FOR ELECTRIC POWER GENERATION AND ENERGY EFFICIENCY

Two factors—inaccurate market prices and the large amount of capital required to build a power plant—have led to a bias in the current mix of power plants in the United States[2] in favor of higher-polluting technologies (see Figure 2-2).

Inflation Calculator for energy can be used to translate that 2005 price to $.0956/kWh in 2007 dollars (BLS, 2005, 2007).

[2]The same is largely true in other countries around the world as well.

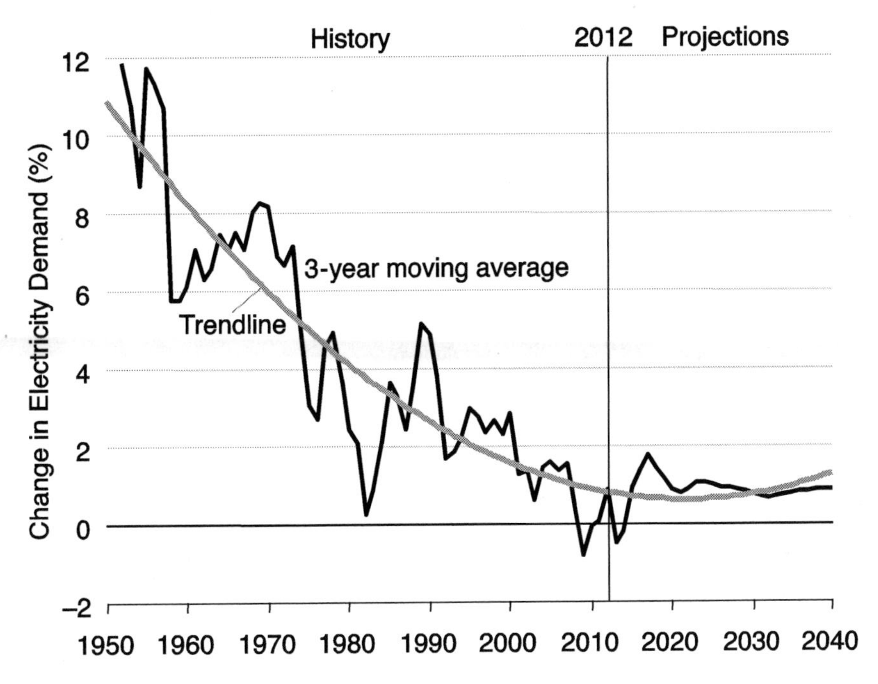

FIGURE 2-1 Growth in electricity demand, with projections to 2040.
SOURCE: EIA, 2014a.

The first factor is that delivered electricity prices do not incorporate the full cost of the harms from the pollution caused by power plants. Because the cost of power plant pollution is not built into the cost of construction, power producers have tended to build more of these plants than they otherwise would have done. And because the delivered price of electricity also does not incorporate the full costs of pollution, end-users consume more electricity from these sources than they otherwise would.

The second factor is that power plants are expensive to construct, requiring large amounts of up-front capital. Such high costs take many decades to fully amortize. Once these costs have been fully amortized, the cost of operating a plant decreases and operating profits increase. Firms may thus have a strong financial incentive to keep a plant operating as long as possible, depending on how the state regulator sets retail rates for electricity (see Chapter 6 for more detail on the ratemaking process). Therefore, retirement of currently operating, higher-polluting plants might be unlikely even if the current price of electricity were to be corrected to include the costs of pollution. Thus power-generating assets are typically kept in operation for 40-50 years, and often even longer. It is important to note this fact when considering the long-term impact of new power plants; choices made today can have pollution consequences for decades.

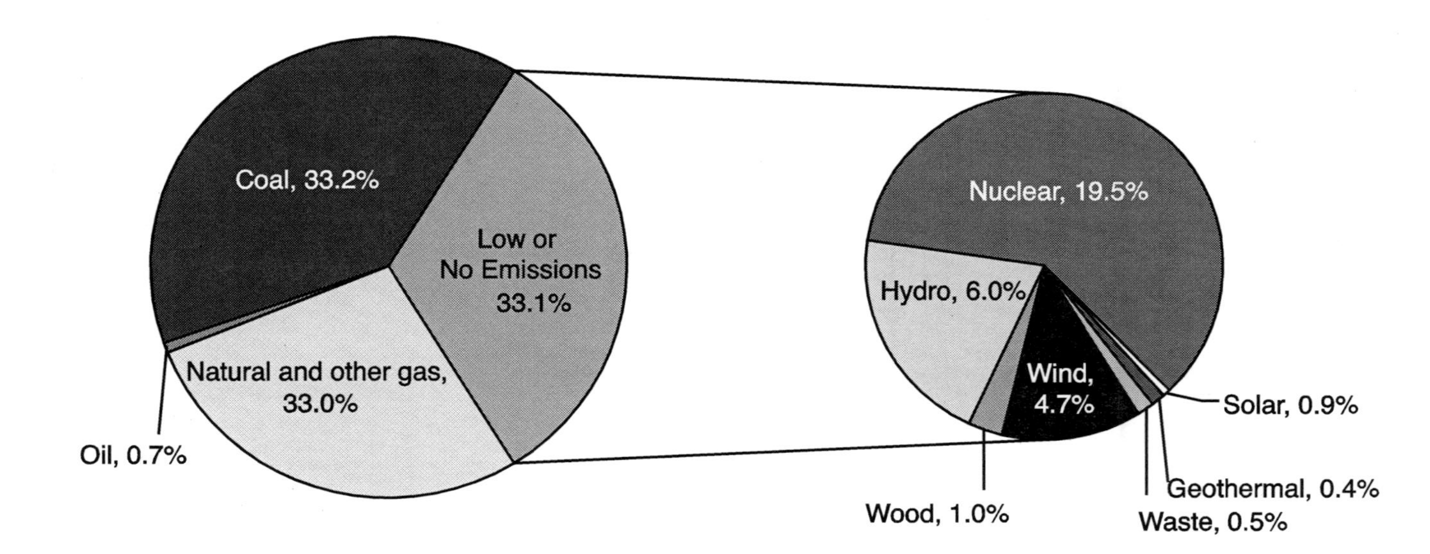

FIGURE 2-2 Percentage of current U.S. net electricity generation by primary fuel source, 2015.
SOURCE: EIA, 2015g, Table 7.2a.

When the financial incentive to keep power plants operating is combined with low market prices for fossil fuels and other factors, it comes as no surprise that most new plants built over the past 30 years have been powered by fossil fuels. As seen in Figure 2-3, from 1989 to 2011, more fossil fuel plants were built than any other type. This figure also shows the likely impact of policies on new plant builds. The Clean Air Act amendments of 1990 created an SO_2 trading system, effectively a price on SO_2, to help diminish the impacts of acid rain resulting from power plant pollution. Coal-fired plants produce more SO_2 per kWh of generated electricity relative to natural gas-fired plants, so it is not surprising that from 1991 to 2011, most capacity additions were natural gas-fired plants. Figure 2-3 also shows increasing construction of new wind and solar facilities following the increase in tax subsidies for these facilities in 2005.

Looking to the future, most new plants are expected to continue to be predominantly fossil fuel-powered, with these capacity additions being greater than they would be if the market reflected the true costs of pollution. Since the market price does not reflect the full costs of pollution, government policies are required to ensure prices that more accurately reflect actual costs. Given such policies, production and consumption will be closer to its efficient and socially optimal quantities.

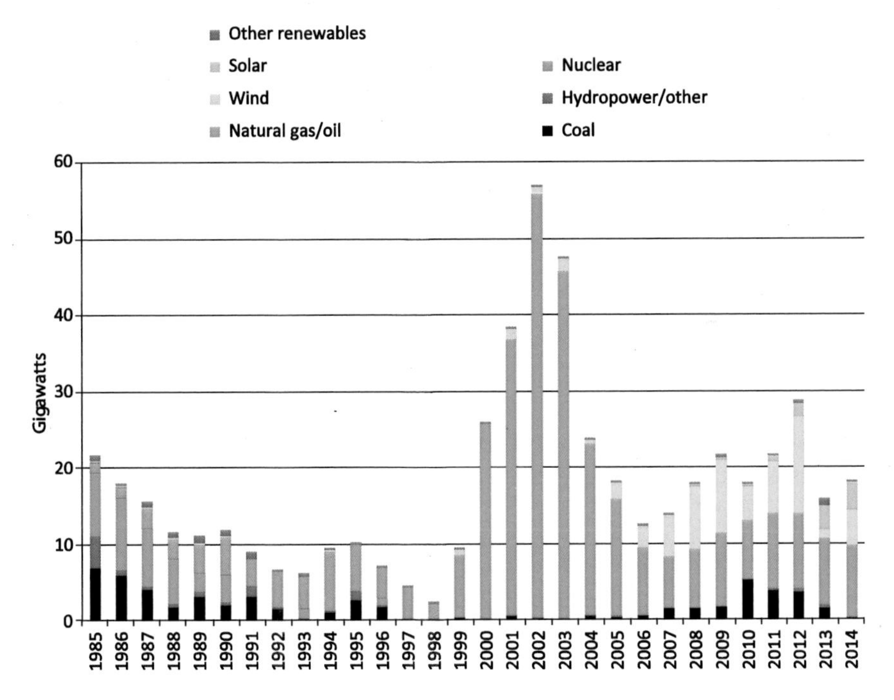

FIGURE 2-3 Additions to U.S. electricity generation capacity, 1985-2014.
SOURCE: EIA, 2016a.

In light of the historic low natural gas prices at the time of this writing (2016), most new plants projected to be built through 2040, like those built in recent decades, are expected to be natural gas-fired. For example, according to the Energy Information Administration's (EIA) projections from its National Energy Modeling System (NEMS)—based on assumptions about future fuel prices and expiring tax subsidies for renewable sources such as wind and solar—new builds may be primarily wind and solar for a few years, but will be predominantly natural gas as the tax subsidies for new wind generating facilities decline through 2019.[3] Notably, EIA and other forecasters expect very few new plants to be powered by nuclear fuel, currently the largest source of nearly emissions-free electricity.

In its *Annual Energy Outlook 2016*, EIA projects total installed electricity generation capacity through 2040 (EIA, 2016a). As seen in Figure 2-4, those projections include an approximately 17 percent increase in total installed capacity between 2016 and 2040, with much of that increase occurring after 2030. The mix of capacity types is expected to change as well. EIA projects that the share of coal will decrease from 29 to 18 percent, mainly before 2020, while that of renewables will increase from 17 to 29 percent. Natural gas is projected to fluctuate slightly until 2020 and then remain stable at 43 percent, and nuclear to decrease from 10 to 8 percent.

IMPACT OF THE MIX OF ELECTRICITY GENERATION SOURCES ON EMISSIONS OVER TIME

As of 2014, emissions from power plants of SO_2 and particulate matter 10 microns or less in size had decreased by 80 percent and of NO_x by 65 percent relative to their levels at the time of the Clean Air Act amendments of 1970 (EPA, 2015). Even at those decreased levels, however, these pollutants are known to cause harms to human health, as discussed earlier (NRC, 2010b). On the other hand, emissions of GHGs due to electric power generation rose by a bit more than 60 percent during the same period (EIA, 2016a, Table 12.6). Meanwhile, CO_2 emissions per megawatt hour (MWh) of electricity produced decreased modestly (Figure 2-5) as a result of improvements in the efficiency of coal plants in the 1950s and 1960s, the growth of nuclear power, and a partial switch from coal to natural gas and wind power in the late 2000s and early 2010s.

---

[3]Chapter 7 provides a more detailed discussion of these tax and other subsidies. Schedule available at http://energy.gov/savings/renewable-electricity-production-tax-credit-ptc (DOE, n.d.-c).

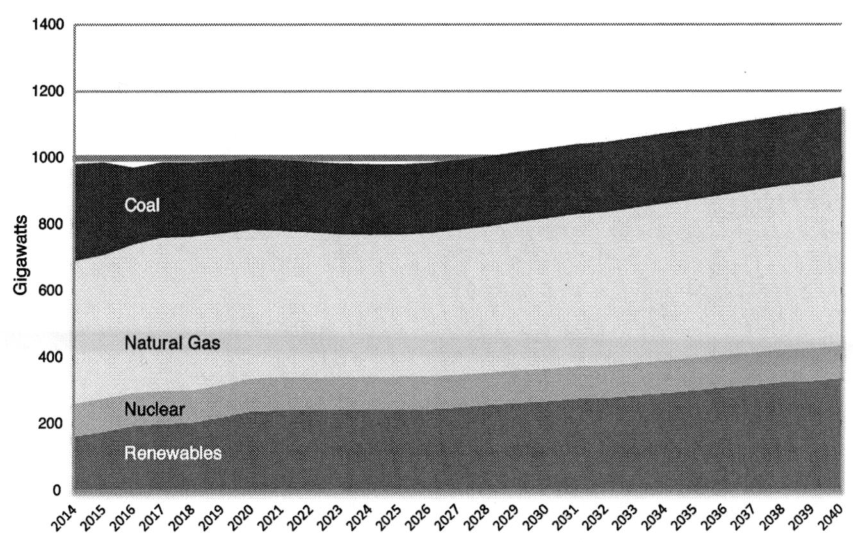

FIGURE 2-4 Total installed U.S. electricity generation capacity, 2014-2040 (projections from 2016 onward).
SOURCE: EIA, 2016a.

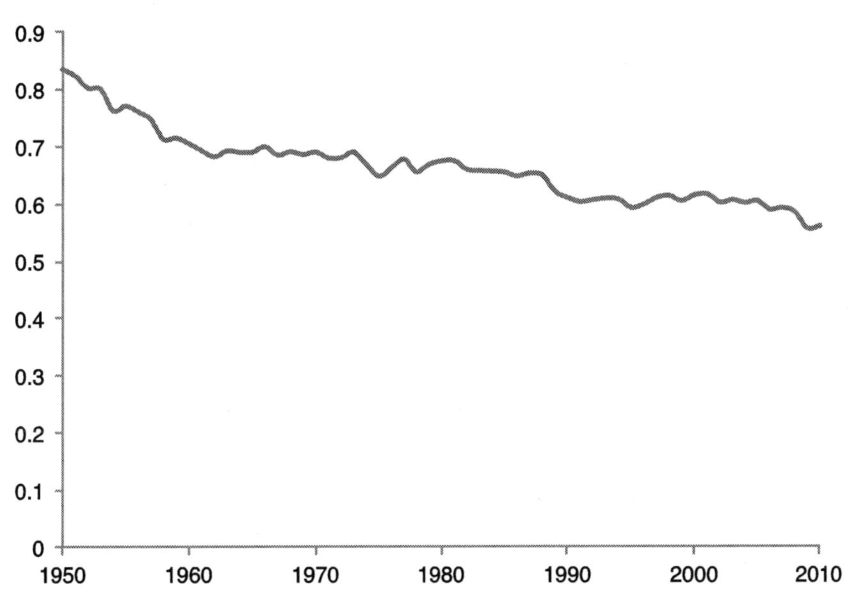

FIGURE 2-5 Emissions of carbon dioxide from electric power generation in metric tons per megawatt hour (MWh), 1950-2010.
SOURCE: EIA, 2015e.

TECHNOLOGY READINESS AND COST OF CURRENTLY AVAILABLE CLEANER TECHNOLOGIES

To understand better the barriers to greater adoption of increasingly clean electric power generation technologies—that is, to understand why power producers are likely to choose to build fossil fuel-powered plants over plants with carbon capture technology or those powered by wind or solar energy—the committee took an in-depth look at the technology readiness and cost of currently available cleaner technologies.

Technology Readiness

A key first step in understanding the barriers to market adoption for low- and no-emission technologies is assessing their readiness to be incorporated into existing infrastructures. Technologies that can readily and easily be incorporated into the existing electric power grid and associated infrastructure are much more likely to be adopted and utilized. There currently exist a wide range of increasingly clean electric power generation technologies that can produce lower or no emissions when used. The committee assessed the technology readiness of the most promising of these technologies in each of the following categories:

- *Renewable power generation*—These technologies focus on the generation of electricity from wind, solar, biomass, geothermal, and hydropower sources. They include, for example, advanced and improved wind turbines, photovoltaic (PV) devices, and enhanced geothermal power generation. The committee also included in its assessment technologies whose deployment would enhance the ability of the grid to host increasing amounts of renewable power production, such as storage technologies (including batteries) since improved storage can support variable power generation from renewables.
- *Advanced fossil fuel power generation*—These technologies focus on improving the pollution control technologies of coal- and natural gas-fired power plants, such as advanced carbon capture and storage. The committee also included water treatment technologies since treating cooling water is a significant obstacle to the construction of new thermal plants (including nuclear plants).
- *Nuclear power generation*—This category includes new and next-generation nuclear technologies and the development of cost-effective technologies that can maximize the use of existing nuclear plants.
- *Electricity transmission and distribution*—This category includes technologies with the potential to reduce losses from and increase the efficiency of the transmission and delivery of electricity to end-users.

As much as 11 percent of all electricity generated is lost during transmission and delivery (Jackson et al., 2015).

- *Efficient electrical technologies for buildings and industry*—This category includes technologies being deployed and developed to reduce building energy needs and energy used in industrial processes.

The detailed assessment of each of these technology categories in Appendix D includes a description of the category; an estimate of the technology readiness level (TRL)[4] of promising technologies in that category in 2016, 2020, and 2035, if estimates were available; and associated technological and commercialization barriers. Table 2-1 summarizes the 2016 TRLs of these technologies.

Cost

To understand impediments to the deployment of increasingly clean energy technologies, the committee reviewed assessments of the economic competitiveness of such technologies in a technology-neutral policy environment.[5] The results of this assessment provide a baseline for evaluating the competitiveness of these technologies and reveal the need for further technological advances.

Developing a levelized cost of electricity (LCOE) is a commonly used method for estimating current and future costs of producing electricity from different generating technologies. While any LCOE must be viewed in light of its assumptions, LCOE estimates can provide a convenient indicator of the relative costs of different technologies and thus a basis for comparing each technology's ability to compete on the basis of its underlying economic performance. Several sources develop estimates of recent and projected future LCOEs for specific technologies.

EIA has a long history of developing LCOE estimates for different electric power generation technologies (EIA, 2014h).[6] These empirical estimates take the location of wind and solar resources into account and incorporate the cost of transmission. EIA also adjusts the LCOE estimates to reflect the relative value

---

[4]A well-accepted method for identifying the readiness of a technology for ultimate dissemination in the marketplace is the TRL taxonomy developed by the National Aeronautics and Space Administration (NASA) as a means of managing its space-related research and development. Further information is provided in Appendix D.

[5]That is, comparing the market prices of technologies absent technology-specific policies that, inter alia, lower prices through subsidies or prevent costs from being incorporated into the market price.

[6]EIA's LCOE estimates include capital costs, fuel costs, fixed and variable operations and maintenance (O&M) costs, financing costs, and an assumed utilization rate for each type of generation technology. EIA also provides information on regional variations in the LCOE for different technologies.

of variable wind and solar generation that cannot follow dispatch instructions easily or at all. The EIA estimates show that the value of wind, which blows more at night, when energy prices are low, can be 12 percent below the unweighted average price of electricity; and the value of solar power, with the sun shining when energy prices are higher, can be 16 percent greater than the unweighted average price of electricity (Schmalensee, 2013). EIA also adjusts its LCOE estimates for wind and solar power based on differences in the avoided costs of electricity derived from sources displaced by each. The agency reports its LCOE estimates based on the factors used in modeling for its *Annual Energy Outlook* publication. In its *Annual Energy Outlook 2016*, EIA provides detailed information on the estimated LCOEs for different electric power generation capacity additions anticipated to enter service in 2020 (EIA, 2016a).[7] Appendix B provides a more complete description of the essential elements used for consistent cost estimates in *Annual Energy Outlook 2016*.

In developing the *Annual Energy Outlook*, EIA must make assumptions regarding future policies. As a result, its LCOE estimates reflect two important policy assumptions that are not technology-neutral. First, EIA adds 3 percent to the weighted average cost of capital for new coal plants as a proxy for anticipated carbon reduction policies (EIA, 2014h). Second, certain technologies, including nonhydro renewables and combined heat and power, are allowed to use a modified accelerated tax depreciation that is not available to other technologies,[8] resulting in substantially lower fixed-charge rates for renewable capital costs (see Appendix B). The committee adjusted EIA's LCOE estimates to eliminate the impact of these two assumptions and enable comparison of supply options on the basis of technology-neutral policies.

The committee then compared the economic competitiveness of different technologies, first incorporating only an estimate of including the harms from criteria pollutants (see Appendix B) and not GHG emissions. For criteria pollutants—SO_2, NO_x, and small particulates—the actual externality costs for a given unit will vary based on its location and differences in the pollutant exposures resulting from its emissions. Conventional coal-fired power generation emits substantially greater quantities of these pollutants relative to other technologies. The committee used Greenstone and Looney's (2012) values for impacts of criteria pollutants based on estimated emission rates from Muller and colleagues (2011). Figure 2-6 compares the projected costs of various electric power generation technologies against the cost of an advanced combined-cycle natural gas plant from this perspective where firms do not directly bear the costs of GHG pollution.

---

[7]EIA also provides LCOE estimates for 2040 that include an assumed learning rate for newer technologies. This assumption contributes to renewables being somewhat more competitive in EIA's forecasts for 2040 relative to those for 2019 (EIA, 2013d).

[8]Section 125 of the Tax Increase Prevention Act of 2014 extended the placed-in-service date for the modified accelerated cost-recovery system (*Tax Increase Prevention Act of 2014*, Public Law 113-295, 113th Congress [December 19, 2014]).

TABLE 2-1 Promising Technologies for Increasingly Clean Electric Power

	Technology Readiness Level[a]								
Technology Category	1	2	3	4	5	6	7	8	9
Renewable Power Generation									
1: Electric energy storage									
2: Hydro and marine hydrokinetic power[b]									
3: Advanced solar photovoltaic power[c]									
4: Advanced concentrating solar power									
5: Advanced solar thermal heating									
6: Advanced biomass power									
7: Engineered/enhanced geothermal systems									
8: Advanced wind turbine technologies									
9: Advanced integration of distributed resources at high percent									
Advanced Fossil Fuel Power Generation									
10: Carbon capture, transport, and storage									
11: Advanced natural gas power and combined heat and power (CHP)[c]									
12: Water and wastewater treatment									
Nuclear Power Generation									
13: Advanced nuclear reactors									
14: Small modular nuclear reactors									
15: Long-term operation of existing nuclear plants									

(Continued)

TABLE 2-1 Continued

Technology Category	Technology Readiness Level[a]								
	1	2	3	4	5	6	7	8	9
Electricity Transmission and Distribution									
16: Advanced high-voltage direct current (HVDC) technologies							■		
17: Reducing electricity use in power systems									■
18: Smart-grid technologies (grid modernization)				■	■	■	■	■	■
19: Increased power flow in transmission systems			■	■	■	■	■	■	■
20: Advanced power electronics						■	■		
Energy Efficiency									
21: Efficient electrical technologies for buildings and industry	■	■	■	■	■	■	■	■	■

[a]Technology readiness levels are shown on a scale of 1 to 9, where 1 is the least ready. Most of the technology categories shown include technologies with varying readiness levels. A shaded box below a TRL number indicates there is at least one technology at that TRL. See Appendix D for more detail.
[b]The committee identified barriers at lower TRLs for hydropower technologies but was unable to make specific level assignments.
[c]For concepts beyond three junctions.

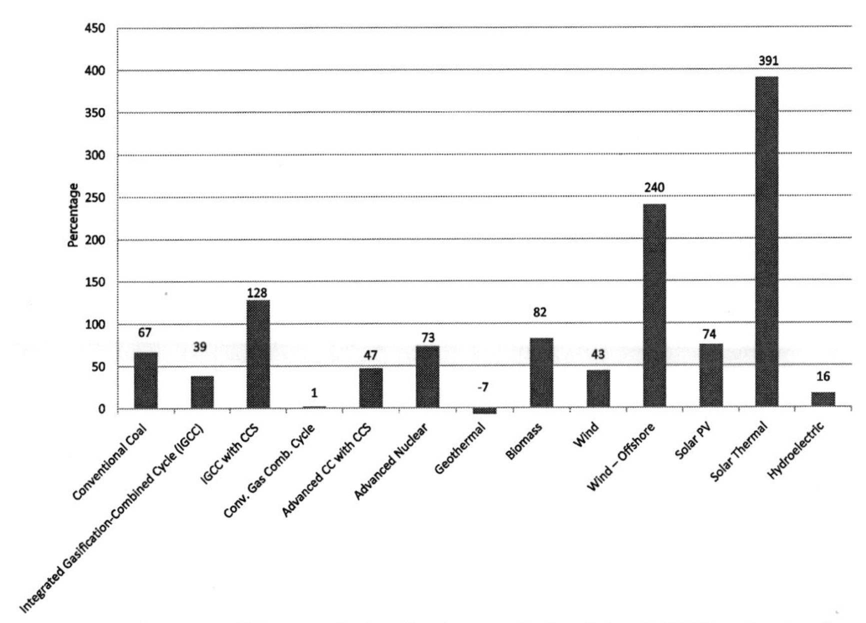

FIGURE 2-6 Percent difference in levelized cost of electricity (LCOE) estimates for electric power generation technologies entering the market in 2022 compared with advanced combined-cycle natural gas power generation when GHG pollution costs are not included.
NOTE: CC = combined cycle; CCS = carbon capture and storage; IGCC = integrated gasification-combined cycle; PV = photovoltaic.
SOURCE: EIA, 2015f, 2016g. Because *Annual Energy Outlook 2016* does not assess conventional coal and IGCC technologies, their values (in 2013 dollars) were sourced from *Annual Energy Outlook 2015* and then converted to 2015 dollars using the Bureau of Economic Analysis' gross domestic product (GDP) implicit price deflator.

These estimates suggest that most increasingly clean power technologies are uncompetitive in the market compared with advanced combined-cycle natural gas power generation unless supported by a technology-specific policy. For example, EIA's benchmark LCOE for onshore wind generation is 43 percent higher than that for an advanced combined-cycle natural gas unit. Without accounting for GHG externalities, wind and solar energy also are often not competitive with new IGCC coal plants (see Appendix B). There may be selected locations and circumstances in which increasingly clean technologies can currently compete with fossil fuel generation. However, EIA's projections suggest that for the United States as a whole, existing low-carbon increasingly clean technologies would not be economically competitive from a market participant's perspective in a technology-neutral policy environment.[9]

---

[9]See Appendix B for a detailed description of the plausible impacts of learning rates and the use of learning rates in NEMS projections.

Including a cost of $15/ton for GHG pollution still shows a similar picture. While the relative costs for fossil fuel electric power generation technology without carbon capture increase, the costs of increasingly clean technologies such as wind, solar, and carbon capture still remain significantly higher than that of advanced combined-cycle natural gas generation. Figure 2-7 compares the projected costs of various electric power generation technologies against the cost of an advanced combined-cycle natural gas plant where firms account for the costs of pollution from GHGs when installing power generation technologies, using EIA's assumption that future carbon abatement policies will add roughly $15/ton to the cost of capital for carbon-intensive technologies.[10] Figure 2-7 shows that wind is still 32 percent more expensive than advanced combined-cycle natural gas generation when the costs of pollution are taken into account. Even if the price charged for the carbon pollution were doubled from $15 to $30/ton of CO_2, which would approximate EIA's estimate of the possible future cost of carbon assuming a 2.5 percent rather than a 3.0 percent discount rate, onshore wind would still remain 23 percent more expensive on average than an advanced combined-cycle natural gas unit.

In addition to EIA, several other groups have begun to produce LCOE estimates for recent historical prices, as well as projections. The National Renewable Energy Laboratory (NREL) has developed its own model that produces a range of high, middle, and low LCOE estimates for essentially the same technologies as those to which the EIA estimates apply. Because of its focus on renewable energy sources, NREL's model characterizes renewable fuel technologies in greater detail relative to NEMS. Like EIA, NREL expresses the caveat that any method of estimating LCOE is subject to high levels of uncertainty and is dependent on modeling assumptions.[11] Also like EIA, NREL generates a range of scenarios when developing its projections. Importantly, current and future cost reduction trajectories are not estimated but are defined as inputs to NREL's model (Sullivan et al., 2015).

NREL's available scenarios do not explicitly include a price on pollution but do account for technology-specific tax policies, making it difficult to

---

[10]By comparison, the Interagency Working Group on Social Cost of Carbon (2015) estimates the social cost of carbon pollution to be $36/ton.

[11]NREL developed and uses the Regional Energy Deployment System (ReEDS) model. Assumptions in this model are intended to account for, inter alia, transmission infrastructure expansion costs, electric system operation costs, cost of capital, busbar costs at the plant gate, costs of transmission spur lines, site-specific construction costs, and projected changes in capacity factor (Sullivan et al., 2015).

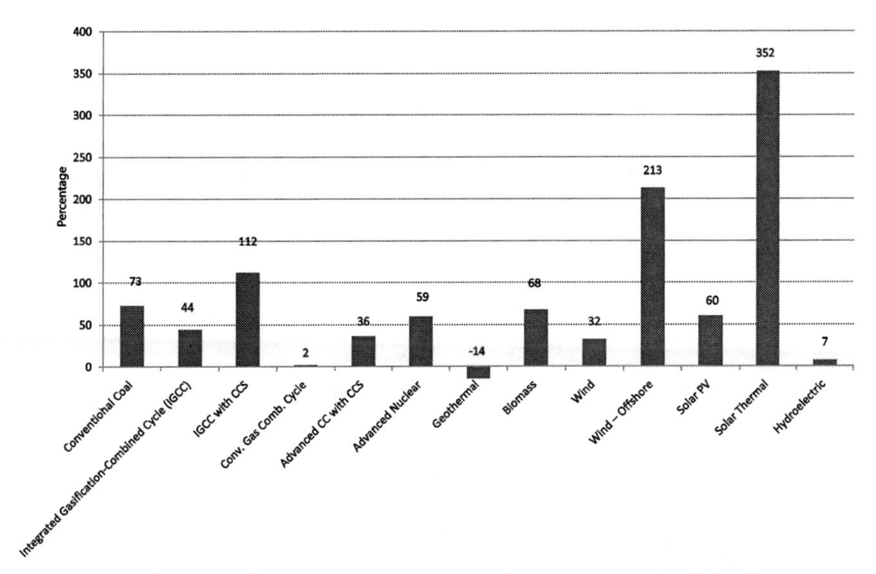

FIGURE 2-7 Percent difference in average levelized cost of electricity (LCOE) estimates for electric power generation technologies entering the market in 2022 compared with advanced combined-cycle natural gas power generation when all pollution costs are internalized.
NOTE: CC = combined cycle; CCS = carbon capture and storage; IGCC = integrated gasification-combined cycle; PV = photovoltaic.
SOURCE: EIA, 2015f, 2016g. Because *Annual Energy Outlook 2016* does not assess conventional coal and IGCC technologies, their values (in 2013 dollars) were sourced from *Annual Energy Outlook 2015* and then converted to 2015 dollars using the Bureau of Economic Analysis' gross domestic product (GDP) implicit price deflator.

develop a useful comparison of unsubsidized technologies at full cost. As with EIA's estimates, the committee adjusted the NREL estimates to eliminate the tax policies for all technologies in order to compare LCOE estimates in a more technology-neutral policy environment. Figure 2-8 compares the projected costs of various electric power generation technologies against the cost of conventional combined-cycle natural gas power generation from the market perspective where firms do not directly bear the costs of pollution.

Like EIA's estimates, these estimates suggest that while the costs of renewable technologies have declined significantly in recent years, absent subsidies or an appropriate price on pollution, increasingly clean technologies often cost more in the marketplace. Based on NREL's estimates, for example,

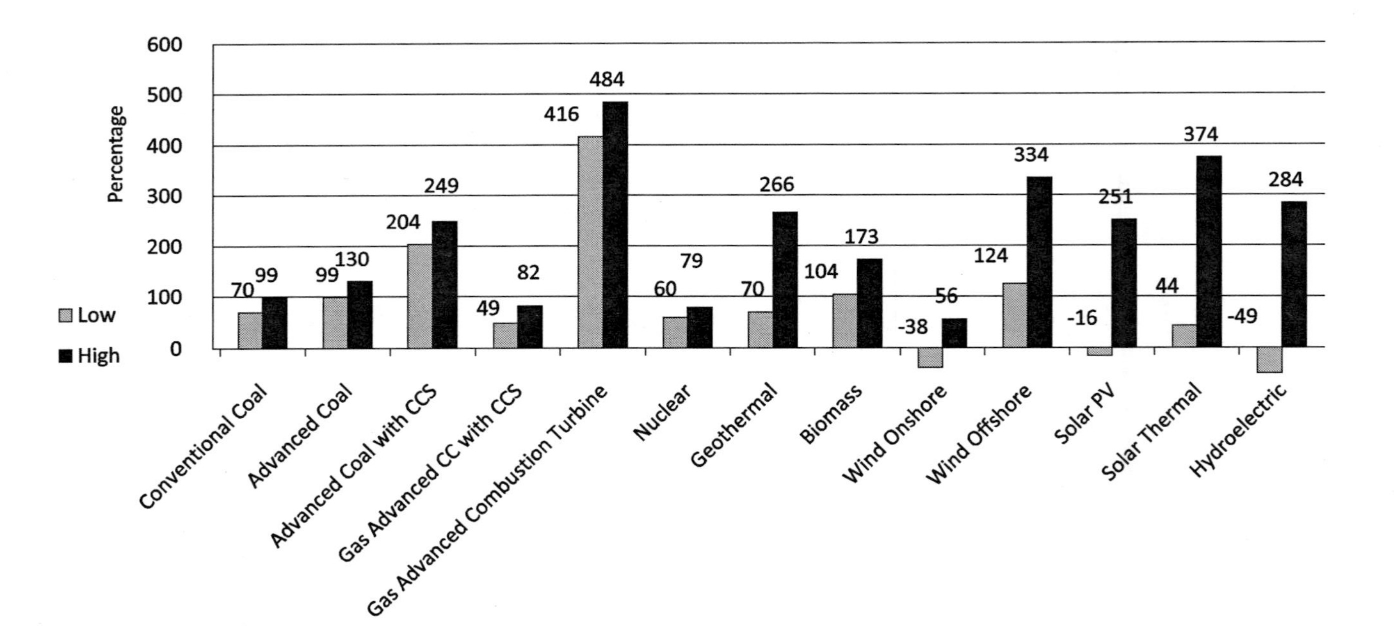

FIGURE 2-8 Percent difference in levelized cost of electricity (LCOE) estimates for electric power generation technologies entering the market in 2020 compared with conventional combined-cycle natural gas power generation when pollution costs are externalized.
NOTE: CC = combined cycle; CCS = carbon capture and storage; IGCC = integrated gasification combined cycle; PV = photovoltaic.
SOURCE: Sullivan et al., 2015.

solar costs may range from as low as 16 percent less than those of natural gas to as great as 251 percent more, while onshore wind may range from 38 percent less to 56 percent more costly.

Financial advisory and asset management firm Lazard has produced a set of LCOE estimates since 2007 using its own parameters. The most recently available set, from November 2015, provides a range of scenarios, including one that is labeled "unsubsidized" (Lazard, 2015). The assumptions stated, though, are less clear than the assumptions and model parameters specified by EIA and NREL. For example, Lazard notes that its estimates do not include factors that could have a potentially significant effect on its estimates, including "capacity value vs. energy value; stranded costs related to distributed generation or otherwise; network upgrade, transmission or congestion costs; integration costs; and costs of complying with various environmental regulations (e.g., carbon emissions offsets, emissions control systems)" (Lazard, 2015, p. 19). Lazard also states that in its LCOE estimates, it does not attempt to account directly for such externalities as the cost to society of pollution. Instead, a cost of carbon abatement is calculated separately. The analysis makes no mention of the lead time for construction or the year the assets are expected to enter service.

Nonetheless, Lazard's estimates show ranges similar to those from EIA and NREL. For example, the estimates for utility scale solar PV range from 49 percent less expensive than conventional gas to 56 percent more costly. Onshore wind is estimated to be as much as 62 percent less to as high as 71 percent more costly. Advanced coal with CO_2 capture and storage is projected to cost 311 percent more than conventional gas. Again, these ranges suggest that progress has been made in improving the cost-competitiveness of increasingly clean technologies, but continued cost declines are still needed. This is especially true once grid upgrade costs, such as the cost of new transmission assets to accommodate additional wind and solar plants, are incorporated into cost estimates.

Data and news provider Bloomberg New Energy Finance (BNEF) publishes an annual *Sustainable Energy in America Factbook* that includes retrospective LCOE estimates for a wide range of electric power generation technologies (BNEF, 2016). One advantage of the BNEF estimates is that the analysts attempt to use data from actual constructed power plants whenever possible. A full analysis of BNEF's estimates, however, was complicated by the opacity of their presentation; estimates are presented as a graphic without underlying figures. BNEF also does not provide details of its model or the assumptions used in any particular scenario except to state that "EIA is source for capex ranges for nuclear and conventional plants" (BNEF, 2016). For example, the estimates are identified as being for "unsubsidized…power generation technologies" (BNEF, 2016, p. 35), but the notes provided do not describe the methodology used to adjust for and eliminate the impact of various subsidies.

Despite these difficulties with its use in the present analysis, the BNEF graphic is illustrative of current costs. For example, in the 2016 factbook presenting estimates for plants entering service in 2015 around the world (BNEF, 2016), it appears that approximately 20 percent of wind plants and fewer than 2 percent of solar PV plants constructed cost less to build than gas- or coal-fired plants. Apparent costs compared with a combined-cycle natural gas plant range tremendously. Onshore wind ranges from 34 percent less to as much as more than 240 percent more expensive. The range for solar PV[12] is 26 percent less to nearly 400 percent more expensive. And for offshore wind, the range is 54 percent to approximately 430 percent more expensive. BNEF's factbook does not provide LCOE estimates for carbon capture-equipped fossil fuel plants.

In addition to estimating a range of costs for various technologies, BNEF estimates "central values" using actual data in its model[13] (BNEF, 2016). Looking at these average estimates, it is clear that, except for small hydropower stations, they all either held steady from the first to the second half of 2015 or declined; small hydro costs apparently increased slightly. The data points reveal that the global average costs for onshore wind-generated electric power are greater than the costs of conventional gas- or coal-generated power in the United States and China, and lower than those in Europe and Australia.[14] Costs for thin-film and stationary crystalline silicon solar PV appear to be close to competitive with those for coal-fired power generation in Australia but higher than the costs for coal-fired generation for the rest of the world and lower than the costs for gas-fired generation anywhere. Tracking crystalline silicon solar PV is estimated to be more expensive than any fossil fuel generation source. Average costs for large-scale hydropower plants are estimated to be greater than those for natural gas-sourced power in the United States, and greater than those for coal-sourced power in China but lower than those for fossil fuel-fired plants in the rest of the world.[15] Offshore wind and concentrating solar power are both estimated to be more expensive than fossil fuel-fired plants anywhere.

Reviewing this evidence and the salient recent literature, it becomes clear that the higher average cost of key increasingly clean electric power generation technologies remains a barrier to their broad deployment (Aldy, 2011).[16]

---

[12]Includes three technologies: crystal silicon without tracking, crystal silicon with tracking, and thin film.

[13]The accompanying notes refer to these as "global central scenarios," explaining that "these central scenarios are made up of a blend of inputs from competitive projects in mature markets" (BNEF, 2016, Slide 35).

[14]BNEF provides estimates for the cost of gas-fired generation only for the United States, China, Europe, and Australia.

[15]The factbook does not define the difference between large and small hydropower plants.

[16]Uncertainty regarding climate change and the possibility of its very large negative impacts have raised questions about the application of cost-benefit analysis and the social

> **Finding 2-1:** On average, unsubsidized increasingly clean electric power generation technologies are estimated to cost between 43 percent and 391 percent more than a new combined-cycle natural gas facility when prices do not account for the costs of pollution.[17]

EIA's *Annual Energy Outlook 2016* No Clean Power Plan (CPP) case assumes that the performance of increasingly clean power generation technologies will continue to improve and that governments will continue other policies favoring those technologies.[18] Applying these assumptions, EIA projects increases in renewable electric power generation through 2040. Renewables start from a current cost disadvantage and power generation market share of approximately 13 percent, but they (including hydroelectric resources) are projected to provide 23 percent of electric power generation in 2040. That penetration rate for renewables is still far below the projected market shares for natural gas and coal of 32 percent and 29 percent of U.S. electric power generation, respectively (EIA, 2016a).[19] The projected increase in renewable electric power generation has only a limited impact on the overall mix of power generation technologies, as reflected in Figure 2-9. In the *Annual Energy Outlook 2016* No CPP case, nuclear power's approximately 20 percent market share is projected to fall to about 15 percent by 2040. As a consequence, with projected growth in demand, EIA's No CPP case projects that CO_2 emissions from the production of electricity could be nearly 4 percent higher in 2040 than they were in 2015.

EIA also conducted one sensitivity analysis that assumes the extension of some policies through 2040 and expansion of other policies meant to decrease CO_2 emissions.[20] Under those assumptions relative to the No CPP case,

---

cost of carbon. For a discussion of these issues, see Weitzman (2009, 2011), Nordhaus (2011), and Pindyck (2011).

[17]Geothermal and hydroelectric power generation costs are exclusive of constraints on capacity increases. The smaller number is associated with wind and the higher number with solar thermal generation.

[18]EIA's No CPP case "assumes that the final CPP rule is permanently voided and is not replaced by other controls on power sector CO_2 emissions." The committee used these projections given the U.S. Supreme Court's stay of the Clean Power Plan in February 2016 (see Martin and Jones, 2016).

[19]This citation refers to data found in Martin and Jones (2016, Table 8).

[20]Specifically, this sensitivity case assumes that tax policies such as the production tax credit extend beyond their current sunset dates and remain in force, while corporate average fuel economy standards, appliance standards, and building codes are expanded beyond current provisions, and the Clean Power Plan is reinstated with tightening regulation of CO_2 emissions starting in 2030.

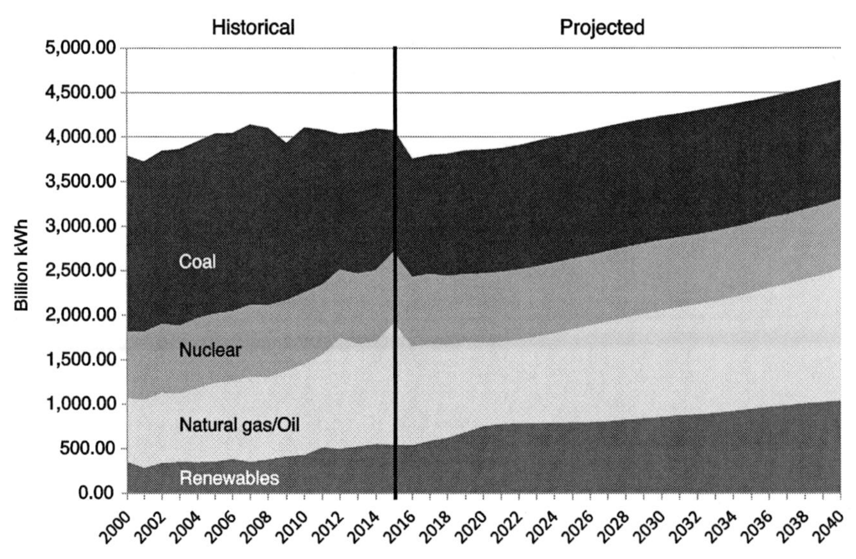

FIGURE 2-9 Electric power generation by fuel (billions of kilowatt hours [kWh]) assuming No Clean Power Plan, 2000-2040.
SOURCE: EIA, 2016a.

renewables are projected to grow to supply 34 percent of electric power generation, the same as natural gas, while coal and nuclear both shrink to 16 percent. These results are shown in Figure 2-10. This case projects CO_2 emissions from the electricity sector to be roughly 30 percent lower in 2040 than in 2015.

Comparison with an earlier, alternative sensitivity analysis is helpful to consider how extending current policies compares with enacting a policy that would incorporate the cost of pollution into the market price of electricity. In its *Annual Energy Outlook 2014*, EIA projected that a significant price on carbon starting at $25 per ton of CO_2 could increase the market share for nuclear power to more than 37 percent and reduce the electricity sector's CO_2 emissions by nearly 80 percent compared with its 2012 emissions.[21] This assumed price on CO_2 emissions was projected to increase the average electricity price for 2040

[21]In sensitivity cases, EIA examined policies that would favor increasingly clean technologies and made modestly more favorable assumptions regarding the cost of renewable electric power generation. The additional cases included those in which the capital cost of nonhydroelectric renewables was assumed to be 20 percent below reference case levels, and a case in which the carbon price is initially set at $25 per ton of CO_2 and increases at a rate of 5 percent per year. In both of these cases, renewable generation was forecast to increase but remain at below a 25 percent market share in 2040 (EIA, 2014a, Appendix B).

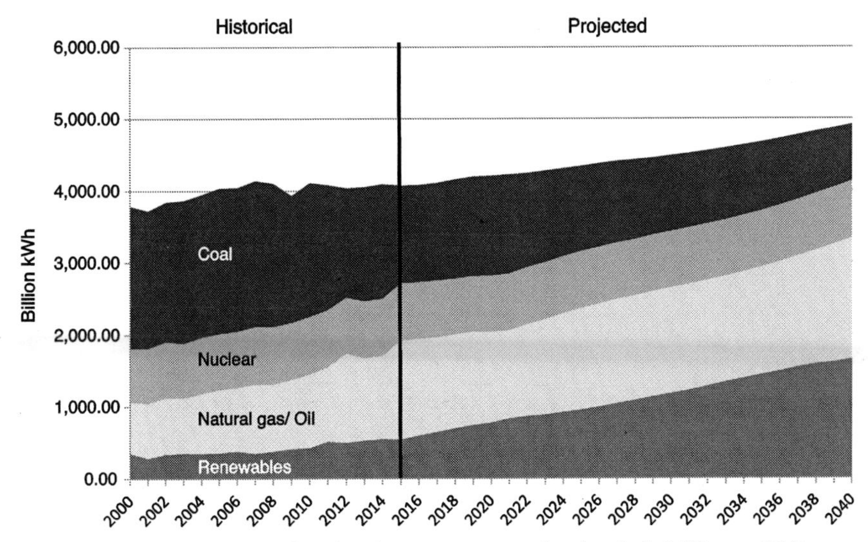

FIGURE 2-10 Projections for electric power generation by fuel (billions of kilowatt hours [kWh]) assuming specific policies are extended and expanded through 2040.
SOURCE: EIA, 2016a.

by 23 percent compared with the reference case and (in constant 2012 dollars) by 39 percent relative to the average 2012 price. Other studies assuming limited improvements in the cost of low-carbon resources have reached similar conclusions: achieving large reductions in the U.S. electricity sector's carbon emissions by incorporating the full costs of pollution into electricity prices would lead to significant increases in prices to ultimate consumers.[22]

These results suggest that major improvements in the cost-competitiveness of low-carbon increasingly clean technologies—improvements that go beyond those assumed in EIA's or NREL's analyses—will be required if those technologies are to be market-competitive globally to a degree that encourages significant displacement of incumbent technologies. These improvements will be essential to achieving long-term reductions in GHGs, such as the reduction called for in the COP21 agreement,[23] without significantly increasing electricity prices.[24]

---

[22]Other assessments of the costs of reducing carbon emissions can be found in Clarke et al. (2009, 2014), Fawcett et al. (2009, 2013), CBO (2009), EIA (2009a), Paltsev et al. (2009), Fischer and Newell (2008), and CCSP (2007).

[23]Under that agreement, the "United States intends to achieve an economy-wide target of reducing its greenhouse gas emissions by 26%-28% below its 2005 level in 2025 and to make best efforts to reduce its emissions by 28%" (United States, 2015).

[24]Regarding the collateral goal of preventing large increases in electricity prices, see Parry et al. (2015, Chapters 6 and 7).

Finding 2-2: Achieving long-term targets for reducing GHG emissions from the electricity sector by 80 percent or more without significantly increasing electricity prices would require significant improvements in the performance of low-carbon increasingly clean technologies.

WILL EXPANDED DEPLOYMENT MAKE INCREASINGLY CLEAN TECHNOLOGIES MORE ECONOMICALLY COMPETITIVE?

Given the above finding that currently available increasingly clean electric power generation technologies are not yet economically competitive compared with conventional, higher-polluting technologies, the committee considered the extent to which policies designed to expand the deployment of cleaner technologies produce meaningful performance improvements and associated cost declines as a result of "learning by doing" (LBD),[25] and the extent to which LBD benefits might offset the difference in the societal costs of low-carbon and conventional resources. The committee found no evidence that increasingly clean technologies could become economically competitive in the near term based primarily on performance improvements achieved through expanded deployment and LBD alone (Gallagher et al., 2012), even though LBD is often assumed to have a material effect on costs. This leads to the conclusion that improving the cost-competitiveness of increasingly clean technologies will require that attention be paid to the larger innovation system.

"Experience curves" are a common component of innovation system models. Simple experience curves have been developed for various technologies and industries in which an historical doubling of technology deployment is associated with coincident reductions in costs or improvements in performance. For newer alternative energy sources, single-factor experience curves typically estimate a 15-20 percent improvement in costs with each doubling of a technology's adoption (McDonald and Schrattenholzer, 2000). This relationship of improvements in cost or performance to deployment is often presented as a "learning rate."[26] However, experience curves simply document historical associations and by themselves provide limited information on the impacts of increased deployment or appropriate public policy choices.

There are significant limits on the inferences that should be drawn from such experience curves. First, an historical "learning rate" does not necessarily imply that a given technology will continue to improve along its historical trend line. Emerging technologies are quite complex and often include both improving

---

[25]For purposes of this discussion, the phenomenon of customers "learning by using" new technologies is included as a subset of LBD.

[26]For a discussion of experience curves and their use in the energy sector, see Junginger et al. (2010).

and relatively mature components. Improving cost or performance often requires solving "problems" across a wide range of TRLs. At some point, progress in improving components is likely to diminish and be offset by other factors, such as increased input prices. Second, long-term forecasts of technology costs are sensitive to small variations in the choice of the underlying historical data (Nemet, 2006; see also NRC, 2010d; Weisenthal et al., 2012). Third, simple experience curves do not reveal what factors led to observed performance improvements. Proponents of the experience curve method acknowledge that it treats the mechanisms of performance improvement as a "black box" (Junginger et al., 2008). Thus, single-factor experience curves cannot help answer the policy question of the optimal balance between public investment in research and development (R&D) and direct support for increased market adoption, such as expenditures for deployment. Fourth, the associations documented in experience curves do not imply a causal relationship (Clarke et al., 2006; Popp et al., 2010). An increase in the adoption of a technology, for example, may reflect cost reductions that were the result of an independent government research program, innovations developed in a different industry, or other external factors. Thus, the "learning ratios" in experience curves would not, by themselves, demonstrate that programs subsidizing larger-scale deployment of increasingly clean technologies led to material performance improvements.

Experience curves often reflect the impact of multiple factors in addition to LBD (Clarke et al., 2006). This point has important public policy implications. The basic mechanisms by which performance improvements may occur have been widely documented (Junginger et al., 2010). Broadly defined, they include the following:

- Learning by (re)searching (LBS)—This is R&D broadly defined. LBS is an intentional and often costly effort to seek out and develop innovations. Its goal is to develop an innovation until it is at or near the stage of large-scale deployment. Typically, R&D is risky and can have large spillover benefits that are not fully captured by the organization sponsoring the research. These knowledge spillovers justify public support, as private entrepreneurs would otherwise underinvest in R&D activities.
- LBD—This is the creation of new information that reduces the cost of future production. LBD is passive (Thompson, 2010)—it is a free by-product of deployment rather than an explicit undertaking with its own costs. LBD can produce spillover effects that, depending on the cost of additional deployment and the rate of learning, may justify some amount of public support.
- Economies of scale—Economies of scale reflect decreasing unit production costs as production at a plant or firm reaches an efficient size. In the energy sector, economies of scale may be relevant at the unit, plant, firm, or industry level (Gillingham and Sweeney, 2010).

According to Borenstein (2012, p. 83), "The distinction between learning-by-doing and economies of scale may seem minor, but the implications for public policy are immense. If one firm can drive down its costs by producing at large scale in its factory or its installation operation, those benefits are highly appropriable by that large firm....Thus, significant economies of scale in any industry, short of creating a natural monopoly, are not generally seen as a basis for government intervention." Private entrepreneurs can be expected to invest in the realization of economies of scale when doing so will produce an economically competitive product.

- Learning by waiting (LBW)—The spillover effects from other industries, technologies, or countries are essentially exogenous—that is, developed on the outside, from the perspective of the firm (Thompson, 2010). The resulting innovations will appear over time and can be exploited by waiting. In some cases, government support may have played a role in the development of the borrowed technologies for another industry; in other cases, government may be able to accelerate technology transfer and the adaptation of technologies developed in other fields, often with limited intervention. However, LBW is the result primarily of innovation that occurs elsewhere and not of accelerating technology deployment. Separating LBD from the impacts of external technological change is difficult, such that estimates of learning rates based on experience curves can easily be biased upwards (Nordhaus, 2014).

Studies examining the factors contributing to performance improvements that coincide with policy-driven deployments of increasingly clean technologies indicate that single-factor experience curves should be viewed with caution. They may overstate the extent to which significant innovation and performance improvements can be achieved through policies focused primarily on expanding deployment. Söderholm and Sundqvist (2007) developed two-factor models examining the impacts of both R&D and LBD on improvements in wind generation in four European countries from 1986 through 2000. They concluded that the problem of omitting such variables as LBS must be taken seriously. After accounting for R&D, their models estimated that LBD was associated with improvements of about 5 percent with every doubling of deployments (Söderholm and Sundqvist, 2007). Other two-factor studies have reached similar conclusions and indicated that learning rates associated with LBS may be higher than those associated with LBD (Jamasb, 2007; Kahouli-Brahmi, 2008; Kobos et al., 2006). In a widely cited analysis, Nemet (2006) further disaggregated the factors contributing to reductions in U.S. PV energy costs from 1975 to 2001. He considered seven different factors, including economies of scale, efficiency improvements, and reductions in material costs. According to Nemet, "Overall, the 'learning' and 'experience' aspects of cumulative production do not appear to have been major factors in enabling firms to reduce the cost of PV" (Nemet, 2006, p. 3226). He goes on to state that "a much broader set of influences than experience alone contributed to the rapid cost reductions" (Nemet, 2006, p. 3230).

Other studies have examined the impacts of deployment subsidies on patent filings and other evidence of innovation. A 2010 study by Swiss researchers, for example, analyzed the effectiveness of "demand-pull" (i.e., deployment) and "technology-push" (i.e., R&D) policies for PV across 15 OECD countries. Their analysis found that "demand pull policies only foster incremental innovation," and the authors cite "anecdotal evidence that in phases of rapid induced market growth such policies even disincentivize non-incremental innovation." They conclude that "only technology-push support is able to incentivize non-incremental innovation" (Peters et al., 2011, p. 2).

A recent analysis of the impact of the German Renewable Energy Sources Act, the so-called Erneuerbare-Energien-Gesetz (EEG), on patents for innovation in renewable energy technologies questions whether German feed-in tariffs for PV, wind, and geothermal energy have led to innovation in these technologies. The authors found statistically significant negative correlations between feed-in tariffs for hydroelectric and biomass generation and innovation in these technologies. They conclude that "empirical data of the German feed-in regulation over the last two decades...do not lend support to the proposition that German feed-in tariffs under the EEG spur innovation." The study found that in the case of PV, which received very high incentives for deployment, "the EEG does not engender innovative output" (Böhringer et al., 2014, p. 15).

An analysis by Nemet (2012) of $1 billion in public investments leading to the deployment of $2 billion in wind generation in California and in contemporaneous performance improvements between 1985 and 2005 found evidence of LBD. However, the LBD benefits were found to diminish with additional deployments (Nemet, 2012). The finding of diminishing benefits from LBD are similar to results from other industries (Arrow, 1962; see also Argote et al., 1990; Benkard, 2000; Darr et al., 1995). They suggest that increasing the scale at which new technologies are deployed cannot be expected to produce a proportionate improvement in performance from LBD because learning rates and their benefits also may moderate as a technology begins to mature. Moreover, Nemet (2012) found that the benefits from LBD also may diminish over time. This may occur because some of the knowledge acquired during deployments may be retained by employees as tacit knowledge and be lost to the firm when they leave, and other lessons learned may become less relevant with changes in technology, demand, or industry structure. Qiu and Anadon (2012) analyzed improvements associated with the Chinese government's wind power concessions during 2003-2007. Chinese wind prices saw reductions during this time. These reductions, however, can be explained largely by economies of scale and other factors. Taking such factors into account, Qiu and Anadon estimated an LBD rate of only 4 percent for each doubling of production.

These results do not suggest that LBD should be discounted entirely as a mechanism of progress (Arrow, 1962). However, caution is necessary in attributing observed performance improvements to deployment and LBD alone. While LBD appears to have a positive impact, learning rates for LBD may be lower than those for LBS. Importantly, when other factors are taken into account, reasonable estimates for LBD learning rates may be in the single digits. Moreover, evidence of diminishing returns suggests that in some cases, most LBD benefits may be gained through more limited deployment initiatives.

Studies examining patent filings and other evidence of innovation suggest that LBD may play a larger role in incremental improvements in technology, while LBS may be more important for fundamental improvements. Thus, the importance of LBD may depend on the stage of a given technology's development. And in some cases, given the iterative nature of the innovation process, LBD may complement LBS investments.

Appendix C provides an analytical approach to evaluating LBD and the relative merits of deployment versus focusing near-term policies on R&D and waiting for the costs of increasingly clean electric power technologies to approach becoming commercially competitive.[27] This appendix contains a quantitative analysis illustrating the analytical model with specified assumptions. While the analysis suggests that LBD is relevant and that a material learning benefit can be associated with deployment, it also suggests that this learning benefit may be too small to offset much of the cost of large-scale deployment of increasingly clean electric power technologies. The analysis indicates that LBS (i.e., investments farther upstream in the innovation system, such as at the R&D stage) is more important in the near term (Nemet and Baker, 2008; NRC, 2010c).

Some increasingly clean electric power technologies will be economically competitive, either generally or in specific applications, independent of any LBD benefits that might result from their deployment. In the near term, it will be advisable to continue to deploy energy-efficient (see Chapter 4) and other increasingly clean technologies that are already cost-effective—in many cases more aggressively. Policies that lead to prices fully incorporating the costs of pollution will aid deployment and reduce emissions by providing more accurate market signals. It is important to note, however, that optimal policies for encouraging innovation in energy technologies rely less on pollution pricing than on directed support of innovation (Acemoglu et al., 2012). Reducing pollution to socially optimal levels by implementing only a pollution price would likely cost more than doing so in tandem with complementary innovation-focused policies (Parry et al., 2015). Thus while pollution pricing is a critical complement to innovation policies, achieving the level of desired pollution abatement will require tailoring policies to promote innovation in energy technologies, to be comprehensive, to address undue barriers to

---

[27]Other relevant U.S. policies to support deployment are discussed in Chapter 7.

innovation in each stage of the innovation process, and to provide significant support for research, development, and deployment (RD&D).

> **Finding 2-3:** Evidence suggests that policies focused disproportionately on subsidizing deployments of increasingly clean technologies will not produce the large, timely, cost-effective improvements in the cost and performance of these technologies required to address pollution problems. Rather, what is required to achieve these improvements in currently available technologies and to create new, as yet unknown breakthrough technologies is a major investment in innovation.

The development of affordable low-carbon increasingly clean electric power generation technologies could position the United States to take more effective measures to address the risks and uncertainties of climate change. In its report *Limiting the Magnitude of Future Climate Change*, the NRC (2010c) identifies "an urgent need for U.S. action to reduce greenhouse gas emissions." Recent assessments of risks and uncertainties associated with climate change are consistent with that conclusion (IPCC, 2013, 2014a,b; Walsh et al., 2014). Multiple reviews and expert panels likewise have concluded that global GHG emissions pose clear risks to U.S. economic prosperity (CEIR, 2007; CBO, 2009; Dell et al., 2012, 2014; Interagency Working Group on Social Cost of Carbon, 2013; Nordhaus, 2013) and national security (CNA Military Advisory Board, 2007, 2014; Defense Science Board Task Force, 2011; DoD, 2014).

The committee both agrees with these prior conclusions and acknowledges the uncertainty inherent in making forecasts for complex climate systems. Yet the existence of uncertainty does not mean that the United States should eschew mitigation measures. Avoiding the potential negative consequences of significant climate change is critical to protecting the nation's economic and security interests. Effective mitigation of climate risks may require a transition to low-carbon energy technologies on a global scale and possibly within a compressed time frame. Significantly reducing the cost and improving the performance of low-carbon energy resources appears both the most efficient and the most likely path to providing options for making an affordable transition to a low-carbon global economy. There is an urgent need for the development of energy technology options that could make the global transition to a low-carbon economy practical, affordable, and timely.

The federal government has taken a number of recent actions to support innovation in electric power generation technologies. For example, the Department of Energy made clear in 2015 that it planned to increase its focus on "crosscutting R&D," including electric power grid modernization, with a primary goal of continuing to decrease the costs of increasingly clean energy technologies (DOE, 2015c). DOE expects it will need "partnerships with university scientists and engineers, researchers at both established and entrepreneurial companies, federal and state agencies, and others" to induce the level and kind of transformational innovation needed (DOE, 2015c, p. iii). The committee finds this a positive development, as it is expected that an increase in support for innovation activities such as R&D will be cost-effective in reducing the costs of increasingly clean electric power generation technologies (Baker et al., 2015). The remainder of this report is aimed at providing guidance on how the Department of Energy and other federal entities, along with state governments and other stakeholders, can take action to support and encourage breakthrough innovation to meet the energy challenge.

CONCLUSION

The committee's review of currently available increasingly clean electric power generation technologies suggests that they are not yet capable of meeting the challenge of supplying reliable electric power at socially acceptable pollution levels at prices that make them competitive in current electric power markets. Policies therefore need to focus on both the improvement of currently available and the development of new increasingly clean energy technologies. The approach of increasing deployment in the hope that LBD will drive down costs and increase performance appears unlikely to succeed at the scale needed to address the pollution challenge adequately. The gains from LBD are too small to expect that expanded deployment will yield the level of innovation needed. While adequately pricing pollution would also help—both to induce additional innovation and to create a level playing field so that prices reflect the full costs of technologies—it also would likely be insufficient absent other policies.

The implication of these findings for increasingly clean energy innovation policy is that the most important priorities are identifying and creating new options, demonstrating the efficacy of these options, and setting the stage for early adoption of those that are most promising. Although policies could be instituted that would enhance the conditions for eventual large-scale take-up and improvements in use, these policies are likely to be expensive and ineffective without a substantial investment in the earlier stages of the innovation process. The emphasis needs to be on developing technologies that can truly compete with incumbent energy sources. Such technologies are not available today, and efforts to create these future technologies need to be expanded and accelerated. A major investment to this end is warranted, with a clear view of the challenges ahead. These challenges create an opportunity and a need for action by governments at all levels, keeping an eye on the prize of expanding the innovation machine.

Recommendation 2-1: The U.S. federal government and state governments should significantly increase their emphasis on supporting innovation in increasingly clean electric power generation technologies.

Recommendation 2-2: Congress should consider an appropriate price on pollution from power production to level the playing field; create consistent market pull; and expand research, development, and commercialization of increasingly clean energy resources and technologies.

3

Supporting and Strengthening the Energy Innovation Process to Expand the Technological Base for Increasingly Clean Electric Power

This chapter addresses the need to expand the technological base for increasingly clean electric power by supporting and strengthening innovation in electricity generation, transmission, and distribution systems, as well as by spurring innovation in the design and integration of distributed energy generation and management systems, including efficient demand technologies. The scale of this challenge underscores the importance of investing in research and development (R&D) to discover and improve transformative innovations. The deployment of viable existing technologies is also important, but as discussed in the previous chapter, will be far from sufficient to meet the global clean energy challenge, particularly in controlling the concentrations of greenhouse gases (GHGs) in the atmosphere to hold the future rise in average global temperatures to less than 2° C over the preindustrial equilibrium in accordance with the agreement among the United States and 194 other nations at COP21(UNFCCC, 2015).

This chapter first provides additional detail regarding the importance of innovation in increasingly clean energy technologies and then looks at the key stages of the energy innovation process, describing some of the main obstacles to accelerated innovation at each stage. Finally, the chapter proposes a set of strategies for overcoming those obstacles. The broad goal is to build an innovation system that is matched to the scale of the challenges confronting the electricity sector (Lester and Hart, 2012). All of the stages in this system—research, development, demonstration, and take-up—need either increased or more flexible support or new mechanisms to address gaps. Policies designed to support innovation need to be informed by recognition that innovation takes place within an interlinked, iterative system; failure to take this system-level view may reduce the effectiveness or increase the cost of individual policies focused on a single stage in the innovation process. In particular, the committee advises against an excessive focus on deployment initiatives at the expense of early R&D.

THE IMPORTANCE OF INNOVATION IN INCREASINGLY CLEAN POWER TECHNOLOGIES

The primary rationales for public support for energy innovation are overcoming market failures and internalizing major externalities (Popp et al., 2010; see also Chapter 2). Government supports innovation through a range of policies, including funding technology research, development, and demonstration; facilitating the availability of capital to small and start-up businesses; providing tax incentives; and protecting intellectual property rights that provide a temporary monopoly to inventors.

General-purpose technologies (GPTs) are innovation technologies that have many potential applications for wide use and are capable of ongoing technological improvement. GPTs also enable innovation in different application sectors and create innovation complementarities that raise returns both to the GPT itself and in various application sectors as the technology improves in response to application-sector requirements. Because of these factors, GPTs tend to produce large knowledge spillovers, a form of market failure that prevents the inventors and firms that invest in innovation from appropriating the full benefits that flow from their R&D expenditures. Electricity is one example of a GPT that contributes to technological dynamism (Clarke et al., 2006). Governments have played a major role in supporting development of GPTs (Bresnahan and Trajtenberg, 1992; Janeway, 2012; Mazzucato, 2011).

STAGES OF THE ENERGY INNOVATION PROCESS[1]

The energy innovation process is a complex network of market and nonmarket institutions and incentives that includes public and private research and educational institutions; individual entrepreneurs and small entrepreneurial firms; large, mature firms; financial intermediaries ranging from large commercial and investment banks to venture capital firms and individual angel investors; local, state, and federal regulatory and standards-setting agencies and legislative units; other government agencies engaged in research, development, or procurement; and innovation users of many different kinds. These institutions and individuals are connected by a set of incentives, regulations, and laws (e.g.,

[1]This section draws on Lester and Hart (2012, Ch. 2).

governing competition, intellectual property protection, environmental protection, building codes, and the behavior of capital markets).

The innovation process rarely starts with a lone inventor experiencing a flash of insight, but more often germinates from collaborations among teams of researchers or among designers, users, manufacturers, and others. Whatever the source, the initial insight is just the first step. If a new idea is to create value, it must be reduced to practice, that is, converted into a product, process, or service that works. It must then be tested by its users to show that it is economically viable and that there is a demand for it. Then, to have real impact, it must be "scaled"—that is, adopted by a significant fraction of the population of potential users. This means that firms must also develop profitable business models for delivering the technology to users. Most innovations continue to be refined even after they have been deployed at scale.

It is helpful to distinguish among the stages that occur in the progression from new idea or concept to large-scale deployment (Figure 3-1). The process is not linear, and important feedback loops connect these stages (Janeway, 2012; Mazzucato, 2011). Yet while the process of technological change has been depicted in other ways that show the continuous interaction among the innovation stages (see, e.g., Rubin, 2005), the activities involved in each stage are distinctly different. To accelerate the flow of energy innovations over a sustained period, all stages must be emphasized.

Option Creation/Proof of Concept

This is the first stage of the innovation process, when new possibilities for products, services, or processes are identified and developed. Option creation is closely associated with R&D, but the two are not synonymous. Advances in fundamental research often yield new insights that are translated into practical applications, but ideas for new products and services frequently arise elsewhere—for example, from observations of user behavior or as a result of conversations among different members of a design team. However, strong investment in R&D is necessary to a healthy innovation system, and contributes not just to the discovery of new possibilities but also to later stages of the innovation process. A key goal is to encourage experimentation with new concepts. Another important goal is to conduct proof-of-concept testing to establish that there are no technical showstoppers that would prevent practical realization of a new concept.

Demonstration

The primary goal at this stage is to enable technology developers, investors, and users to obtain credible information about cost, reliability, safety, and other dimensions of performance under conditions that approximate actual

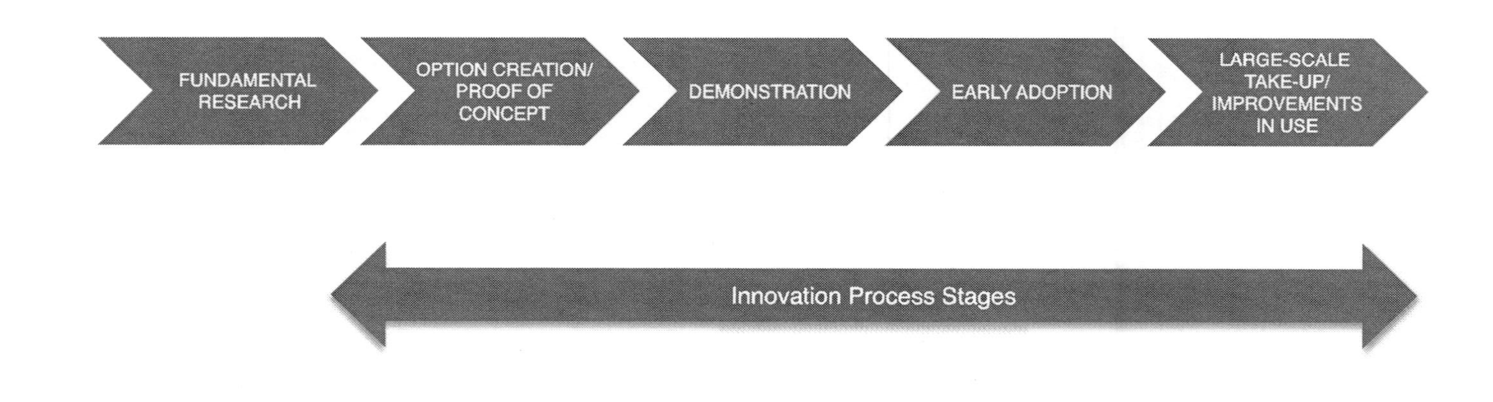

FIGURE 3-1 Stages of the energy innovation process.
SOURCE: Adapted from Lester and Hart, 2012, Figure 2.1, p.33.

conditions of use. In other words, the goal is to reduce technological, regulatory, and business risks to levels that would allow private investment in the first few commercial projects. Achieving this goal entails building, operating, and debugging pilot-scale and then full-scale prototypes, and often also requires proof of system—demonstrating that the new technology is compatible with other technologies with which it must interact, and that it can be integrated effectively into the larger system of which it is part. Other important tasks at this stage may include settling on standards and manufacturing and other infrastructure requirements, and identifying key legal and regulatory barriers that would need to be overcome for widespread use. Private innovators and their investors assume an increasing share of costs and risks in the demonstration stage relative to the option creation stage, and for smaller-scale innovations may assume all of the cost and risk. But for large-scale, complex, *system* innovations—such as central station power plants or systems for carbon capture and storage—that entail high costs, long development times, and, typically, large regulatory uncertainties, private firms are unlikely to move forward with demonstration projects unless public institutions share the costs and risks.

Early Adoption

This stage typically involves the most forward-looking users, or perhaps those with the strongest need to use the innovation. The main goals include market development and early deployment of the various infrastructure elements needed for scale-up, such as manufacturing and distribution capabilities and other key parts of the supply chain, as well as regulatory systems and processes. At this stage, too, early adopters play a key role in learning processes, providing feedback that allows valuable features to be enhanced and practical problems to be addressed. Reliability and affordability also are typically improved at this stage.

Large-Scale Take-up/Improvements in Use

At this stage, the market and regulatory environments settle into more stable and predictable patterns. Nonetheless, designs continue to be refined, production systems and business models continue to be improved, and the behavior of customers comes to be better understood. The cumulative impact of evolutionary improvements to an energy technology or system, which may continue over a period of decades, often greatly exceeds the performance gains achieved when the technology is first brought to market.

OBSTACLES TO ACCELERATED INNOVATION

Obstacles must be overcome at each stage of the innovation process. Some of these obstacles are referred to colloquially as "valleys of death." They include (1) the technological "valley of death" as new concepts move from laboratory research to proof of concept, and (2) the commercialization "valley of death" as innovations move from the demonstration stage into the marketplace or early adoption stage (Figure 3-2). Figure 3-3 lists specific obstacles that hinder the progress of innovations and may be major contributors to the "valleys of death" phenomenon.

Inadequate and uncertain funding for R&D is one such obstacle. A lack of market pull at the proof-of-concept stage may be another. In addition, high capital costs (an issue for many increasingly clean electric power technologies) and free-rider or spillover effects discourage private investment and may necessitate public/private partnerships. Obstacles further downstream can include the complexity of siting for demonstration projects, inadequate standards for scale-up and demonstration of new technologies, and a lack of vehicles for financing precompetitive pilot and demonstration projects. Regulatory review also can delay utility investments and create uncertainty regarding cost recovery for utility R&D expenditures, pilot projects, and first-of-a-kind investments. The time required for demonstration of capital-intensive technologies in a regulated, risk-averse industry and a slow pace of commercial adoption undermine the value of time-limited intellectual property rights, and create gaps between the risks and time frames acceptable to venture equity investors and the availability of project debt models. The committee notes in particular that even with a pollution price providing greater market pull, these obstacles remain because they are structural features of the regulated market for power technologies.

At the early-adoption stage, unpredictable market forces and a lack of alignment between federal and state standards are common obstacles, as are risk aversion and institutional barriers among utilities and utility regulators. Lack of incentives for early adoption among major customers also plays a role. At the level of large-scale take-up/improvements in use, a market failure of particular importance is a lack of full life-cycle costs, including the costs of carbon emissions and other externalities. Still another obstacle is the absence of real-time pricing, which would help match retail prices with production costs, as discussed in detail in Chapter 6. The failure of markets to provide entrepreneurs with the expectation of an opportunity to capture the full value of increasingly clean technologies, including environmental benefits, depresses activity at the large-scale take-up stage, as well as at earlier stages in the innovation process.

Additionally, unique obstacles to innovation in the electric power industry arise from the challenges of incentivizing a regulated local distribution company or a vertically integrated utility that is the sole generator, seller, or distributor of

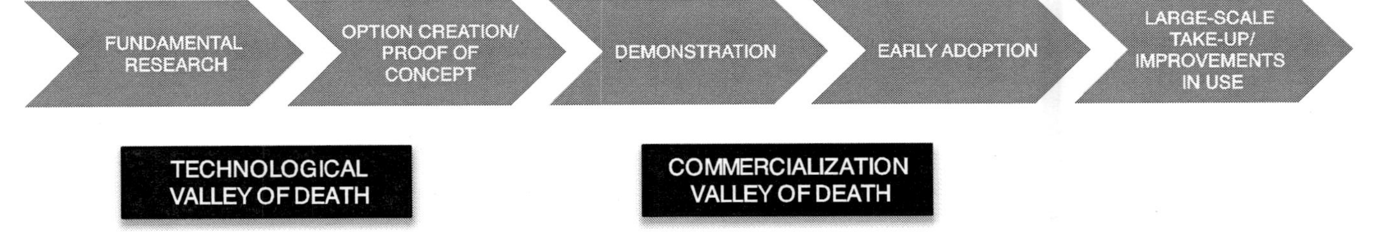

FIGURE 3-2 Stages of the innovation process and valleys of death.
SOURCE: Adapted from Lester and Hart, 2012, Figure 2.1, p.33.

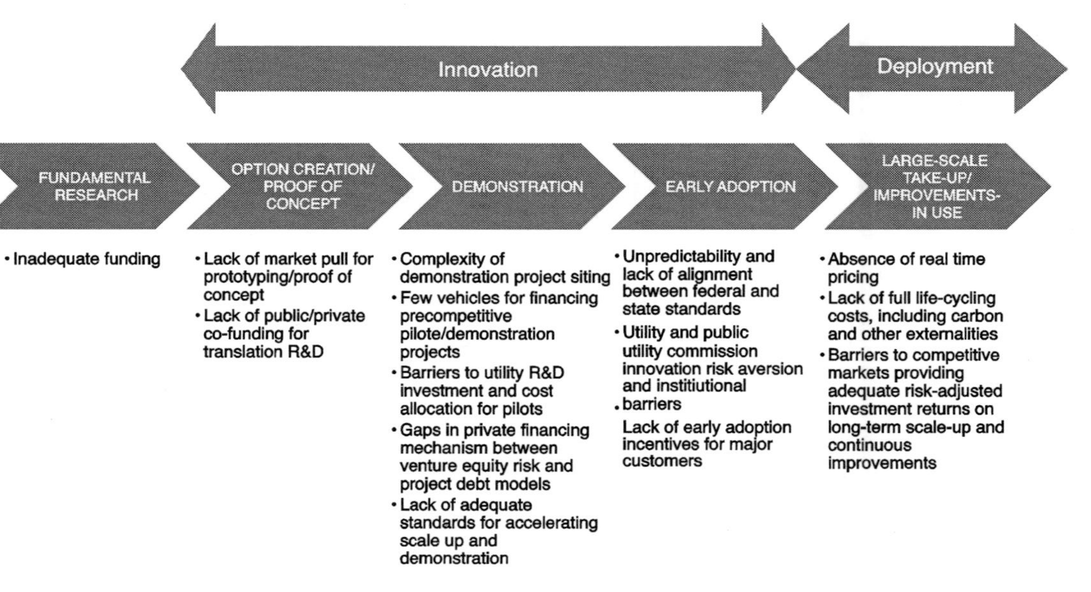

FIGURE 3-3 Stages of the innovation process and key obstacles to acceleration.
SOURCE: Adapted from Lester and Hart, 2012, Figure 2.1, p.33.

electricity to customers, issues discussed in greater detail in Chapter 6. Four such obstacles are particularly important:

- Obstacles to entry into the electric power industry, such as regulatory barriers, limit the development of new business models and the paths for introduction of new technology.
- Cost-of-service regulation, a widely used regulatory model, provides little incentive for a utility to innovate, as any savings tend to be passed on to customers, and the utility receives little or no reward for improvements in service beyond the minimum service quality standards imposed by regulators.
- Utilities may face first-mover risks, as costs may be disallowed if an innovation fails to perform as expected, and a utility may be criticized for not adopting a successful innovation more broadly or rapidly.
- While firms in competitive markets can rapidly innovate, learn, and, if necessary, redirect their efforts, a regulated utility may need to cycle through a lengthy regulatory review process and justify changes from previously approved practices (Malkin and Centolella, 2014).

Removing obstacles to innovation in the utility sector may require changes in utility regulation and in utility business models. Moreover, other regulatory policies and unresolved legal issues present additional obstacles to the development of specific technologies. For example, the Nuclear Regulatory Commission has focused on the licensing and regulation of light water reactors, but has not developed an adequate framework for licensing other types of advanced reactors. Similarly, the availability of low-cost carbon capture and storage (CCS) technologies has been projected to have a large impact on the long-term cost of policies designed to stabilize the atmospheric concentration of GHGs (Krey et al., 2014), but legal and regulatory uncertainty regarding larger-scale applications of CCS are an additional obstacle to innovation in these technologies. These technology-specific issues are taken up more fully in Chapter 5.

Finding 3-1: Market failures and nonmarket barriers for increasingly clean power technologies exist at all stages of the innovation process.

STRATEGIES FOR OVERCOMING THE OBSTACLES TO ACCELERATED INNOVATION

Some possible strategies for overcoming the obstacles at each stage of the innovation process are shown in Figure 3-4. The figure combines solutions for

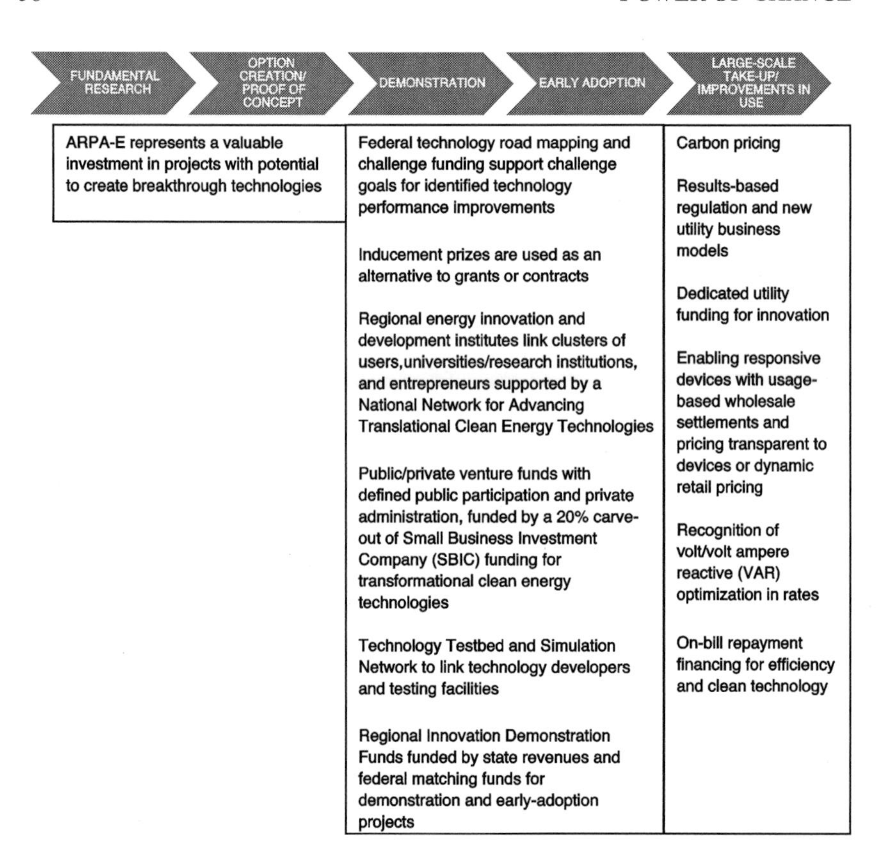

FIGURE 3-4 Obstacles at specific stages of the innovation system and candidate solutions.
SOURCE: Adapted from Lester and Hart, 2012, Figure 2.1, p.33.

the demonstration and early-adoption stages, as they are difficult to distinguish. While the proposed strategies are not comprehensive, they were selected to address the major obstacles in the innovation system while leveraging federal, state, regional, and private-sector capabilities and models with demonstrated applicability.

It is important to emphasize that public support needs to address all stages of the energy innovation process, not just fundamental research and improvements in use (Lester and Hart, 2012). Support for early-stage research will increase the rate at which new options are created but will have much less impact on the intermediate stages of the process, in which many of the greatest obstacles to innovation arise. Some strategies call for federal leadership to address gaps in the innovation system and enable new breakthroughs to advance to become market solutions, while others call for federal support for state and regional initiatives.

The case for strengthening the local and regional dimension of innovation policy is bolstered by the well-documented importance of local and regional innovation systems to economic development. Geographic proximity facilitates interactions among researchers, entrepreneurs, investors, potential customers, and others, and the development of local and regional entrepreneurial ecosystems is an important part of removing obstacles to innovation (Lerner, 2010). Moreover, as the dominant pattern of innovation has shifted away from the old model of closed, in-house corporate research laboratories toward more open innovation networks encompassing multiple companies specializing in different stages of the value chain, as well as universities and other public research institutions, proximity has become even more important to the innovation process (Ketels and Memedovic, 2008).

Although government policies cannot create these innovation networks, they can support their development in various ways. State and local governments and multistate entities are today supporting energy technology innovation through economic development and utility regulation policies, as well as through workforce development programs and programs designed to link entrepreneurs with local universities and research institutions (Lester and Hart, 2012). Early-stage cleantech companies need help developing their products or services; developing and proving their business models and strategies; building their teams; leveraging appropriate mentors and advisors; finding and connecting with customers and partners; and attracting capital with which to pilot, scale, and commercialize their technologies. Successful start-ups tend to cluster, as the concentration of these resources contributes significantly to venture development and a greater percentage of successful ventures. Some of the most cost-effective innovation acceleration mechanisms involve local, state, and regional governments, nonprofits, and public/private partnerships focused on building the necessary connections across a regional ecosystem to leverage regional innovation assets and economic competitiveness (Porter, 2001).

Strategies That Address Obstacles in the Early Stages: R&D and Option Creation/Proof of Concept

The Advanced Research Projects Agency-Energy (ARPA-E) is critical to the innovation pipeline because it was chartered by Congress specifically to address obstacles and market imperfections in the earlier innovation stages. The development of increasingly clean, low-carbon technologies that are both globally scalable and affordable will require exploration of a broad range of potentially transformational technologies. ARPA-E illustrates a governmental commitment to a focus on transformational innovation. The emphasis is on new technologies that go well beyond incremental improvements to provide potentially transformational breakthroughs. ARPA-E's goal is not to avoid risk, but to recognize risk and manage it to maximize the chances of big successes.[2]

Comparisons are often drawn between ARPA-E and the Defense Advanced Research Projects Agency (DARPA), after which it was modeled. However, ARPA-E and DARPA differ in three important ways:

- DARPA is funded at a level roughly 10 times higher than that of ARPA-E.
- While the Department of Defense is a likely customer for successful DARPA projects, the Department of Energy (DOE) generally is not a customer for ARPA-E projects. Finding funding and early-adoption customers is thus a more significant challenge in energy markets, especially at the proof-of-concept stage.
- At a point when the Department of Defense can support the continued development of technologies, including higher-cost technologies, that address national security risks, energy technologies must demonstrate an ability to achieve near-term commercial viability to attract private capital.[3]

Recommendation 3-1: DOE should direct funds to a broader portfolio of projects than will ultimately prove viable and should tolerate the inevitable failure of some experiments, while at the same time winnowing at each stage of the innovation process.

In addition to being essential to limit costs, downselecting at each stage would provide opportunities to identify at earlier stages of the innovation process technologies that are unlikely to succeed commercially (in their current

---

[2]Congress established ARPA-E with a broad mission "to overcome the long-term and high-risk technological barriers in the development of energy technologies." In addition, Congress established goals for ARPA-E: "(A) to enhance the economic and energy security of the United States through the development of energy technologies that result in—

(i) reductions of imports of energy from foreign sources;

(ii) reductions of energy-related emissions, including greenhouse gases; and

(iii) improvement in the energy efficiency of all economic sectors; and

(B) to ensure that the United States maintains a technological lead in developing and deploying advanced energy technologies." 42 U.S.C.S. 149 §§16538(b) and (c) (2016).

[3]The committee concluded that early indicators suggest ARPA-E generally is poised to produce a positive public return on public dollars invested. At the time this report went to press, however, another committee of the National Academies of Sciences, Engineering, and Medicine was close to completing a full evaluation of ARPA-E. The reader is referred to the report of that committee (expected to be released in late 2016) for a more detailed assessment of ARPA-E.

form). The most important objective would not be to avoid failure, but to ensure that failure is recognized, understood, and addressed without delay. This could be accomplished by ending funding for projects that failed to meet preset cost and performance improvement targets.

The committee recognizes that implementation of this recommendation, as well as others in this chapter, would require additional spending on innovation programs in a budget-constrained political environment. In some cases, existing funding could be redirected for this purpose, and budget commitments could be shifted from supporting the deployment of existing technologies or incremental improvements to existing options that would remain too expensive to the development of technologies that showed promise for becoming economically competitive. However, decisions regarding the sources for any additional funding would be the product of a political process that is beyond the scope of this study.

Still, there are actions that Congress and the Executive Branch could take to help ensure the accountability of public entities in a way that would sustain support for public participation in the early stages of innovation in energy technology—for example, creating firewalls between elected officials and program administrators to limit political influence and the perception of political influence on funding. Also helpful would be to give agencies legal authority to establish personnel policies that would ensure professionalism in program administration. For example, both DARPA and ARPA-E are allowed to hire key technical personnel on time-limited terms and empower them to suggest what projects to support. Thus both agencies are able to attract highly qualified personnel, often from preeminent research institutions and other notable organizations. Additionally, high levels of transparency (including reporting of successes and failures), proactive communications, and independent program reviews provide information needed to assess performance and maintain accountability.

Strategies That Address Obstacles in the Intermediate Stages: Demonstration and Early Adoption

While much policy attention is focused on the early and late stages of the innovation process, some of the most significant barriers to innovation occur at the intermediate stages of demonstration and early adoption. As innovations approach the point of commercialization, their capital requirements typically increase significantly, as does the importance of engagement with markets, customers, and private investors. As discussed previously, much of this activity takes place most effectively in local and regional energy markets and innovation systems, and especially at these intermediate stages of the innovation process, the federal government needs to augment its own leadership roles with support for regional, state, and local innovation initiatives.

Federal Sector-Specific Technology Road Mapping and Challenge Funding

Federal sector-specific road mapping and challenge funding developed with specific technology development milestones have been used effectively to drive private-sector innovation and investment, as well as DOE programs and grants in the proof-of-concept and demonstration stages. In other industries, projects with a clearly defined mission have been most successful at achieving the desired outcomes (Janeway, 2012).

DOE has used its expertise to analyze the technology readiness of specific energy technology categories (see Chapter 2 and Appendix D), and to develop road maps that consider targets for spurring innovation at the component and supply chain levels to meet levelized cost of electricity goals for each specific technology. These analyses and road mapping efforts are not aimed at addressing the new, disruptive breakthrough ideas that ARPA-E performers may consider. Rather, they are more focused on specific sectors that have understood product architectures and components, and in which improvements across every component of a deployed solution (including advances that reduce the cost of deployment) can be targeted for a significant combined improvement and competitiveness in energy markets.

An excellent example of this model is DOE's SunShot Initiative, whose mission is to make solar energy fully cost-competitive with traditional energy sources by 2020. According to DOE, "The SunShot Initiative aims to reduce the total installed cost of solar energy systems to $0.06 per kilowatt-hour (kWh) by 2020. Today, SunShot is about 70% of its way toward achieving the program's goal, halfway into the program's ten year timeline. Since SunShot's launch in 2011, the average price per kWh of a utility-scale photovoltaic (PV) project has dropped from about $.21 to $.11" (DOE, 2016a). Additionally, SunShot is funding research into next-generation solar technologies with "the potential to dramatically lower costs and/or increase efficiencies of PV module[s] beyond the SunShot targets of $0.50/W and 20%, respectively" (DOE, 2014d). SunShot has developed a detailed technology road map for every major PV component. The program is organized around a series of challenge solicitations that consider competitive proposals from companies, laboratories, and universities for R&D grants targeted to achieving these specific road map milestones. Amounts for most SunShot grants are between $100,000 and several million dollars, with varying degrees of matching funds required. In addition to funding, the SunShot Initiative includes working groups and conferences that are well attended by researchers and innovators across the solar PV sector, and serves as an accelerator of competitive ideas in all areas of solar-related innovation.

DOE's Quadrennial Energy Review and other DOE research already provide major portions of the research and road map planning necessary to consider initiatives similar to SunShot for other energy sectors and technologies. DOE could compile these research insights and develop road map challenge initiatives to align its own program areas and programs supported by the national laboratories with a clearly defined mission to support the timely development of affordable, scalable technologies that could effectively mitigate potential GHG impacts. In some areas, this might require redirecting DOE and national laboratory R&D programs toward the achievement of more ambitious cost and performance objectives. The funding for these challenges could be provided by a pooling of current R&D funds, including Small Business Innovation Research (SBIR) funds from DOE and other agencies, such as the Department of Defense, the National Science Foundation, the Department of Agriculture, and others, that are customers of or have expertise in the markets for those specific technologies. DOE could use road mapping and challenge funding to set targets and funding priorities consistent with a clearly defined mission for the timely development of technologies that could enable an affordable global transition to low-carbon energy resources.

Inducement Prizes

Inducement prizes are another way to accelerate certain types of innovation, and could be a valuable addition to the federal, state, regional, nonprofit, and private-sector increasingly clean energy innovation toolkit. Inducement prizes are particularly relevant at the proof-of-concept and early demonstration stages, and in cases when efforts such as sector-specific road mapping or the Quadrennial Energy Review have determined that a goal may require a stepwise change in performance or a novel integration of components and ideas from both known and unknown areas. DARPA has used comparable grand challenges to accelerate major advances in such fields as autonomous vehicle control. These prizes would be intended to incentivize new ideas at these early and middle stages, but prior to full-scale demonstration. If supported by expert judgment, prizes also could be used to help ensure the deployment of market-ready advanced technology at an appropriate price point.

Prizes can be an alternative or a supplement to grant funding and might be used in place of additional, less narrowly targeted inducements to promote learning by doing and learning by searching (discussed in Chapter 2). A variety of criteria have been used to determine the suitability of inducement prizes (NRC, 2007). Prizes tend to be most appropriate when the objective is clear,[4]

---

[4]In the demonstration and commercialization/early-adoption stages, contestants are demonstrating a proof of system (combinations of technologies that together represent a new system or application) that with definable additional steps could be commercialized and brought to market. At a stage at which the potential for commercial applications becomes apparent, prize contests are more likely to provide reputational benefits to successful participants and elicit third-party financing for contest participants. For prizes linked to commercialization and deployment, the prize (e.g., credible advance market commitments, deployment incentives with specified qualification requirements, or intellectual property rights buyouts) can ensure reasonable pricing and achieve such other conditions as may be necessary to facilitate broad adoption.

but when the path to achieving that objective is not. In such cases, the competition will invite alternative approaches, and the contest objective will serve as a surrogate for success in a competitive market. For the prize to work, however, clear criteria for victory must be established, and it must be possible to measure performance. The contest objective needs to be achievable in a reasonable amount of time (e.g., 2 to 10 years). In addition, prizes are most suitable when there are many potential contestants who could produce a winning solution—for example, when the prize captures the imagination of the public or within a field and can attract the participation of contestants or teams from diverse backgrounds, ideally including those that ordinarily might not participate in research grants or contracts.

Prizes have grown in consideration, authorization, and use in recent years. In response to a request from the National Economic Council, a National Academy of Engineering workshop assessed the potential value of federally sponsored prizes and contests in advancing science and technology in the public interest. The workshop's steering committee recommended that "Congress encourage federal agencies to experiment more extensively with inducement prize contests in science and technology" (NRC, 1999, p. 1). This recommendation reflected the following views of the steering committee:

> When compared with traditional research grants and procurement contracts, inducement prizes appear to have several comparative strengths which may be advantageous in the pursuit of particular scientific and technological objectives. Specifically, these include:
>
> - The ability of prize contests to attract a broader spectrum of ideas and participants by reducing the costs and other bureaucratic barriers to participation by individuals or firms;
> - The ability of federal agencies to shift more of the risk for achieving or striving toward a prize objective from the agency proper to the contestants;
> - The potential of prize contests for leveraging the financial resources of sponsors; and
> - The capacity of prizes for educating, inspiring, and occasionally mobilizing the public with respect to particular scientific, technological, and societal objectives. (NRC, 1999, p. 1)

The steering committee viewed inducement prizes as "a complement to the primary instruments of direct federal support of research and innovation—peer-reviewed grants and procurement contracts" (NRC, 1999, p. 1).

A 2007 National Research Council study on inducement prizes at the National Science Foundation (NSF) produced this finding:

> Inducement prize contests are clearly not well suited to all research and innovation objectives. But through the staging of competitions they are thought to have in many circumstances the virtue of focusing multiple group and individual efforts and resources on a scientifically or socially worthwhile goal without specifying how the goal is to be accomplished and by paying a fixed purse only to the contestant with the best or first solution. Inducement prize contests with low administrative barriers to entry can attract a diverse range of talent and stimulate interest in the enterprise well beyond the participant pool. (NRC, 2007, p. 1)

The National Research Council concluded that "…an ambitious program of innovation inducement prize contests will be a sound investment in strengthening the infrastructure for U.S. innovation" (NRC, 2007, p. 2). Further, it found that the area of "low carbon energy systems," among others, "has potential to yield one or more worthy prize contests" (NRC, 2007, p. 6).

An expansion of the number and size of government, private-sector, and public/private-sponsored prize contests has occurred over the last decade. Purses increased from $74 million in prize competitions with awards of more than $100,000 in 1997 to $315 million in such competitions in 2007 (McKinsey & Company, 2009). Prizes in the category of climate and environment increased from $6 million to $77 million, in science and engineering from $18 million to $88 million, and in aviation and space from $12 million to $88 million. Together these categories went from accounting for fewer than half of large prize competitions in 1997 to 80 percent by 2007 (McKinsey & Company, 2009).

In addition, there has been a recent expansion of federal authority to use inducement prizes, translating to valuable experience that can continue to be tapped in appropriate situations:

- Section 1008 of the Energy Policy Act of 2005 (EPACT) gives the secretary of energy authority to award cash prizes of $10 million for "breakthrough achievements in research, development, demonstration, and commercial application" that are related to DOE's mission. It also gives the secretary of energy authority to award "Freedom Prizes" of $5 million for innovations that reduce dependence on foreign oil.
- Section 105 of the America COMPETES Reauthorization Act of 2010 gives all federal agencies broad authority to conduct prize competitions and includes provisions for different aspects of prize

design, implementation, and oversight. In particular, this act authorizes the use of prizes for one or more of the following:
   - find solutions to well-defined problems;
   - identify and promote broad ideas and practices, and attract attention to them;
   - promote participation to change the behavior of contestants or develop their skills; and
   - stimulate innovations with the potential to advance agencies' missions.
- The America COMPETES Reauthorization Act of 2010 also allows agencies to accept funds for cash prizes from other federal agencies and the private sector; allows agencies to enter into agreements with private, nonprofit entities to administer a prize competition; and requires reporting of prize activity for each fiscal year.

Recommendation 3-2: The federal government, including DOE, should continue to expand the appropriate use of inducement prizes as a complement to patents, grants, procurement contracts, and other types of support for energy innovation.

An inducement prize for increasingly clean power and energy-efficiency technologies should follow certain criteria, consistent with the findings and recommendations of previous National Research Council studies on inducement prices (NRC, 2007). First, these prizes should take advantage of the development of additional energy technology road maps and the Quadrennial Energy Review led by DOE. In addition, sponsors, including DOE, should consult with experts, affected parties, and categories of potential participants in choosing prize topics and objectives. DOE should consult with experts regarding circumstances in which deployment prizes should be used to reduce economic welfare losses from monopoly pricing of patent rights or to supplement undervalued patent rights for low-carbon and other increasingly clean energy technologies, as well as the most appropriate designs for such deployment-related prizes (targeted deployment prizes, advance market commitments, cost/pricing conditions, or intellectual property rights buyouts).

Given the recent growth in energy-related prize competitions and the National Research Council's prior recommendations, DOE could undertake efforts to evaluate such competitions as was laid out for NSF in 2007 (NRC, 2007). These could include considering whether the desired technology might have been developed more quickly or a more effective version of the technology might have been developed if the structure of the competition had been different. The ongoing learning and experience from these efforts should be applied to prize competitions across DOE. DOE could also appoint a coordinator of innovation prizes in the office of the under secretary—a step comparable to what the National Research Council previously recommended for NSF—to manage the administration of prize competitions in conjunction with applicable program offices (NRC, 2007, p. 24). This office could coordinate and support consultation experts, affected parties, and potential contest participants; administer or contract for the efficient administration of prizes; and coordinate the evaluation of prize contests to identify lessons learned that could be used to improve future competitions.

In addition, given the global nature of climate change, the development of increasingly clean energy options should have a global component. The commitment of 20 nations to Mission Innovation[5] during the December 2015 United Nations Framework on Climate Change 21st Annual Conference of Parties (COP21) is aimed at accelerating global clean energy innovation, with the objective of making clean energy widely affordable.[6] Each of the participating countries will seek to increase governmental and/or state-directed clean energy R&D investment over 5 years. New investments will be focused on transformational clean energy innovations that can be scaled to address varying economic and energy market conditions.

These national commitments are linked to a private initiative—the Breakthrough Energy Coalition—supported by more than 20 institutional and wealthy individual investors. The Breakthrough Coalition was developed to "add the skills and resources of leading investors with experience in driving innovation from the lab to the marketplace." Its development was based on a recognition that "in the current business environment, the risk-reward balance for early-stage investing in potentially transformative energy systems is unlikely to meet the market tests of traditional angel or VC [venture capital] investors," and that "even the most promising ideas face daunting commercialization challenges and a nearly impassable Valley of Death between promising concept and viable product, which neither government funding nor conventional private investment can bridge" (Breakthrough Energy Coalition, n.d.). The coalition is creating a network of private capital to accelerate early investments in a broad range of reliable, affordable energy technologies that do not produce carbon.[7]

---

[5]See mission-innovation.net.

[6]The participating countries account for more than 80 percent of current clean energy R&D and include the 12 largest national economies (based on 2015 gross domestic product). The countries making the initial pledges to Mission Innovation are Australia, Brazil, Canada, Chile, China, Denmark, France, Germany, India, Indonesia, Italy, Japan, Republic of Korea, Mexico, Norway, Saudi Arabia, Sweden, the United Arab Emirates, the United Kingdom, and the United States.

[7]For additional information on inducement prizes, see Adler (2011), Brunt et al. (2011), Dalberg Global Development Advisors (2013), Davis and Davis (2004), Kay (2011, 2013), Kremer (1998), Newell and Wilson (2005), Nicholas (2011, 2013), Williams (2012); see also https://www.challenge.gov.

Finding 3-2: The development of affordable, reliable, widely available increasingly clean energy technologies that can be rapidly deployed in both developed and developing economies will be enhanced by public/private collaborations and international partnerships.

Regional Energy Innovation and Development Institutes (REIDIs)

Electricity markets are regional, and different regions have differing energy resources, fuel- and technology-specific R&D capabilities, and regulatory and market structures that create varying incentives and opportunities for new increasingly clean energy technologies. Public/private partnerships to accelerate new market development and evolve regulations for new entrants are being formed in clusters and regions. The United States has a significant number of emerging increasingly clean energy clusters, as well as regional initiatives designed to connect the region's innovation resources with early-stage ventures. Federal policy for energy innovation can take advantage of the strengths of these regional differences in innovation conditions, capabilities, and priorities.

A local, state, or regional public/private partnership—what the committee refers to as a regional energy innovation and development institute (REIDI)—could be created to help spur the development of both early-stage innovations and innovations that show appropriate promise. This type of regional institute structure would complement federal innovation agencies and programs such as ARPA-E, SunShot, and DARPA. It would extend support after proof of concept through a technology's optimization, iterative prototyping, piloting, testing, and readiness for commercial demonstration. It would help accelerate the movement of technologies through the middle stages of the innovation process by developing institutional capabilities specifically tailored to the earlier-detailed obstacles to development commonly faced by energy technologies. And with input from potential users, it should be able to address potential barriers to market adoption.

The committee estimates that an optimal annual budget for a REIDI would range from $2 million to $40 million and would be linked to scale (whether it covered a metropolitan area, a state, or a multistate region), scope (whether it had a focus on a small or large number of technologies or markets), and stage of development (whether it was a nascent organization focusing on early-state proof of concept and business validation or had the capability and partnerships to accelerate innovations through development to reach commercial demonstration readiness). The amount of funding support provided to an individual project would depend on the innovation stage. An appropriate scale for early start-ups would be $50,000 to $500,000, while more advanced ventures might need between $500,000 and $5 million, even if they already had private funding, to demonstrate commercial potential quickly.

A network of these regional institutes would facilitate access where capabilities already exist. Where capabilities may not yet exist to meet anticipated needs, networked institutes could help identify likely development needs for promising technologies and fund or plan and create the support capabilities, physical infrastructure (where applicable), and translational relationships often needed for four activities that can accelerate innovation in energy technology:

- simulating;
- testing;
- accelerating (or paralleling) the development of standards and specifications for related physical, information, and/or control architectures and implementation or integration templates; and
- certifying products using appropriate proof-of-system test protocols.

As new energy technologies move beyond laboratory research to prototype development and beyond, they require funding, services, expertise, and market connections to develop a commercial prototype product and prove its basic market viability. Individual institutes could develop general capabilities to expedite the movement of technologies through the middle stages of the innovation process. If national laboratories participated, they could support a translational approach to accelerating development by providing core capabilities for some regional institutes in partnership with other institutions and potential customers. A recently initiated example of national laboratory involvement in promoting clean energy innovation is Cyclotron Road, housed at Lawrence Berkeley National Laboratory. Where the regional institutes created new simulation and testing capabilities, they would also be participants in the Technology Test Bed and Simulation Network described below, along with other organizations. Some of the activities supported by the regional institutes, such as standards development, would be coordinated at the national level through the proposed National Network for Advancing Translational Clean Energy Technologies (NNATCET), also detailed later.

While ARPA-E is an important funder for technical development and derisking for some potential breakthrough technologies, ARPA-E by itself is not in a position to address many of the business-related risks and is not designed to support achievement of all the milestones for commercialization of promising innovations. However, some of the resources most actively and successfully supporting the acceleration and development of innovation at this early stage tend to come together in the country's regional innovation clusters,[8] presenting

---

[8]The U.S. Cluster Mapping Project, led by Professor Michael Porter, Institute for Strategy and Competitiveness, Harvard Business School, and built with funding support from the Department of Commerce's Economic Development Administration (EDA), lists hundreds of organizations in one or several categories addressing early-stage energy

an important opportunity for the federal government to follow and build on state and regional initiatives.

REIDIs would be energy-specific venture development organizations (VDOs) that would add several capabilities specific to the energy innovation system and its needs. As defined by the Department of Commerce,[9] a VDO is a "business-driven, public or nonprofit organization that promotes regional growth by providing a flexible portfolio of services, including: assisting in the creation of high-growth companies; providing expert business assistance to those companies; facilitating or making direct financial investments; and, speeding the commercialization of technology." VDOs sit at the center of a regional network of universities, laboratories, technology development organizations, incubators, accelerators, state programs, entrepreneurial networks, industry organizations, private capital communities, and other partners. REIDIs, as energy-specific VDOs, would bring together and apply the regions' energy innovation capabilities to develop and accelerate those projects and ventures with the most promising new energy technologies through early-stage proof of concept, pilot, and commercial readiness.[10] There are dozens of REIDI-like organizations (or nascent efforts to form such organizations) across the United States. Most of them are modestly funded at several hundred thousand to the low millions of dollars per year, and most are only a few years old.

REIDIs and their partners would in some ways complement ARPA-E, but could become involved with promising innovation projects at the earlier formative stages, and could remain engaged beyond ARPA-E early technical support to help innovations reach viability for commercial demonstration. Modest levels of public support provided when a promising technology is completing a proof-of-concept demonstration or further testing can have a significant impact on the subsequent ability to access venture capital and generate revenue. Small high-tech firms have been able to compete for limited awards from DOE's SBIR program. A recent working paper suggests that a small grant of $150,000 approximately doubles the recipient firm's chance of subsequently obtaining venture capital, leads on average to the award of an additional 1.5 patents within 3 years, and increases the probability that the firm will earn revenue and either be acquired or transition to an initial public offering. At this stage, modest public support can reduce uncertainty and risk

---

innovationdevelopment(http://clustermapping.us/organization-type/cluster-organizations-and-initiatives;http://clustermapping.us/organization-type/innovation-and-entre-preneur-ship-centers; and http://clustermapping.us/search/site/Regional%20Energy%20Innovation).

[9]VDOs are defined by the EDA Initiative on Regional Innovation, http://regionalinnovation.org/content.cfm?article=fundamental-characteristics.

[10]VDOs and the partners they assemble are represented by the EDA support to the Regional Innovation Acceleration Network, http://regionalinnovation.org/content.cfm?article=about-rian.

without crowding out private capital (Howell, 2015).[11] By diversifying the evaluation of promising innovation projects that are in their formative stages and connecting such projects to potential partners, the modest financial support available from REIDIs could leverage greater access to candidate projects and complement DOE's SBIR program.

Beyond funding, REIDIs would serve an important role in the middle stages of innovation because they would focus on the following:

- Innovation acceleration—providing support and modest innovation funding and services to promising projects; supporting technology development, but with an additional focus on leveraging local resources of mentors, customers, investors, entrepreneurs, teams, and early-adoption market connections to help new innovations prove their business and economic value.
- Market and cluster research—focused on regional market and cluster potential, and seeking to connect cluster and market needs to innovators and innovation concepts to rapid market feedback.
- Access to technical resources—including developing and supporting access to a regional network of test beds and simulation modeling laboratories, and coordinating the leveraging and growth of test resources with the recommended national Technology Test Bed and Simulation Network (as described below).
- Ecosystem development—programs designed to develop and leverage regional innovation resources (including mentors; experienced entrepreneurs; customers; partners; R&D facilities, including national laboratories; test sites; capital providers; educators; and team members); initiatives to invest in regional assets for incubation, acceleration, R&D, business development, mentoring, and education.
- Policy and regulatory alignment—initiatives to change the policy and regulatory structures to eliminate obstacles and implement market signals for emerging categories of increasingly clean technologies.
- Smart deployment—initiatives to stimulate market demand, siting processes, customer and innovator connections, business development connections, and early-adoption customers for emerging increasingly clean technologies (including the public sector as customer).

While the best examples of existing REIDI-like entities have benefited from collaboration and modest, pilot funding from DOE (2011b),[12] the

---

[11]Professor Howell's results came from analyzing outcomes for small firms receiving SBIR Phase I grants.

[12]DOE's Innovation Ecosystem Development Initiative awarded more than $5 million to research institutions and universities across the nation to "nurture and mentor clean energy entrepreneurs."

Department of Commerce (EDA, 2011),[13] and the Small Business Administration (Regional Innovation Cluster program),[14] these regional efforts are still relatively disconnected from federal partnerships, and their funding and operating levels are below what is required for sustainability. Additionally, many other regions of the country have the potential to house regional increasingly clean energy innovation clusters, but have lacked a model and formative support for initial programs and partnerships.

The most recent example of federal recognition of and support for REIDI-like entities (albeit at very modest funding levels) is the announcement of DOE's National Incubator Initiative for Clean Energy. DOE competitively selected three regional partnerships from the Midwest, Texas, and California to receive federal support "to run innovative programs with commercialization services for startups including mentorship, business development, capital access, and testing and demonstration" (DOE, 2014c). DOE also made funds available to sponsor two institutes of the National Network for Manufacturing Innovation. As of this writing (2016), DOE has announced that it will provide $70 million to support an institute that will "enable the development and widespread deployment of key industrial platform technologies that will dramatically reduce life-cycle energy consumption and carbon emissions associated with industrial-scale materials production and processing through the development of technologies for reuse, recycling, and remanufacturing of materials" (DOE, 2016b). DOE also has established another institute, PowerAmerica, to support the development, demonstration, and deployment of advanced power electronics.[15] PowerAmerica[16] primarily creates technology road maps and funds demonstrations, with a stated aim of improving device performance and reducing the perceived risk of adoption by industry.

As these examples show, REIDI-like entities may be the lead organizations supporting projects at the earliest postresearch stage to help projects reach proof of concept and market need. These entities can partner with sources of private capital, corporate investors, and others that often invest after proof of concept. The innovation resources and partnerships they assemble can help ventures with private funding to reduce the capital and time requirements for prototyping, testing, business/market/economic assessments, connections to potential customers and markets (leveraging the regional clustering of utilities, engineering and construction firms, energy service companies, manufacturers,

---

[13]The i6 Green Challenge in 2011 made $12 million available to six teams across the country "with the most innovative ideas to drive technology commercialization and entrepreneurship in support of a green innovation economy" and new jobs.

[14]For the locations of the 58 federally funded clusters, see http://www.sba.gov/sba-clusters.

[15]Power electronics can reduce losses of electricity during its transmission and distribution, enable greater grid penetration of intermittent increasingly clean power technologies, and increase the energy efficiency of semiconductors.

[16] See www.poweramericainstitute.org.

etc.), and other venture development milestones to reach commercial demonstration readiness. Many investors at the venture and similar stages lack the technical capability to assess which energy technologies hold the greatest potential. Because of the technical expertise they offer, REIDI-like entities can help private investors do exactly this and thereby lower the technical risk.

Given the regional diversity of the U.S. economy, these entities can have a variety of organizational forms and priority areas of focus. They can be a single nonprofit, or more of a partnership, network, or consortium that brings together many of the incubator, accelerator state program, academic, laboratory, business, entrepreneurial, capital, market, and other regional energy innovation resources in their region. They may be focused on specific targeted technology and market intersections, based on the characteristics and assets of their regional economies. Alternatively, in regions such as the Northeast and California where research, industry, markets, and expertise cover many technology and market segments, they may be more broadly based.

> **Finding 3-3:** REIDIs could help sustain the development of promising technologies and ameliorate funding gaps associated with achieving intermediate milestones as technologies move toward commercialization, including gaps that are not covered by federal programs such as ARPA-E.

The funding for these REIDIs could come from an equal match of federal and regional funds, with the regional funds derived from state and private-sector sources, including potential allocation of electricity sector systems benefit charges or other funds allocated to accelerating increasingly clean energy innovation. The latter funds could include new electricity system charges similar to the Network Innovation Competition funding allocation that is a key part of the new U.K. regulatory model RIIO (Revenue set to deliver strong Incentives, Innovation and Outputs) (see the section on "Dedicated Innovation Budgets and Roles for Utilities" in Chapter 6). Federal funding might over time require more than a 1:1 regional match to encourage multiple regional funding partnerships. However, federal funds would need to be flexible enough to support the majority of the operating capital required to launch new REIDIs. Federal funds would also need to be flexible enough to support a REIDI's general operations, enabling the majority of regional funds to be deployed for innovation programs and support for promising entrepreneurial ventures.

The combined budgets for an assortment of REIDIs spread across the United States might eventually reach $250 million, with scaling to this level over a 5- to 10-year period. Modest federal funding support would be critical to incentivize states, regions, state regulators, and private companies to come together to provide matching regional funds for these institutes. The federal government could consider creating a dedicated office—likely within DOE—to help coordinate and provide support to REIDIs and a mechanism for sharing best practices across the institutes through the proposed NNATCET.

The NNATCET could be a joint operation across DOE, reporting directly to the secretary of energy and shared across the major technology offices, such as Electricity Delivery and Energy Reliability, Energy Efficiency and Renewable Energy, Fossil Energy, and Nuclear Energy, plus the Office of Science. Its annual budget might start at $50 million and grow over 5-10 years to reach $150 million. Approximately $125 million would be allocated as matching funds to the REIDIs, with the remaining $25 million supporting NNATCET operations, development and sharing of best practices, related events and programs, and the flexible ability to seed regional innovation initiatives with the potential to become regionally supported REIDIs. The NNATCET also would have the flexibility to support new initiatives designed to address gaps in the REIDI network.

At a minimum, the NNATCET would act as a source of structured support for the REIDIs, verifying their quality and output and providing financial support for their operating capabilities; investments in their regions' innovation resources; and specific energy technology projects at the prototype, pilot, demonstration, and field test stages. The NNATCET might also be the logical home for a broader energy innovation-enabling initiative that could facilitate the movement of technologies from laboratory to market by supporting dispersed components of the larger energy innovation system. In this capacity, the NNATCET would additionally promote collaboration and resource sharing among the regional organizations to facilitate knowledge transfer and guard against unnecessary redundancy, offer a source of streamlined support for navigating regulatory processes and for updating relevant regulations and market policies, and provide a checkpoint for the dissemination of all federal financial support to these organizations.

Public/Private Venture Funds

Although venture capital accounts for a modest fraction of total investments in increasingly clean energy technology, it plays a critical role in the innovation system. Venture capital contributes to the development of prototype and pilot-scale technologies and new business ventures. In 2001, venture capital investment in clean energy technology totaled on the order of $500 million, representing about 1 percent of total venture capital investments in U.S. companies. By 2011, investments in U.S. cleantech companies had peaked at about $7.5 billion, or 25 percent of total venture capital investment in U.S. companies. Since that peak, venture capital investments in increasingly clean energy technologies have declined, although they may have stabilized in 2015 at around half the 2011 peak (BNEF, 2016; Clean Edge, 2014).

Observers of venture capital and equity investment markets explain this decrease by pointing to the mismatch between the business model of venture capital funds and the needs of increasingly clean energy entrepreneurs. Venture capitalists invest in early-stage companies and plan to exit after 4-7 years, whereas energy technologies typically require much more time before an investor can exit (Schwienbacher, 2008). Similarly, venture capitalists typically can make investments at the level of $500,000 for seed funding to about $2.5 million for initial investment in growth companies. The capital needs for energy technologies, however, can be significantly greater just to complete proof of concept. Moreover, many energy technologies require significant capital to evaluate technical performance and thus reduce technical risk. Venture capital firms, on the other hand, are generally much better suited to assessing market opportunity and operational risks for a given venture (Madison Park Group, n.d.).[17]

It is not surprising, then, that venture capital typically will not be the lead source of financing for demonstration and early-adoption activities. At these stages, project debt and other demonstration financing mechanisms are needed, including the Regional Innovation Demonstration Funds proposed in this report (discussed in the next section). However, venture capital can serve as the lead source of capital in the early stage of innovation if providers of new venture capital funds are able to adjust their risk profile, their capital return expectations, and the life of their fund to be appropriate for cleantech early-stage opportunities. In this section, the committee discusses the potential benefits of an expansion of the Small Business Administration's (SBA) Small Business Investment Company (SBIC) program for cleantech early-stage public-private venture funds to address this issue.

Through the SBIC program, the federal government has been a provider of matching funds to catalyze new privately managed investment funds for important sectors with private-sector capital gaps. The SBIC program has been in existence since 1958. Since then, SBA has licensed more than 2,100 SBIC funds that have invested more than $67 billion in total, roughly 64 percent being private capital, in American small businesses (SBA, 2016c).[18] These include early investments in information technology companies such as HP, Apple, and Intel. However, there have been very few SBIC-licensed funds for early-stage cleantech innovation.

The topic of structured public participation in privately administered venture capital funds is addressed by Lerner (2009, 2010). He focuses on how to stimulate venture capital formation generally, but his analysis may provide useful lessons for energy innovation specifically. Lerner argues that governments can incentivize private investment at stages typically attractive to

---

[17]According to the venture capitalists and other energy innovation financiers who attended the committee's February 28, 2014, and April 8, 2014, workshops, their experience was that the venture capital business model's available capital and timelines for investment were mismatched, often greatly, with the needs of firms developing increasingly clean energy technologies.

[18]SBICs have made more than 166,000 investments.

venture capital. Common features he identifies as salient to attracting private capital include

- government matching funds provided to an investment fund;
- a buy-back or similar right whereby the private investors can buy out the government's investment at a predetermined interest rate; and
- no government involvement in running the fund, with the government acting instead as a "limited partner" of sorts.

Lerner underscores that the use of matching funds to determine where public subsidies should go is a central feature of public participation in privately administered venture capital funds, one also used by SBIC. While there are many examples of state-supported venture capital funds and a leveraging of local and regional public/private capital structures, the success of SBIC funds in addressing early-stage capital gaps offers a valuable model for the establishment of a range of new cleantech early-stage venture funds to address the option creation/proof-of-concept stage.

More recently, the SBIC program has added a new category of qualified SBIC funds focused on "impact investments." SBA has defined eligible impact investment categories as including start-ups through a linkage with the goals of the Start-Up America Initiative (SBA, 2016a), and has separately defined clean energy as a "sector-based impact investment" area to encourage the use of existing and new SBIC funds to address these investment areas (SBA, 2016b). To help close the important gap in early-stage venture capital for increasingly clean energy start-ups developing new innovations, SBA could be directed to set a goal of creating $1 billion in new venture capital funds focused on early-stage increasingly clean energy technologies.

A fund focused on clean energy could be created if SBA were to phase in a "carve-out" of current SBIC funding allocations, aiming for 20 percent of current SBIC commitments, or $440 million in federal funds out of the current $2.2 billion/year SBIC allocation. Doing so would allow SBA to license dedicated cleantech funds that would be focused on directing significant portions of capital allocation to early-stage innovation, that would have lives of longer than 10 years to better match cleantech commercialization timelines, and that would involve fund managers and private capital limited partners with demonstrated expertise in cleantech. Further, SBA could review its regulations defining allowable clean energy sector-based impact investments to ensure that they would support the development of increasingly clean energy technologies, including early-stage start-ups, that could result in the timely development of affordable, scalable technologies capable of effectively mitigating potential

GHG impacts.[19] Given that the current distribution of private cleantech venture capital is weighted toward later-stage investments, this allocation of existing early-stage SBIC funds could have a significant impact in increasing the availability of early-stage capital for cleantech ventures.

In addition to filling funding gaps, the committee notes that market entry is a key factor in spurring innovation (Lockwood, 2013). Therefore, public policy needs to encourage new entrants into energy markets, including both new firms and established firms from other sectors (Lester and Hart, 2012). Historically, entry and innovation have been encouraged by government support for and sponsorship of competitions for funding for research at the technological frontier that have open specifications any supplier could meet.

Technology Test Bed and Simulation Network

Increasingly clean energy innovations would benefit substantially from a national Technology Test Bed and Simulation Network. One of the most significant challenges for developers of new cleantech innovations is to find partners and resources for effectively testing their innovations, or to avoid some of the cost and time of expensive testing with appropriate simulation systems.[20] Opportunities to connect new innovations to uniquely relevant test sites and simulation modeling resources could provide value by lowering the cost and time requirements for testing and development, as well as by confirming the expertise of validated external assessments. The needs of innovators vary widely—from simulation modeling and testing laboratories, to materials and component laboratory testing, to small pilot testing, and eventually to full-scale demonstration projects. A network of test beds and simulation laboratories would be instrumental in accelerating the development of increasingly clean energy technologies.

Such a network would provide streamlined identification of and access to new and existing federal, state, regional, and private testing resources; simulation modeling and testing laboratories; and preconfigured test sites. This network would also have funds with which to analyze gaps in the national network, and run solicitations and provide partial funding for the development of new test bed and simulation resources that could cost-effectively accelerate the testing and development of important technology categories.

This network would also work closely with the proposed Federal Technology Road Mapping and Challenge Fund (discussed above) to align test and simulation assets with those projects. The specific road map milestones and

---

[19]Relevant energy savings activities are defined in detail in the "Definitions" section of the SBIC program regulations (13 CFR 107.50), http://www.gpo.gov/fdsys/pkg/CFR-2013-title13-vol1/pdf/CFR-2013-title13-vol1-sec107-50.pdf.

[20]This need for testing and demonstration is particularly acute for nuclear power and carbon capture and storage technologies. See Chapter 5 for a more detailed discussion of the challenge.

challenge grant opportunities would benefit from linkage to test sites that were preconfigured for cost-effective testing (and possibly certification) of technology performance. These include current DOE-funded test beds, such as the National Wind Technology Center and the Pacific Northwest National Laboratory (PNNL) Smart Grid Test Bed, as well as national laboratory user facilities, many of which have relevant capabilities but can be difficult to access. They also include other relevant federal resources from the General Services Administration and, most important, from the significant Department of Defense facilities in the United States and potentially worldwide. This is an area for international partnering to leverage assets across the globe. DOE partnerships already exist with China, Canada, Europe, and various developing economies. In some instances, federal support would be required for the development of new simulation capabilities and testing facilities.

To make it easier for companies with new innovations to find and access this Technology Test Bed and Simulation Network, as well as for new test sites to market their capabilities, DOE could fund the network to manage a portal and clearinghouse for individual test beds and to connect the regional test bed networks across the country, including those associated with the proposed REIDIs.

Finding 3-4: Developers of technologies in the demonstration and early-adoption stages face technological issues in determining how to translate their proofs of concept into commercial products:

- Data from larger test beds and iteration with advanced simulation models can accelerate the resolution of these issues.
- Different technologies will have different simulation and testing requirements. In some cases, these capabilities may already exist or could be developed with modest government support.
- In other cases, such as the development of nuclear test beds (see Chapter 5 for details), substantial government investments could be required.

Regional Innovation Demonstration Funds (RIDFs)

The financing of energy technology demonstration and early postdemonstration projects is challenging. Full-scale demonstrations of innovative central station electricity-generating technologies and new kinds of manufacturing facilities for distributed technologies are commonly billion-dollar-scale projects, and typically carry significant technology, market, and regulatory risks. Even relatively small-scale innovations frequently require large, expensive projects to demonstrate the system-level impacts of their deployment at infrastructural scale. Venture equity funds are structured to finance high-risk technology development activities, but not major, billion-dollar-scale projects. Project financiers are structured to finance large assets but not to take on the risks of technology scale-up. Regulated electric utilities devote a tiny fraction of their revenues to R&D—far less each year than the cost of even a single billion-dollar project. State regulators are focused on keeping short-term electricity costs down and tend to discourage investments in new technologies, even those that promise to stabilize and reduce power costs in the longer run, if the initial cost is higher than that of incumbent technologies.

Attempts to fill this financing gap with federal funding alone face significant obstacles. Some initiatives, such as proposals to establish a federal infrastructure bank, are not designed to address projects with significant technology risk. Other, more targeted proposals, such as those to create a federal Clean Energy Deployment Administration or a federal demonstration corporation, have not progressed. DOE's loan guarantee programs are valuable resources for energy innovators, but loan authorities are modest compared with the scale of the need. Several states have launched "green banks" or clean energy financing authorities, drawing on a range of funding sources that include federal and state grants, bond issues, on-bill repayment mechanisms, and state ratepayer surcharges. For the most part, however, these initiatives are focused on financing the deployment of proven, commercially available technologies with low technology risk.

The committee proposes a new, decentralized strategy for financing energy technology demonstration, early adoption, and scale-up projects, with an enhanced role for states and regions and a new kind of partnership among the federal government, the states, private innovators, and investors. The public funds would be drawn primarily from state-level electric power system public benefit charges or from state and regional carbon mitigation programs, such as the California cap-and-trade program and the Northeast Regional Greenhouse Gas Initiative. The state funds would be augmented by supplementary federal grants to incentivize the creation of regional funding pools and partnerships (Lester and Hart, 2015). The new funding mechanism would specifically target projects designed to demonstrate the performance of potentially transformative energy technologies at commercial scale.

The creation of a network of Regional Innovation Demonstration Funds (RIDFs), staffed by experienced professional technology and project investors, would help reduce the costs and risks and increase the volume of private financing for the intermediate stages of the energy innovation process. The governors of RIDF member states would appoint the members of the RIDF governing board. This arrangement would create new opportunities for regional differences in energy innovation needs and preferences to be expressed at the demonstration stage and would give states a direct stake in innovation outcomes.

RIDFs would be staffed by experienced professional technology and project investors and would manage the regionally aggregated funds. RIDFs could provide multiyear grants to selected projects. These grants would augment private investments in first-of-a-kind commercial-scale demonstrations and "next few" post-demonstration projects with significant technology and/or regulatory risk.

To receive RIDF funds, projects would first have to be certified by an independent Energy Innovation Board comprising individuals publicly acknowledged as authorities in the fields of energy and environmental science, engineering, economics, manufacturing, and business management. Members would be appointed by the secretary of energy. The board also would be able to hire consultants with special expertise to assist on specific matters. The board's role would be to certify that a project would contribute to the public goal of creating cost-competitive, scalable technology options for reducing GHG emissions. Specifically, certification would be based on the potential of the project technology to achieve significant reductions in carbon emissions at a declining unit cost over time and at delivered energy costs competitive with those of high-carbon incumbent energy systems. Certification would be granted only for a limited period, and would be withdrawn at the end of that period if progress proved too slow or if public support were no longer necessary. The Energy Innovation Board would also make recommendations to the federal grant-making authority (probably DOE) regarding incentive payments linked to the board's annual evaluations of the overall performance of the RIDF project portfolios.

Projects precertified by the Energy Innovation Board as contributing to the public interest would seek grants from the RIDFs to augment the private financing assembled by the project team. Project teams could include technology vendors, power generators, transmission and distribution utilities, third-party energy service providers, and national laboratories and universities. Proposers would seek RIDF funding not as their primary source of finance but as a means of lowering the costs and risks of their own investments. The RIDFs would evaluate project proposals partly against standard commercial and financial criteria, including the strength of the project team, the quality of project management, and the extent of self-funding by the proposers. Most important would be the potential of the proposed project to contribute to the reduction of carbon emissions. The most attractive projects would be those with the greatest potential to stimulate major future reductions in carbon emissions while also delivering affordable, secure, and reliable energy services.

Examples of such projects could include demonstrations of integrated carbon capture, transportation and storage systems at full-scale coal- and gas-fired power plants and in different geologies, small modular light water or advanced nuclear reactor projects, grid-scale electricity storage integrated with utility-scale solar or wind systems, and next-generation offshore wind projects. Other eligible projects would include demonstrations of advanced grid infrastructure technologies; community-scale demonstrations of grid-integrated distributed battery storage using electric vehicles; and test beds for next-generation distribution systems with advanced demand-management technologies, microgrids, distributed generation, and dynamic and differentiated pricing schemes.

Projects selected by the RIDFs would receive direct multiyear grants, with end-year funding tied to performance. Alternatively, RIDF funds could be used for customer rebates, subsidized loan programs, credit support for power purchase agreements (PPAs), or other arrangements designed to promote user engagement with the new technology. As a condition of making a grant, the RIDF would acquire a modest equity position in the project, whose ultimate value would depend on the outcome of the project and the subsequent market potential of the project technology.

Over time, a national network of RIDFs might emerge. Certified projects could be proposed to one or more RIDFs for funding, providing more opportunities for new entrants to gain support for their ideas. The RIDFs could operate independently or could co-invest with each other. Over time, some specialization of the RIDFs in technology areas of particular interest to their regions might occur (e.g., offshore wind in the Northeast, nuclear in the Southeast, carbon capture in the Midwest).

RIDFs would likely first be established in parts of the country where there is already a strong commitment to innovation and to interstate collaboration, and where there is existing state-level funding. In these locations, federal matching funds would create incentives for additional state funding. Elsewhere, federal funds would incentivize the introduction of state funding and the creation of new regional partnerships.

The federal matching funds would also be used to encourage effective RIDF investing by rewarding RIDFs whose project portfolios were ranked highly by the proposed independent federal Energy Innovation Board. The board would conduct annual reviews of RIDF portfolios, ranking most highly those combining strong representation of high-potential projects with prompt winnowing of failing projects.

Distribution of the federal matching funds to the RIDFs could be administered by DOE or by a separate, dedicated agency.[21] The committee estimates that at steady state, an RIDF network covering half the country could

[21]Such an agency would have some similarities to a proposal made some years ago to establish a federal Energy Technology Corporation to manage and select technology demonstration projects (Deutch, 2011). In this case, however, the federal agency would not be selecting and managing specific projects, but providing funding at the portfolio level to regional entities that would in turn be providing grants to privately managed demonstration projects. The structure proposed here is also similar to the proposal for Regional Innovation Investment Boards, State Energy Innovation Trusts, and a Federal "Gatekeeper" that would certify projects presented to the regional boards (Lester and Hart, 2012).

lead to the deployment of up to $13 billion/year in public and private funds for demonstration and early post-demonstration projects, with federal cost-matching outlays to the RIDFs accounting for as much as $2 billion/year of these expenditures (see below).

Public benefit charges would be one potential source of state funds for RIDFs. Today, about 30 states have implemented power system public benefit charges, with the revenue being used primarily to fund energy-efficiency and renewable energy projects and low-income assistance and weatherization programs. The charges range from less than 0.05 mills per kilowatt hour (kWh) in North Carolina to nearly 5 mills per kWh in California. Together, these charges produce revenues of $3.5-4 billion per year, and the average increase in electricity costs in the affected states is 2.1 percent (DOE, 2010).

Initially, only a few states might be willing to redirect existing public benefit charges to innovation financing or to implement new surcharges for this purpose. Over time, encouraged by federal matching funds, more states would likely participate, and some states might opt to use funds from other sources, such as state or regional carbon cap-and-trade or taxation schemes.[22] A dedicated 1 mill per kWh electricity surcharge (adding about 1 percent to the average U.S. retail price) applied to U.S. retail electricity sales would generate roughly $3.7 billion per year, and might leverage up to twice that amount in private investment funds. A steady, predictable funding stream of more than $10 billion per year in public and private funding dedicated to financing demonstration and "next few" post-demonstration projects—enough to launch several new such projects each year—would be large enough to have a major impact on the nation's energy innovation challenge. The magnitude of the needed federal funding is uncertain, but assuming that 50 cents of federal matching funds would be required to induce each new dollar of state funding, the federal funding requirement might start at, say, $200 million/year and would eventually grow to about $1.8 billion/year for an RIDF network covering half the country and deploying a total of $13 billion/year in public and private funds.

The regionally based public funding mechanism proposed here would have several advantages over current practice. It would create a large, dedicated funding stream for a critical part of the U.S. energy innovation system—full-scale demonstration and early-adoption projects—that has to date been chronically underresourced. RIDF grant making would not have the stop-and-go pattern that is typical of the annual federal appropriations process, and would generate the steady, predictable supplementary funding that is needed for multiyear private project investment commitments. Putting RIDF project selection decisions in the hands of technology investment professionals would make public funding responsive to market needs and the latest technological information, while the public interest would continue to be strongly represented

---

[22]If eventually implemented, Environmental Protection Agency 111(d) regulations might encourage the introduction of more such schemes at the state and regional levels.

by the Energy Innovation Board. The new mechanism would also create opportunities for the expression of regional differences in energy innovation needs and preferences at the demonstration project selection stage, and would give states a direct stake in innovation outcomes. Finally, the mechanism would introduce multiple levels of competition into the selection of demonstration projects.

> **Finding 3-5:** Regional efforts that leverage regional energy markets and initiatives by states, universities, entrepreneurs, industry, and others can complement federal actions to help bridge funding and commercialization gaps.
>
> **Finding 3-6:** Funding and commercialization gaps for innovations in energy technologies tend to be most acute in, and most closely associated with, the early to intermediate innovation stages.
>
> **Recommendation 3-3: The federal government should provide funding and expertise to leverage regional opportunities and expertise in order to spur innovation in increasingly clean energy technology.**

Examples of this funding and expertise include a number of examples discussed throughout this chapter. For example, two key strategies for addressing obstacles at the proof-of-concept and demonstration stages of increasingly clean energy technology innovation include establishing a network for advancing translational clean energy technologies to support the proposed REIDIs and allocating additional funds within the SBIC program to create new venture capital funds focused on long-term investment in early-stage increasingly clean energy technologies. Additionally, two key strategies for addressing obstacles at the intermediate stages of increasingly clean energy technology innovation include linking technology test beds and simulation laboratories into a network and providing expertise and matching funds for regionally based, competitive public/private funds that would invest in demonstration projects.

Solutions That Address Barriers at the Final Stage: Large-Scale Take-up/Improvements in Use

Carbon Pricing

A national policy on carbon pricing would primarily have the effect of accelerating improvements in those technologies and business models that are already well developed, and would have less effect on early-stage development.

Absent a price on carbon and with uncertainty about the timing and magnitude of action, innovators have little incentive to invest in technology for reducing GHG emissions. Following passage of the Clean Air Act of 1970 and its amendments, innovative activity spiked (Rubin, 2014; Taylor et al., 2006; see also Acemoglu et al., 2014).[23] Those innovations lowered the cost of reducing sulfur dioxide emissions from coal-burning electric power plants by creating a market for scrubber technology. Regulations on GHG emissions and other pollutants would likely lead to similar innovations and lower the cost of well-developed increasingly clean technologies (see Chapter 2).

Regulatory Changes

These strategies—which include results-based regulation and new utility business models, dedicated utility funding for innovation, enabling responsive devices, recognition of volt/volt ampere reactive (VAR) optimization in rates, and on-bill repayment financing for energy-efficiency and increasingly clean energy technology—are discussed in detail in Chapter 6. In addition to the regulatory issues discussed in Chapter 6, siting and permitting remains a difficult obstacle at the regional, state, county, and municipal levels. New policies are needed to incorporate best practices for siting and streamlined review and permitting processes. While this issue applies broadly to the deployment of all increasingly clean energy technologies, DOE's SunShot Initiative, discussed earlier in this chapter, may offer a model for addressing these challenges. One of the key goals of that program is to reduce nonhardware, balance-of-system costs for solar systems. DOE cites these costs as representing up to 64 percent of total installed system prices, including customer acquisition, finance and contracts, permits, interconnection, and inspection; installation and performance; and operations and maintenance (DOE, 2014d).

Further Application of Principles and Analytics

Chapter 5 explores how the above recommendations and principles could be implemented in real-world energy systems that face significant market imperfections and other obstacles to full development and utilization. The systems in question, notably nuclear generation, carbon capture, and large-scale renewables, each manifest their own technological, market, and regulatory complexities within the general context illustrated above. Such attention to the development of necessary details, on the ground as it were, is necessary to build a complete picture of an increasingly clean power system.

---

[23]However, Acemoglu and his colleagues find that "Though it is intuitive to expect that carbon taxes should do most of the work in the optimal allocation—because they both reduce current emissions and encourage R&D directed to clean technologies—we find a major role for both carbon taxes and research subsidies" (Acemoglu et al., 2014, p. 2).

CONCLUSION

Expanding and advancing the portfolio of increasingly clean energy technology options requires analyzing and developing strategies for overcoming obstacles to innovation from R&D through pilot testing, demonstration, and deployment. Innovation has played an important role in providing the United States with secure and affordable energy resources, and governments play an important role in enabling various aspects of the nation's energy innovation system. Given the highly distributed nature of the U.S. electricity system, markets, resources, and innovation resources, advancing increasingly clean energy options through the innovation stages discussed in this chapter also requires insights into the federal government's opportunities to lead or to enable and leverage the efforts of states, regions, the private sector, and public/private partnerships.

The energy innovation system is a complex network of market and nonmarket institutions and incentives that includes public and private research and educational institutions; individual entrepreneurs and small entrepreneurial firms; large, mature firms; financial intermediaries ranging from large commercial and investment banks to venture capital firms and individual angel investors; local, state, and federal regulatory and standards-setting agencies and legislative units; other government agencies engaged in research, development, or procurement; and innovation users of many different kinds. This chapter has considered the obstacles that must be overcome at each stage of the innovation process, and arrived at findings and recommendations for strengthening the nation's critically important energy innovation system. The most important priorities are identifying and creating new options, developing and demonstrating the efficacy of these options, and setting the stage for early adoption of those that are most promising. Before discussing some specific obstacles facing certain technologies (nuclear power, capture and storage of carbon from fossil fuel generators, and renewable technologies) in Chapter 5, the committee turns in Chapter 4 to the potential environmental benefits of expansion of energy-efficiency measures.

4

The Role of Energy Efficiency in Increasingly Clean Electricity

While the rest of this report examines the prospects for increasing the supply of electricity from increasingly clean sources, this chapter deals with economically efficient methods of reducing electricity usage through energy efficiency.[1] Energy-efficiency measures offer the promise of reducing energy use and saving money on electricity bills, as well as reducing negative environmental externalities associated with the production of electricity. These measures can provide pricing and usage transparency to allow residential, commercial, and industrial customers greater control of their energy choices. Appropriate federal and state policies can promote the development of more energy-efficient buildings and products and strengthen incentives for consumers, businesses, and industrial customers to pursue cost-effective energy-efficiency measures and to make investments that will provide future energy-efficiency improvements.

The improvements in energy efficiency achieved over the past 30 years can be attributed to a variety of factors, including technological progress and pressures on households and businesses to cut their spending on energy. In addition to those market forces, there is evidence that policies and programs designed to improve energy efficiency, such as energy-efficiency standards, funding for research and development (R&D), educational and informational efforts, and financial incentives to accelerate the development and adoption of energy-efficiency measures, have contributed to the improvement in energy efficiency experienced in OECD countries (Geller and Attali, 2005; citing

[1]This chapter focuses on improving the efficiency of energy consumption. An improvement in energy efficiency occurs when there is a reduction in the energy inputs required to provide a given unit of energy services (e.g., lighting, cooling, heat, or drive power) to a specific end-user of such services. However, improvements in energy efficiency can occur with simultaneous increases in the use of energy services and in overall energy use.

Bosseboeuf and Richard, 1997; Gillingham et al., 2004; NRC, 2001, among others).

Despite the large potential to reduce spending on electricity, however, market imperfections can lead to underinvestment in energy efficiency. One such market failure is attributable to unpriced pollution from the production of electricity, discussed in Chapter 2. But even if electricity is priced to include environmental externalities, there may be other market failures that lead to underinvestment in energy efficiency. For instance, consumers often have inaccurate or incomplete information about their energy use, price, or the net savings from energy-efficient investments. Asymmetric information or misaligned incentives, such as between landlord and tenant, may also reduce energy-efficient purchases. Improving information available to consumers through energy labeling, together with advanced metering infrastructure and customer systems, could increase cost-effective energy efficiency (NAS et al., 2010, NRC, 2011).

In addition to market failures, electricity consumers do not always behave in an economically rational way to minimize costs or maximize profits or benefits, and policies designed to reduce market failures may not affect behavior. Consumers may be reluctant to make new purchases of energy-efficient appliances because of inertia, risk aversion, or uncertainty about other characteristics of the appliances (see Wilson and Dowlatabadi [2007] for a broad range of influences on consumer behavior). For these reasons, firms cannot completely capture the benefits of their investments in energy-efficiency innovations. This fact, coupled with the knowledge spillovers that impede realizing the full benefits that flow from their R&D (see Chapter 3), results in firms underinvesting in innovation in these technologies.

One challenge for policy makers, then, is to ensure that the benefits of energy-efficiency measures are greater than their costs. The savings from energy-efficient investments are difficult to calculate, however, and estimates derived from currently available methods have large uncertainty. Consequently, this chapter presents no estimates of the costs per kilowatt hour (kWh) of energy-efficiency measures. Improving the accuracy of savings estimates and measuring actual savings could aid in evaluating the effectiveness of existing and the design of future policies.[2] Also beneficial would be understanding barriers that contribute to the gap between actual savings and costs and consumers' product use, and developing behavioral models of how gains from energy efficiency can be realized (Allcott and Greenstone, 2012).

This chapter details potential electricity savings through energy efficiency, barriers to achieving the development and adoption of cost-effective energy-

---

[2]It should be noted that there are other, unobservable costs and benefits to energy-efficiency investments that make their net benefits difficult to measure even if energy savings are correctly assessed (Allcott and Greenstone, 2012, p. 5).

efficiency technologies, and policies of the federal and state governments designed to address those barriers.

POTENTIAL ELECTRICITY SAVINGS THROUGH ENERGY EFFICIENCY

Americans today spend almost $400 billion annually on electricity to power their homes, offices, and factories, a large share of which is used in residential and commercial buildings (more than two-thirds of all the nation's electricity use and 40 percent of total energy consumption) (EIA, 2015g [data as of April 2016]; see also Austin, 2012). On average across the United States, energy usage in buildings contributes about 41 percent of total U.S. carbon dioxide emissions (DOE, n.d.-a). And in some medium to large U.S. municipalities, building energy use can be responsible for 50 to 75 percent of citywide carbon emissions—much higher than the national average (City Energy Project, 2014, slide 9).[3]

As described in detail below, evidence suggests that energy-efficiency measures have been effective at reducing energy consumption. Moreover, there is potential for improvements in energy efficiency in the future. According to a previous Academies study, "Energy-efficient technologies for residences and commercial buildings, transportation, and industry exist today, or are expected to be developed in the normal course of business that could potentially save 30 percent of the energy used in the U.S. economy, while also saving money" (NAS et al., 2010, p. 278).[4]

Over the past 40 years, ongoing technological change, structural changes in the economy, changes in energy prices, and improved efficiency of energy use have contributed to the decline seen in energy intensity (units of energy as a share of per capita gross domestic product [GDP]) (NAS et al., 2010). Indeed, research suggests that three-quarters of the decline in U.S. energy intensity from 1970 to 2001 is attributable to improvements in the use of energy (Huntington, 2009; Levinson, 2015[5]; Metcalf, 2008). All told, total energy used per dollar of goods produced is down, as is spending on energy services (from lighting to refrigeration) (EIA, 2016h). In sum, efficiency measures in the electricity sector

---

[3]Using data from individual cities, it was found that buildings in the following cities contributed 50 to 75 percent of citywide carbon dioxide emissions, based on the cities' individual greenhouse gas inventories: Dallas, Philadelphia, Minneapolis, Chicago, Washington (DC), New York City, and Salt Lake City.

[4]Note that there is considerable uncertainty around the magnitude of energy-efficiency savings, particularly the costs of those improvements, discussed in detail in the rest of this chapter.

[5]Levinson shows that 90 percent of pollution reduction in the manufacturing sector is due to changes in production technique rather in than manufacturing composition.

have reduced energy consumption and could continue to save energy in the coming decades, while also helping to reduce pollution (EIA, 2016h).

According to the Energy Information Administration (EIA), the adoption of energy-efficiency measures such as more efficient equipment, better insulation, and improved windows has contributed to a decline in residential energy use, as have migration patterns within the United States as more people have moved to states that are warmer in the winter (EIA, 2013b). Despite being 30 percent larger in size, newer houses use only 2 percent more electricity than those built before 2000 (EIA, 2009b, Forms EIA-457 A and C-G). Improvements in data on energy use due to technological advances in information collection, storage, and access also have contributed to improvements in energy efficiency in all sectors.

BARRIERS TO THE DEVELOPMENT AND ADOPTION OF COST-EFFECTIVE ENERGY-EFFICIENCY TECHNOLOGIES

There is a literature investigating whether and why firms and consumers leave profitable or cost-effective energy-efficiency investments on the table. This "energy-efficiency gap" literature compares actual energy savings with the costs of energy-efficiency investments to see whether there are really unexploited energy-efficiency opportunities (Allcott and Greenstone, 2012; Jaffe and Stavins, 1994; Jaffe et al., 2004, Gillingham et al., 2009). If there is no energy-efficiency gap or the gap is negative, it means that the costs of the energy-efficiency measure outweigh the benefits.

Some researchers have questioned the existence or size of the energy-efficiency gap, offering evidence that predicted savings from certain energy-efficiency programs are overstated (Allcott and Greenstone, 2012; Davis et al., 2014; Dubin et al., 1986; Fowlie et al., 2015; Geller and Atalli, 2005; Gillingham and Palmer, 2014; Gillingham et al., 2009; Houde and Aldy, 2014; Jacobsen and Kotchen, 2013; Levinson, 2014a; Metcalf and Hassett, 1999).[6] These studies highlight the need for rigorous, randomized evaluations of programs to ensure that those who do and do not receive a program or treatment are statistically identical. Allcott and Greenstone (2012) note that results on a small group of volunteers rather than a randomly selected group may be biased. Ramos and colleagues (2015, p. S19) note that "experimental methodologies with rigorous design and the use of large-scale random samples are extensively being employed to study novel aspects of energy efficiency in buildings."

---

[6]Other studies have found that models have overestimated the costs of energy-efficiency standards because of technological changes (Dale et al., 2009; Taylor et al., 2015). In addition, in a recent study, Kotchen (2016) finds that engineering forecasts do not significantly overestimate realized savings of energy efficiency, and Auffhammer and colleagues (2008) find that average savings and costs may be consistent with average utility-reported savings on demand-side management programs.

Much of the difference between energy savings predicted by current engineering models and actual energy savings may be attributable to how users adopt and use technology. Differences in consumers' behavior may arise from differences in their values (including how much they discount future payments relative to the present) and their abilities to process or calculate savings from the available information. Another cause may be consumers' increased use of an appliance when the costs of its use decline because of energy efficiency—termed the "rebound effect" (see, e.g., Gillingham et al., 2009, 2015). Recent empirical evidence suggests that the rebound effect in energy efficiency is small (Davis, 2008; Dumagan and Mount, 1993; Fowlie et al., 2015; but see Davis et al., 2014), but more analysis of this phenomenon is needed to understand how large it may be (see, e.g., Chitnis and Sorrell, 2015 [suggesting that including indirect income effects makes the rebound effect larger relative to relying only on price effects]). Finally, energy-efficiency investments may have hidden costs or large transaction costs that are not accounted for in engineering models (Fowlie et al., 2015).

To evaluate the cost-effectiveness of policies aimed at overcoming barriers to investment in energy-efficiency measures, it will be necessary to have better information on actual energy savings and market adoption rates of different technologies. Opportunities to collect such information have been enhanced by advances in communication and information technology. Analysts have hypothesized that a number of market imperfections, discussed below, inhibit customers from adopting energy-efficiency measures even when their benefits outweigh their costs, and the United States and Europe have adopted a number of measures designed to overcome these imperfections.

Economic, institutional, and political barriers to the development and adoption of energy-efficiency measures include

- incorrect energy prices (prices that do not reflect the full societal cost of energy, including costs of pollution), which reduces incentives to invest in energy efficiency (see Chapter 2);
- inadequate and imperfect information and—as anyone who has rushed to replace a broken water heater, furnace, or refrigerator knows—often insufficient time to make good energy-efficiency decisions;
- so-called "split incentives," denoting cases in which decisions about energy efficiency often are made by those who do not pay the utility bills, and therefore will neither reap the benefits of improved efficiency nor bear the cost of poor efficiency;
- capital market failures, whereby customers may lack access to—or face competing demands for—the funds needed to make structural improvements or replace major pieces of equipment to improve efficiency;

- behavioral constraints such as inertia, limited attention, or heuristics that may lead consumers to make less than perfectly rational decisions; and
- knowledge spillovers that cannot be captured by the manufacturer and lead to underinvestment in new energy-efficiency technologies.

Potential policy solutions to these barriers are listed in Table 4-1 and discussed briefly in the subsections below. More detailed discussion of these solutions is provided in the section that follows on energy-efficiency policies in the United States.

Incorrect Energy Prices

If the externalities associated with electricity production and use are not reflected in their costs, incentives to invest in energy efficiency will be blunted. Since prices will not reflect the full societal cost of producing electricity, public intervention may be warranted to take prices to their correct levels. Potential solutions include a carbon price, other energy taxes, and carbon-trading instruments. Additional benefits of a price on pollution are addressed in Chapter 2.

TABLE 4-1 Market and Nonmarket Barriers to the Development and Adoption of Energy-Efficiency Measures and Potential Policy Solutions

Barriers	Potential Policy
Incorrect energy prices	Carbon price Other energy taxes Carbon-trading instruments
Inadequate and imperfect information	EnergyGuide/Energy Star™/Leadership in Energy and Environmental Design (LEED) labels Comparative bills/home energy audits
Split incentives	Appliance standards Building codes
Capital market failures	Third-party energy-service providers On-bill repayment programs
Behavioral constraints	EnergyGuide/Energy Star/LEED labels Comparative bills/home energy audits Appliance standards Building codes
Knowledge spillovers	Investments by the federal government R&D tax credits Inducement prizes

While a carbon price would help spur investments in energy efficiency, other barriers, discussed below, may inhibit the adoption of cost-effective energy-efficiency measures (Ryan et al., 2011). Moreover, the effectiveness of increased electricity prices in inducing conservation is limited by the very low measured price elasticity of demand for electricity, especially in the short term. Measures of the short-term price elasticity of demand for electricity range from 0.14 to 0.44 in absolute value for residential customers, meaning that a 10 percent rise in the price of electricity will reduce the quantity of electricity consumed by only 1.4 to 4.4 percent (Gillingham et al., 2009, Table 1). Over the longer term, these efforts would be more fruitful. The long-term price elasticity of demand for electricity is estimated to be between 0.32 and 1.89 (values greater than 1.0 mean that a 10 percent rise in the price of electricity would be expected to lead to a decline in electricity demand of more than 10 percent).

> **Finding 4-1:** A more accurate price for electricity that includes its full social costs might help spur energy efficiency in the long run; however, higher electricity prices may be insufficient to overcome all market and behavioral failures that inhibit the adoption of energy-efficiency measures.

Another issue related to the price of electricity is that the bills of many consumers reflect the average cost of electricity, which may be lower than the incremental cost of producing electricity at peak times. Consumers then have less incentive to reduce their consumption during those peaks. Measures intended to move toward real-time pricing for electricity (or time-of-day pricing)—for example, by encouraging the adoption of advanced metering infrastructure or "smart meters"—could help reduce peaks in demand and perhaps overall electricity usage as well (Cappers et al., 2016).[7] Because electricity produced during peak periods is usually generated from fossil fuels, reductions in peak demand could reduce pollution. In addition, evidence indicates that some form of time-of-day pricing would reduce overall demand for electricity.

Previous estimates of reductions in energy use induced by the adoption of smart meters range from 5.7 to 17 percent.[8] The 2009 American Recovery and Reinvestment Act (ARRA) included $3.4 billion for the Smart Grid Investment Grant program, which promoted investments in smarter grid technologies, tools, and techniques, including funding for advanced metering infrastructure. Ten utilities were funded to undertake consumer behavior studies. A number of such

---

[7]A full estimate of the impacts of time-varying rates on different consumers requires analysis of how changes in the aggregate load curve impact investment requirements and market prices.

[8]Houde and colleagues (2013) find a 5.7 percent drop in energy use only for the first month; Faruqui and colleagues (2010) find average energy use reduced by 7 percent; Gans and colleagues (2013) report that energy use in Ireland dropped by 11-17 percent.

studies will be conducted to investigate consumer acceptance, retention, and response to time-based rates (Cappers et al., 2016). The first consumer behavior study looked at differences in opt-in versus opt-out time-of-use metering programs, finding that many customers who defaulted to a time-of-use rate appeared to be better off, although some "inattentive" consumers who failed to opt out may have incurred higher electricity bills because of time-of-use pricing (Cappers et al., 2016).[9]

While utilities have reasons other than energy efficiency to deploy advanced metering infrastructure (addressed in more detail in Chapter 6), it is unclear whether the costs of deploying smart meters and the necessary communication infrastructure are great enough to dwarf the energy-efficiency savings to residential customers associated with these investments at this point in time (see Ramos et al., 2015 [citing Conchado and Linares, 2012]).

Inadequate and Imperfect Information and Split Incentives

Consumers do not directly observe the amount of electricity that is being used to wash clothes or dishes, keep a house at a certain temperature, or provide adequate lighting (see Ramos et al. [2015] and Gillingham et al. [2009] for more detailed analysis of such information failures). Additionally, it can be difficult for consumers to translate energy use into its cost, especially over a period of time, or to determine the savings derived from an energy-efficient device. Additionally, as discussed below, there may be cases in which the purchaser of an appliance is not the same as the person who uses it and pays the electricity bill, which may eliminate incentives for investments in energy efficiency.

A number of policies have been instituted to address the information failures leading to an energy-efficiency gap. These policies address energy use in buildings (for lighting, space heating, and cooling), as well as energy use by appliances. They include energy certificates (labeling such Energy Star™, EnergyGuide, or Leadership in Energy and Environmental Design [LEED] labels) and comparative bills/home energy audits whereby consumers receive personalized information on their electricity use compared with that of their neighbors and receive tips for reducing their electricity consumption (Allcott and Rogers, 2014).

As noted above, another informational failure relative to energy efficiency arises when the purchaser of the appliance is different from the user who is responsible for the electricity bill. Studies to date provide support that such split incentives exist (Davis, 2010; Gillingham et al., 2012). One example is set-top boxes that are owned by cable or telephone companies and are rented to

---

[9]Suggesting that utilities could use focus groups or other forms of market research to make these customers aware of the transition to time-of-use pricing, better understand their options, and more easily navigate the opt-out process if they do not want to make the transition.

consumers.[10] Another is the landlord/tenant situation, where the landlord is responsible for the purchase of refrigerators, heat pumps, water heaters, and other appliances, but the electricity bill is paid by the tenant.[11] In the United States, 37 percent of households are in rental housing, and the vast majority (86.4 percent) of those renters are responsible for paying their own electricity bill (Census Bureau, 2014). There is evidence of split incentives causing differences in behavior around energy efficiency: renters were found to be 1 to 10 percentage points less likely than owners to have Energy Star appliances (Davis, 2010). Potential policy solutions to the problem of split incentives include appliance standards and building codes.

Capital Market Failures

Energy-efficiency measures generally require a large up-front investment, which may be returned through a stream of smaller energy payments in the future. If some potential purchasers are unable to obtain credit, underinvestment in energy efficiency may result. As yet, there is little evidence regarding the size of this particular market failure (Gillingham et al., 2009).

Potential solutions to this barrier lie with third-party energy-service providers who pay the capital cost of an investment and receive a share of the resulting savings as payment, and on-bill repayment programs whereby the investments are recovered through charges on utility bills. Much more work is needed to identify and design appropriate policies for addressing failures to adopt energy-efficient technologies because of imperfect capital markets.

Behavioral Constraints

There are a variety of reasons why consumers, even when they have the correct information about energy prices and the net benefits of adopting an energy-efficient technology, may still choose not to adopt such technologies. For example, people put off decisions or find decision making difficult, or must weigh other characteristics (such as location, price, or size) in choosing a house (Allcott and Mullainathan, 2010; Gillingham et al., 2009). Consumers also may fear hidden costs or high transaction costs. In some cases, consumers may find qualitative attributes of the new technology less desirable than those of the existing technology (such as the inability to use a dimmer switch with compact fluorescent lighting) (Broderick, 2007). Hidden or high transaction costs that

---

[10]The Federal Communication Commission (FCC) has issued a notice of proposed rulemaking (NPRM) that would allow consumers to purchase set-top boxes in the commercial market. See FCC NPRM MB Docket No. 16-42 and CS Docket No. 97-80, adopted February 18, 2016.

[11]A comparable split can occur in commercial buildings if the specification and operation of the heating, ventilation, and air conditioning (HVAC) system are controlled by the building owner.

make adoption of an energy-efficient product or service more difficult also may affect consumers' purchasing decisions. An example of these costs is found in weatherization assistance programs, which require residents to be available for at least two site visits (Fowlie et al., 2015). Consumers also may resist purchases that have high up-front costs, even when they recognize that the benefits in the long run are positive.

Another important factor is a fragmented retail market and a stovepiped energy policy landscape. Consumers cannot go to a single source to manage their energy requirements efficiently. Electricity and fuel supplies, distribution services, lighting, appliances, heating and cooling, building shell improvements, and control systems often are provided by different vendors, none of which necessarily have an incentive to optimize the consumer's overall energy usage. The advent of advanced metering infrastructure, platform markets, data analytics, and "intelligent efficiency" providers that leverage increased data availability could help change the fragmented efficiency landscape. This is a potentially positive opportunity that is still in its early stage of development.

Any interventions designed to provide behavioral incentives on energy efficiency need to be scalable, to incorporate careful impact evaluation protocols that include control group comparisons and randomized field trials, and to have observable and measurable outcomes (Allcott and Mullainathan, 2010). One example is comparative bills or home energy audits, discussed above. A study of the effectiveness of these programs found that the monthly reports received by consumers provided cues that induced energy conservation, which persisted even after the program was terminated (albeit with some backsliding) (Allcott and Rogers, 2014). This finding suggests that further research on behavioral interventions could lead to positive economic benefits (Allcott and Mullainathan, 2010). Other potential policy solutions to behavioral constraints include EnergyGuide/Energy Star/LEED labels. In addition to residential and commercial customers, industrial customers may benefit from policies targeted to changing behavior (Gosnell et al., 2016). As with the problem of split incentives, moreover, appliance standards and building codes hold potential for addressing behavioral constraints.

Knowledge Spillovers

As discussed in Chapter 3, electricity is an example of a general-purpose technology—an innovation technology that contributes to technological dynamism. Such technologies are subject to large knowledge spillovers—a form of market failure that prevents investors in innovation from realizing the full benefits resulting from their R&D investments. This problem can be addressed through government investments, tax credits for R&D, or inducement prizes (discussed in detail in Chapter 3).

ENERGY-EFFICIENCY POLICIES IN THE UNITED STATES[12]

The federal government can help overcome many of the obstacles to efficiency discussed above through indirect programs such as energy labeling, appliance standards, and building codes. The federal government also plays an important role in spurring innovation in energy efficiency through the R&D programs of the Department of Energy (DOE). In addition, it is poised to lead by example through direct efforts to promote energy efficiency in the 500,000 buildings it owns or operates. State and local governments also offer incentives to utilities to encourage retail customers to adopt energy-saving measures, among other incentives.

Energy Labeling and Certificates

Energy labels on appliances provide information about the energy savings that can be realized from adopting more energy-efficient appliances or equipment or assure consumers that a product is more efficient than the average appliance on the market. Energy labeling represents an inexpensive source of information on the operating costs of different appliances. While willingness to pay for labeled appliances varies across consumers, appliances, and states, evidence suggests that consumers may value the information provided by these labels (Ramos et al., 2015).[13] Evidence also indicates that consumers may trust labeling by the government more than that by appliance manufacturers or other private parties (Banerjee and Solomon, 2003; but see GAO, 2010 [showing that Energy Star's certification process is vulnerable to fraud and abuse]).

EnergyGuide labels, administered by the Federal Trade Commission (FTC), are mandatory energy usage labels.[14] They apply to certain consumer products, such as clothes washers, refrigerators, freezers, televisions, water heaters, dishwashers, air conditioners, and boilers. EnergyGuide labels inform

---

[12]In addition to the policies to spur energy-efficiency deployment described here, the Qualified Energy Efficiency investment tax credit, Internal Revenue Code (IRC) Section 25C, provides a 10 percent credit for the purchase of qualified energy-efficiency improvements to existing homes. The maximum credit for a taxpayer is $500, and no more than $200 of the credit can be attributed to exterior windows (in 2009 and 2010, the maximum credit was $1,500) (IRS, n.d.; NRC, 2013c). The credit is allowed for qualifying property in service through December 31, 2013. Additionally, IRC Section 25D allows for a credit for qualified expenditures made by a taxpayer for residential energy-efficient property placed in service before January 1, 2017.

[13]However, see Newell and Siikamäki (2013) (showing that the impact of Energy Star certification may be due to a perceived endorsement of a model rather than the information provided) and Houde and Aldy (2014) (finding that net energy savings are small when a labeling program promotes products that have a high market share).

[14]The EnergyGuide program was established in 1979 (FTC, n.d.).

consumers of a product's projected energy use, efficiency, and/or cost, based on DOE test procedures.

Energy Star, administered by the Environmental Protection Agency (EPA), is a voluntary label applied to more efficient heating, ventilation, and air conditioning (HVAC) equipment, lighting, home electronics, office equipment, and other appliances. The label is based on whether the appliance exceeds federal minimum efficiency standards by a certain percentage, which varies over time "depending on proportion of certified products offered on the market, the market shares, and the availability of new technologies" (Houde, 2014). In addition, certain state rebate programs for appliances are tied to Energy Star certification (Houde, 2014). Energy Star products are currently based on self-certification by manufacturers. In the absence of verification, products that do not meet the standard may end up on the market with the Energy Star label (GAO, 2010).

Although Energy Star was designed to supplement the EnergyGuide program, a consumer seeking to purchase a certain appliance that is covered by both programs would be confronted with various logos and labels containing different information using different formats. Thus the programs would be more effective if the relevant agencies attempted to harmonize their approaches and present the consumer with a common, consistent performance label wherever possible. Furthermore, while the Energy Star label indicates that a product meets a single, category-specific energy-efficiency benchmark, experience from other countries indicates that multitiered labeling by categories, such as the graded approach in the European Union, work well (CLASP, 2005).

Although some evidence suggests that too much information overwhelms consumers, other evidence indicates that better information leads to better outcomes. In one study, consumers were presented with information on how their state's energy usage and prices were higher or lower than the national average used on EnergyGuide labels (Davis and Metcalf, 2014). Those consumers whose state's energy usage and prices were higher than the national average were more likely to make a more energy-efficient purchase relative to consumers from states with lower energy prices and use.

While these energy labeling programs may be effective at reducing energy use in a cost-effective way, there is evidence that the state and utility energy rebates associated with Energy Star products may not be cost-effective (Alberini and Towe, 2015). Most consumers purchase a new appliance when their old one is no longer working, and it is unclear how many consumers would have purchased a more energy-efficient appliance even without the rebate (the "free rider" problem)[15] (Alberini and Towe, 2015; Boomhower and Davis, 2014 Houde and Aldy, 2014; Malm, 1996 [finding that 73-92 percent of program participants were free riders]). While replacing appliances with more energy-

[15]Free riders are program participants who would have participated without any intervention.

efficient versions does reduce energy use, it may be that energy-efficiency standards (discussed below) would be a cost-effective option (Alberini and Towe, 2015). Where products have a high market share, there is evidence that a labeling program (especially one combined with rebates) produces small net benefits because of the free rider problem (Houde and Aldy, 2014). In the case of a product with a high market share, one study found that 73 to 92 percent of program participants would have purchased the product without any incentives or labels (Houde and Aldy, 2014).

There are also energy labeling programs for construction of new buildings, including LEED[16] and the Energy Star certification. Researchers have found that commercial buildings certified under these two programs in the United States have higher rents, higher selling prices, and higher occupancy rates relative to uncertified buildings (Ramos et al., 2015) [see Table 2 for a literature survey]). The literature on European buildings reports more varied results, with some researchers finding no effects of certification. Moreover, the actual energy efficiency of certified buildings remains subject to debate (compare Kahn et al. [2013] and Newsham et al. [2009]).

Research on the impact of energy labels on the residential sector has yielded mixed results (Ramos et al., 2015; Walls et al., 2016). While some studies have shown that homes with LEED or Energy Star certification do use less energy than noncertified homes, the reduction in energy use is not always reflected in their selling price (see, e.g., Walls et al., 2016; but see Ramos et al., 2015, p. S21 [stating that "the market has not yet been able to generate enough data to estimate the effect of introducing certificates on energy demand, neither at the aggregate nor at the disaggregate level"]). In addition, with both commercial and residential buildings, it is unclear whether certified buildings sell for higher prices because of the perceived energy savings, or they are perceived as having higher-quality building materials or better designs (Gillingham et al., 2009; Ramos et al., 2015). It may be that labels provide no additional information about net savings to the consumer, but a simple endorsement of a product may improve customer confidence in the product (Brounen and Kok, 2011; Newell and Siikamäki, 2013). In addition, certification at the state or local level may confound the perceived impact of LEED or Energy Star certification. Local certification may go beyond energy efficiency to include water efficiency, landscaping choices, and building materials, and the coexistence of such certifications makes it difficult to determine which of them is affecting consumers' willingness to pay for a certified home (Walls et al., 2016).

---

[16]LEED certification is administered by the U.S. Green Building Council.

Appliance Standards

DOE announces and implements minimum efficiency performance standards (MEPS)[17] for a variety of residential appliances, including central air conditioners and heat pumps, clothes washers and dryers, major kitchen appliances, and room air conditioners (Office of Energy Efficiency and Renewable Energy [EERE], DOE, current rulemakings and notices). The standards do not cover some classes of appliances, such as computer and battery backup systems, portable ovens, portable air conditioners, set-top boxes, and televisions (although some states have mandatory standards for these appliances). According to EERE, these standards apply to more than 50 categories of products that in the aggregate cover about 90 percent of home energy use, about 60 percent of commercial building energy use, and almost 30 percent of industrial energy use (DOE, n.d.-b).

MEPS work to remove the least efficient appliances from the marketplace, and most researchers agree that these standards appear to have improved consumer welfare (Houde and Spurlock, 2015; Taylor et al., 2015; but see Gayer and Viscusi, 2013 [arguing that consumer welfare is increased only if consumers are not behaving rationally]). The energy efficiency of many appliances covered by these standards has increased substantially, and many consumers are choosing to buy products that exceed the standards (Taylor et al., 2015). The result has been an average energy efficiency of purchased products that exceeds the MEPS requirements (Taylor et al., 2015). While many of the products have not seen price increases, unregulated quality dimensions (e.g., performance, capacity, noise) and product diversity have improved even as the standards have become more stringent (Houde and Spurlock, 2015 [citing Dale et al., 2009; Spurlock, 2013; Allcott and Taubinsky, 2015, to show mixed results on price impacts]). Reliability also has not been harmed by appliance standards; the rate of significant repairs over 5 years of product ownership generally declined from the time appliances were first subject to federal MEPS (Taylor et al., 2015).

These standards have led to much smaller increases in appliance prices than expected ex ante using engineering models, and in fact led to only modest increases (Dale et al., 2009; Houde and Spurlock, 2015; Taylor et al., 2015). Refrigerators are a widely cited example of an appliance whose energy efficiency rose simultaneously with declines in prices. Figure 4-1 shows the

[17]In the 1970s and 1980s, appliance standards were used in states such as California, New York, and Florida. A federal program that included energy targets was established in the Energy Policy and Conservation Act of 1975, although the federal minimum standards did not preempt state-level standards until the passage of the National Appliance Energy Conservation Act of 1987. New categories for federal standards were added in the Energy Policy Act of 1992, the Energy Policy Act of 2005, and the Energy Independence and Security Act of 2007. The Energy Independence and Security Act of 2007 also required that DOE maintain a schedule for regularly reviewing and revising all standards (DOE, n.d.-b).

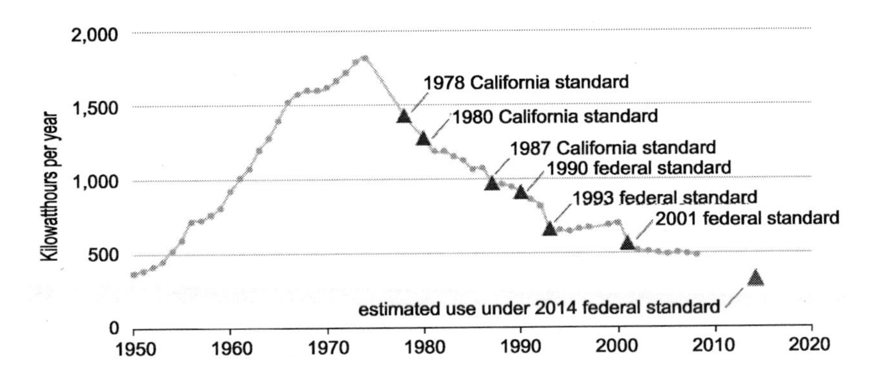

FIGURE 4-1 Annual energy use of a new refrigerator, 1950-2008.
SOURCE: EIA, 2013e.

decline in the energy used by a new refrigerator from the mid-1970s through 2008 along with the enactment of energy efficiency standards. Simultaneous increases in product quality and diversity together with no or little change in prices may at first appear counterintuitive. It may be, however, that stringent standards led manufacturers to compete on other quality dimensions because they all had to meet a certain minimum efficiency standard (Houde and Spurlock, 2015).

One unanswered research question is how great an impact appliance standards have on inducing technological innovation to meet the standards. It is likely that standards did spur manufacturers to innovate to ensure that their products would exceed the minimum standards, although how much of that innovation was due to the standards, to increasing energy prices, or to exogenous R&D efforts is unclear (see Newell et al., 1999).[18]

Legal requirements mandate that the standards be reviewed at least every 6 years,[19] and test procedures must be reviewed at least every 7 years to determine whether updates are warranted.[20] DOE also evaluates new product categories for standards as opportunities for energy efficiency emerge. Other countries use a different type of appliance standard. The Top Runner approach used in Japan is somewhat analogous to the CAFE (Corporate Average Fuel Economy) standards for automobiles in the United States in that it uses a market-weighted average[21] rather than a minimum allowable efficiency. Top

[18]Taylor and colleagues (2015) state that is unlikely that MEPS are technologically forcing because of the rulemaking process.
[19]United States Code, 42 U.S.C. 6295 (m).
[20]United States Code, 42 U.S.C. 6293 (b).
[21]A market-weighted average is calculated for each machinery or equipment category for each of the companies that manufacture (or import) machinery or equipment covered by

Runner sets a high efficiency level for a future date, and the market must meet that weighted-average efficiency.

Efficiency standards and policies can be made more stringent, which may spur technical innovation and market competitiveness. Furthermore, appliance standards may help address the problem of split incentives since a landlord has no choice but to provide appliances with better energy efficiency (Gillingham et al., 2012). More research is needed on the connections and interactions between efficiency standards and other policies designed to spur innovation in energy efficiency, especially any interaction between standards and policies to internalize prices on pollution.[22]

Recommendation 4-1: DOE should on an ongoing basis set new standards for home appliances and commercial equipment at the maximum levels that are technologically feasible and economically justified.

Given the increasing share of electricity going to television and electronics, DOE might consider expanding appliance standards to include these "nontraditional" appliances that are not subject to MEPS. The committee also recognizes that energy efficiency is not a societal value that trumps all other values. Indeed, in promulgating these standards, cost-effectiveness to consumers, economic impacts on manufacturers, impacts on product performance, impacts on competition, and other factors explicitly cited in the law need to be taken into account.[23]

Building Codes and Retrofits

Building codes (and other efforts to improve new construction) can have a positive effect on improving overall energy efficiency. Building codes are the primary policy instrument for influencing the energy efficiency of newly built or renovated buildings and can provide information on best practices to the multiplicity of builders and contractors across the United States. National model energy codes exist (see below), but building codes are determined by state and local governments.[24]

---

the Top Runner program. This average is based on the volume of each of the products shipped and their efficiency (METI Agency for Natural Resources and Energy, 2010, p. 26).

[22]In the case of sulfur dioxide, research suggests that direct measures for reducing the pollutant encouraged only cost-reducing innovations, while economic incentives such as pricing the pollutant encouraged both cost-reducing and emission-reducing innovations (Harrington and Morgenstern, 2004).

[23]The Energy Policy and Conservation Act of 1975, Section 325(o)(2)(B).

[24]http://www.energy.gov/eere/buildings/articles/energy-codes-101-what-are-they-and-what-doe-s-role.

As noted earlier, energy-efficiency measures and migration to states that are warmer in the winter both have contributed to a decline in residential energy use. Despite a large increase in the number of electronic devices and appliances used in households, there has been only a small increase in overall residential electricity demand (EIA, 2013b). According to the most recent survey, newer homes (built between 2000 and 2009) used only 37.1 million Btu (MMBtu)/ft^2 in 2009 for space heating and cooling, compared with 51.6 MMBtu/ft^2 for homes built before 1940 (EIA, 2009b, Table CE2.1).

With respect to rental properties, because builders or property owners generally choose the windows, amount of insulation, and other energy-efficient aspects of the physical property, the electricity payer may have few choices for investing in energy efficiency—the phenomenon of split incentives discussed earlier. While landlords may be able to capture some of the value of energy efficiency through higher rents, renters may value other housing characteristics, such as location, size, number of bathrooms, and other factors, more than energy efficiency, even if their electricity bill is higher in a less energy-efficient rental unit. One study found owner-occupied homes to be 12 to 20 percent more likely to have insulation than rental units, even after controlling for observable characteristics of property, occupant, and neighborhood (Gillingham et al., 2012).

Although evidence shows that energy use for heating and cooling buildings has decreased over time, it is unclear how much of that decline is due to building codes versus improvements in building materials that would have occurred in the absence of the codes. Research on the link between building codes and improved efficiency is mixed, and more work is needed to determine whether such programs are cost-effective (Horowitz and Haeri, 1990; Jacobsen and Kotchen, 2013; Kotchen, 2016; Levinson, 2014b).

The national model energy codes—the American Society of Heating, Refrigerating and Air-Conditioning Engineers (ASHRAE)[25] code and the Interventional Energy Conservation Code® (IECC)—have historically addressed only construction and renovation; there are no requirements covering ongoing operations, such as that energy use be measured; that building operators be trained to operate energy systems properly; or that energy systems be regularly tuned, as is the case for automobiles. Buildings seldom operate the way they were designed to operate, and even if commissioned properly, seldom continue to perform optimally. Lack of controls or inability to use existing building automation properly may be leading to a substantial increase in energy use (Mills, 2011; Piette et al., 2012). Sensors and communication and computational capabilities can help unlock a substantial portion of these operational savings.

---

[25]ASHRAE was formed by the 1959 merger of the American Society of Heating and Air-Conditioning Engineers, founded in 1894, and the American Society of Refrigerating Engineers, founded in 1904 (https://www.ashrae.org/about-ashrae; International Energy Conservation Code, http://www.iccsafe.org/codes-tech-support/codes/2015-i-codes/iecc). The IECC was created by the International Code Council in 2000.

To enable enhanced energy performance in commercial buildings, the key is to install sensors that measure building energy consumption, to link those sensors to a microprocessor, and to apply software to analyze the data and provide insight into opportunities for energy savings. Advanced data analytics supported by advanced or interval meter infrastructure also can assist significantly. For example, IBM's sustainability efforts through its Smarter Planet platform rely on such an information technology-based approach (Hudson, 2012).

> **Recommendation 4-2: DOE's Office of Building Technologies should continue to partner with ASHRAE and the developers of the IECC in the development of these model codes, and should support them in efforts to compile best practices for improving the energy efficiency and operation of building systems.**

Attempts to reduce energy use in residential dwellings need to address not just new construction and renovations—the target of building codes—but also retrofits to the existing housing stock. Houses have a long life—residential buildings have a typical service life of 60 to 100 years (Jaffe et al., 2004)—and building standards for new construction have no effect on the existing housing stock. However, there is little evidence that retrofits have improved energy efficiency. Studies of energy-efficiency retrofits during the 1970s and 1980s have found that the actual energy savings achieved were far below the engineering estimates (see Geller and Attali [2005] for a review of the literature). More recently, a study of the weatherization assistance program in Michigan found that, even accounting for the social benefit of reduced energy use, the rate of return on weatherization was negative and that the up-front costs were twice as great as the energy savings (Fowlie et al., 2015).

Better information on actual energy use is needed before any assessment of a government energy-efficiency program can be carried out. One way to achieve that goal is to improve engineering models so they more closely resemble actual use. Energy-efficiency policies could also be improved by field studies, conducted by both DOE and the state public utility commissions, relying on state-of-the-art evaluation approaches. DOE has already solicited ideas for improving the accuracy of its audits, issuing a request for information (RFI) on means of improving savings prediction methods for residential energy-efficiency upgrades. DOE expressed its interest in information on the current state of the art of savings prediction methods, forthcoming advances that could improve the accuracy and/or reduce the costs of these methods, and the potential market implications of improved methods. In addition, DOE asked for information on "other factors that may influence the accuracy of modelled predictions such as improved field guidance, technician training and certification, or benchmarked models to a nationally accepted DOE prediction model such as EnergyPlus."

Energy Audits

A home energy audit, also known as a home energy assessment, is one way for a homeowner to obtain personalized information on home energy consumption and measures that can be taken to reduce it (Ramos et al., 2015). Costs for home energy audits can range from $300 to $500 (Palmer et al., 2013 [average cost is $349][26]); however, some local governments and utilities provide subsidies to pay for these audits (Alberini and Towe, 2015 [analyzing free energy audits in Maryland]).

Despite the potential for bridging informational and behavioral failures, research on the cost-effectiveness of energy audits is limited (Alberini and Towe, 2015; Frondel and Vance, 2012; Palmer et al., 2015; Ramos et al., 2015). A study of free audits in Maryland found that residential energy audits reduce energy usage by about 5 percent, with the cost per ton of carbon dioxide emissions abated ranging from just under $50/ton to nearly $70/ton (Alberini and Towe, 2015).[27] Even when homeowners pay for audits, they may not follow through on the resulting information (Palmer et al., 2015). Moreover, only about 4 percent of U.S. homeowners have had an energy audit (Palmer and Walls, 2015). Thus while these audits offer the promise of reducing energy use, additional research is needed to ensure that this promise is realized and to illuminate how information policies can be made more effective.

The Role of Innovation

Chapter 3 provides a detailed discussion of innovation, and notes the potential underinvestment in R&D because of knowledge spillovers. Federal funding for R&D for energy conservation and end uses is quite small, amounting to less than $1 billion in fiscal year (FY) 2013 and accounting for about 20 percent of federal spending on energy conservation and end uses when energy assistance programs are excluded (EIA, 2015c, Table 1).[28] As discussed in Chapter 3, without government support for energy innovation, unpriced environmental externalities may reduce incentives for private-sector innovation in new technologies, including those that reduce energy use (Gillingham et al., 2009 [citing Goulder and Schneider, 1999; Jaffe et al., 2005; Schneider and Goulder, 1997).

One important research question has to do with the link between energy-efficiency policies, such as energy labels or standards, and technological change. In other words, do energy-efficiency policies induce greater investments in technological change to improve energy efficiency? While more research on this

---

[26]http://energy.gov/articles/askenergysaver-home-energy-audits.

[27]The authors caution about selection bias in comparing energy use between homeowners who undergo an audit and those who choose not to do so.

[28]This excludes direct expenditures on the Low-Income Home Energy Assistance Program, which assists low-income households with their energy bills.

question is needed, preliminary findings show that a country's commitment to increasing energy efficiency, along with funding for energy-efficiency research, does lead to a higher probability of innovation (Verdolini and Galeotti, 2011). Other researchers have found that increased regulatory stringency leads to more spending on R&D; however, no relationship has been found between increased costs to comply with more stringent requirements and innovation (as measured by successful patent applications) (Jaffe and Palmer, 1997).

One example of DOE's R&D program is a joint effort with industry that has been ongoing for more than a decade according to a road map, focused on the development and implementation of improved light-emitting diodes (LEDs). The program has funded fundamental science, core technology research, product development, manufacturing R&D, and commercialization support, and has included prizes and test procedures and standards. The program has resulted in LEDs that use 84 percent less energy than standard incandescent bulbs while giving off the same amount of light and lasting 25 times longer (25,000 hours versus 1,000 hours). According to EIA, the efficiency of LED bulbs (light output per unit of energy consumed) has risen even as the costs per bulb have fallen dramatically since 2010 (from nearly $70/bulb in 2010 to $10/bulb in 2014), and costs are predicted to continue to decline through 2020 (EIA, 2014g).

Other examples of DOE R&D include a number of programs supported by the Advanced Research Projects Agency-Energy (ARPA-E), including DELTA (Delivering Efficient Local Thermal Amenities), aimed at reducing the costs of heating and cooling buildings through the development of technologies such as on-body wearable devices and more options for maintaining occupants' comfort within a building; BEETIT (Building Energy Efficiency Through Innovative Thermodevices), focused on enhancing energy efficiency, reducing greenhouse gas emissions, and reducing consumer costs for cooling commercial buildings; and SWITCHES (Strategies for Wide-Bandgap, Inexpensive Transistors for Controlling High-Efficiency Systems), aimed at developing next-generation power switches that could increase the efficiency of appliances and lighting.[29] More recently, ARPA-E announced a new program to develop innovative window coats and window panes that could improve the efficiency of existing single-pane windows (ARPA-E, 2016).

In addition to research on technologies for improving energy efficiency, DOE's Quadrennial Technology Review acknowledges the need for further behavioral research, noting that estimates of the energy-savings potential from research on consumers' decision making in the residential building sector are substantial (DOE, 2015b, Section 10.2.4).

R&D drives a much longer-term process that ultimately brings much-improved energy-efficiency technologies to the market. Accordingly, R&D needs to focus on both basic and market-oriented research.

---

[29]ARPA-E (www.arpa-e.energy.gov).

Recommendation 4-3: DOE should increase its investments in innovative energy-efficiency technologies; improve its ability to forecast energy savings from these technologies; and, in conjunction with other agencies, obtain data with which to develop behavioral interventions for improving energy efficiency.

Energy-Efficiency Technologies in the Industrial Sector

The industrial sector accounts for just under a third of the energy use and carbon dioxide emissions[30] in the United States (EIA, 2016a, Table 2) (and about 54 percent globally [EIA, 2016b]). Of course, the energy efficiency of industrial buildings can be increased through many of the same approaches for commercial buildings discussed earlier in this chapter. More important, changes in industrial processes have reduced energy use in the past (see, e.g., Levinson, 2015 [showing that reductions in emissions in the manufacturing sector are driven by changes in technique rather than industry composition]) and offer significant promise for improving energy efficiency in the future. However, there are barriers that may be inhibiting those investments. New equipment is considered an asset and is subject to different accounting treatment from that applied to energy bills or other operations and maintenance (O&M) expenses. Thus managers may be reluctant to make large investments in new equipment if the payoff of reduced O&M expenses is credited to another part of the company. Uncertainty about future energy prices is an additional barrier to investments in industrial energy efficiency. Policies to reduce that uncertainty—including uncertainty about a price on carbon—could increase action by industry to reduce energy use.

In addition to improvements in industrial buildings, more energy-efficient production processes could help reduce electricity consumption. Efforts to improve information on energy use or change the behavior of workers to improve energy efficiency could prove valuable in the industrial sector as well as in the residential and commercial sectors (see, e.g., Gosnell et al., 2016 [showing that performance information, personal targets, and prosocial incentives induced pilots to improve their fuel efficiency on flights]).

The efficiency of industrial processes has improved, leading to much lower energy use in this sector, due partly to strategic energy management programs and partly to sectoral changes. Motor systems (e.g., pumps, fans, air compressors, and motor-driven industrial processes) can be improved by sizing correctly to load, using speed control, improving maintenance, enhancing the efficiency of underlying motor drives and associated components, and other

[30] Assuming that carbon dioxide emissions are proportional to energy use.

measures. Equipment standards for motor drives, along with guidance on the other efficiency measures, can help boost motor system efficiency significantly.

Another potential energy-efficiency measure is for industrial customers to produce heat and power simultaneously from the energy they have at their disposal; this cogeneration is known as combined heat and power (CHP). Waste heat from electrical processes or from the generation of electricity on site is recycled to produce additional electricity and steam that can be used to warm buildings or assist in industrial processes. According to the most recent survey, fewer than 10 percent of industrial establishments use cogeneration technology (EIA, 2010, Table 8.2).[31] While approximately 85 gigawatts (GW) of CHP is installed in the United States, an additional 130 GW could be produced by the commercial/institutional and industrial sectors (DOE and EPA, 2012). The technical potential of CHP could rise further to about 200 GW if the industrial sector sized its CHP systems to sell power, although there are a number of market and policy barriers to realizing this potential (DOE and EPA, 2012).

Encouraging companies to institute a management structure focused on energy savings—strategic energy management—also could help realize the potential of operational/behavioral energy efficiency. In addition, encouraging equipment-based energy efficiency on an ongoing basis could help incentivize companies to invest in cost-effective energy-efficiency measures.

Direct Federal Government Efforts to Promote Energy Efficiency

Positive spillovers, such as reducing the cost of production by producing more or learning by doing, may occur in energy efficiency. In addition, there may be a "free driver" effect whereby the first customer provides information to later customers about the quality of a new energy-efficient technology or appliance. These effects may justify direct federal government efforts to adopt energy-efficient technologies, although more research is needed to determine the magnitude of these effects.

The federal government owns or operates more than 500,000 buildings across the country, comprising more than 3 billion square feet of total floor space and accounting for 2.2 percent of all building energy consumption in the United States (DOE, 2011a). State and local government buildings account for nearly 10 percent of the U.S. total. The federal government spends roughly $7 billion each year on energy to heat, cool, light, and power federal buildings (WBDG, n.d.).[32] State and local governments spend another $30-$40 billion per year on building-related energy consumption. Improving the energy efficiency of this building stock could reduce spending on electricity without impairing the services government currently provides.

---

[31]However, 96 percent of industrial establishments that produce electricity on site produce that electricity through cogeneration (EIA, 2010, Table 11.4).

[32]Note that the federal government spends $20 billion annually on energy, but a large fraction of that amount goes to nonbuilding energy use.

The Federal Energy Management Program (FEMP) provides expertise, training, and other services to help federal agencies improve energy efficiency and increase the use of renewable energy (Sissine, 2015). It pursues those goals through project financing, technical guidance and assistance, and planning and evaluation. Last year, the White House issued an executive order on planning for federal sustainability in the next decade, which ordered the head of each federal agency to promote building energy conservation, efficiency, and management (White House, 2015). The federal government can continue to lead efforts to promote energy efficiency by doing the following:

- **Continuing to lead in developing procurement practices for appliances and equipment that take life-cycle cost into account.** Federal agencies are required to purchase Energy Star or FEMP-designated appliances and equipment where those products are available (FEMP, n.d.). However, the incremental benefits of using life-cycle costing versus the additional administrative burden need to be assessed. A previous National Research Council report on solid-state lighting includes the recommendation that the Office of Management and Budget develop criteria for determining life-cycle costs and including social costs in evaluations of energy purchases, and for incorporating this methodology into agency procurements (NRC, 2013b). The President's Council of Advisors on Science and Technology made a similar recommendation (PCAST, 2010). Applying life-cycle costing ought not to be highly challenging; DOE has extensive expertise in this regard through the National Appliance Energy Conservation Act (NAECA) appliance and equipment energy-efficiency standards program (DOE, 2014b), which could serve as a template for other federal efforts.
- **Evaluating the benefits of improving the energy efficiency of the Department of Housing and Urban Development's (HUD) 1.2 million units of public housing.** Existing accounting and procurement rules make it difficult to improve the energy efficiency of public housing because energy savings cannot be credited against capital expenditures. These rules need to be restructured. In addition, much of HUD's building stock consists of large housing blocks built in the 1950s through the 1970s that are overdue for major capital upgrades. Similar blocks of midcentury social housing in Europe have been rewrapped with highly insulating facades and retrofitted with efficient equipment to achieve energy savings in excess of 50 percent. Additional benefits of such measures include increased resilience, reduced costs of maintaining the building envelopes, and reduced occupant health issues due to drafts and condensation. HUD could partner with local housing authorities and others to pilot rewrappings of midcentury housing blocks, including the development of scalable

financing models based on life-cycle assessments of costs and savings. To the extent that existing rules inhibit the energy efficiency of housing on military bases, those rules could be restructured accordingly.

- **Taking the lead on contracting for services that gives third parties incentives to invest in energy efficiency.** Various statutes and regulations authorize federal agencies to enter into contracts for their utility services, including electricity, hot water, and steam. The General Services Administration (GSA) is typically the lead federal contracting agency. Performance contracts such as energy savings performance contracts (ESPCs) or utility energy service contracts (UESCs), currently in use in the Department of Defense and other parts of the government, could serve as a model for the federal government's providing incentives for the private sector to invest in energy efficiency, with the cost of the investment paid off through reduced energy or O&M costs achieved over the life of the contract.

Current budgetary rules may inhibit the federal government from fully investing in energy-efficiency technologies. A typical ESPC provides payment from the government to the vendor until the vendor's costs have been covered and the contract expires. During the contract period, the government retains only a small share of the savings; after the contract expires, the government keeps all savings from the energy-efficiency investment. The Congressional Budget Office has identified two issues that may inhibit the government from engaging in these investments, even if they pay off over time (CBO, 2015). First, the additional spending for an ESPC is considered mandatory spending, while the potential future savings from the contract are considered discretionary spending. Second, while much of the additional spending for an ESPC falls within the 10-year cost estimate used by the Congressional Budget Office, much of the savings occurs over a much longer period of time.

Recommendation 4-4: The federal government should lead by example and ensure that federal facilities, including leased space, capture all cost-effective energy-efficiency opportunities, such as setting and achieving energy-efficiency targets, using performance contracting to achieve energy-efficiency gains, and procuring energy-efficient appliances and equipment.

State and Local Efforts to Improve Energy Efficiency

In addition to actions taken at the federal level, states are adopting important energy-efficiency measures. Because the retail market for electricity is regulated at the state level (for investor-owned utilities), states can play an important role in providing appropriate incentives for utilities to adopt energy-efficiency measures, including utility and public benefit programs and policies (ACEEE, 2012). In addition, states provide building energy codes, policies encouraging CHP systems, and state government-led initiatives around energy efficiency. Twenty-four states have an energy-efficiency resource standard (EERS) of some kind; 16 of these states have achieved electricity sales reductions of nearly 1 percent or greater, while 6 have achieved savings of greater than 1.5 percent, on an annual basis (ACEEE, 2012). While higher electricity prices do influence and can assist in the uptake of energy-efficiency measures, the evidence suggests that state policies play a similar role (ACEEE, 2012; EIA, 2014f, Table 5.6.A).

Some energy-efficiency programs—such as advanced metering infrastructure and comparative energy bills—may be run more efficiently through utilities because of access to information or lower transaction costs. However, state programs designed to encourage reductions in electricity consumption that are led by the utilities (demand-side management [DSM] and efficiency programs) are hampered by the fact that many utilities recover their fixed costs through volumetric rates that are charged on a per kWh basis. Because revenues are tied to the volume of electricity sold, utilities have little incentive to take measures that would reduce consumption. Allowing utilities to recover their fixed costs through a fixed charge is one way to remove disincentives for utilities to invest in DSM programs, although there are also other mechanisms that effectively decouple revenues from rates to remove the disincentive for utilities to invest in energy efficiency (see, e.g., Arimura et al., 2011).[33] Spending on DSM programs by utilities peaked in 1993, with combined customer incentives and other costs of about $4.5 billion in 2014 for all customer classes (EIA, 2014e, Table 10.6).[34] The cost-effectiveness of DSM programs is still being debated, with savings ranging from 1 cent to more than 20 cents per kWh saved (Arimura et al., 2011 [citing Gillingham et al., 2006, for a literature survey].

Cities and local governments are also beginning to play leadership roles in deploying energy-efficiency measures. City leaders are looking for ways to reduce the substantial amount of energy waste in local buildings as one step toward making their communities more resilient and sustainable.

One caveat to keep in mind is that the demand for energy efficiency varies across states, in large part because of differences in electricity prices and usage. Great variation also exists within states as a result of heterogeneity among consumers due to differences in preferences, incomes, discount rates, and other factors. Small changes, such as a small change in interest rates, may make a

---

[33]Additional programs include performance incentive programs and lost revenue recovery mechanisms.
[34]http://www.eia.gov/electricity/annual/html/epa_10_06.html.

difference between a consumer's buying a more expensive but more energy-efficient appliance versus a less expensive, less energy-efficient appliance.

CONCLUSION

Energy-efficiency measures targeting electricity usage appear to have contributed to improvements in energy productivity during the past 30 years, likely saving many tens of billions of dollars. In particular, the committee considered evidence for an energy-efficiency gap—the difference between projected savings from avoided energy use due to energy-efficiency measures and the actual savings realized. To the extent that such a gap exists, there may be several barriers to higher utilization of energy-efficiency measures.

First, electricity prices do not, for the most part, incorporate the costs of pollution. Second, even if prices are corrected to include the costs of pollution, other market imperfections may limit consumers' purchases. Information about energy use and price is not always readily available to consumers, and even when it is, consumers may be unable to translate price information into actual costs (or into actual savings in the case of energy efficiency). Additionally, consumers may be reluctant to make new purchases out of inertia, limited attention or heuristics. Moreover, the effectiveness of increased electricity prices in inducing conservation is limited by the very low measured price elasticity of demand for electricity, especially in the short term.

The committee notes that energy efficiency is also an area in which innovation is critical. DOE has several R&D programs under way. The solid-state (LED) lighting program is a joint program with industrial partners that has utilized a road map and produced sizable increases in performance and cost declines. ARPA-E has one program aimed at reducing the costs of heating and cooling buildings and another focused on developing next-generation power switches that could increase the efficiency of appliances and lighting. DOE should be encouraged and supported to create more R&D programs that can ultimately bring much-improved energy-efficiency technologies to the market.

Recommendation 4-5: The federal government, state and local governments, and the private sector should take steps to remove barriers to, provide targeted support for, and place a high priority on the development and deployment of all cost-effective energy-efficiency measures.

5

Addressing the Unique Challenges to the Development and Deployment of Nuclear Power, Carbon Capture and Storage, and Renewable Fuel Power Technologies

To avoid the most harmful effects of rising greenhouse gas (GHG) emissions and criteria air pollutants while also meeting the growing demand for affordable, reliable, and secure energy supplies, the world must embark on a rapid transition to cleaner, low-carbon energy systems. Unfortunately, the current portfolio of cleaner power generation technologies is generally not price-competitive with established technologies such as natural gas- and coal-fired power plants, in part because pollution externalities are not fully incorporated into energy prices, limiting the incentives for the deployment of clean energy technologies (see Chapter 2). Indeed, from 2000 to 2014, more new electricity capacity additions were fired by natural gas than any other single fuel source, with the notable exception of 2012, when wind accounted for the most new installed capacity (EIA, 2016c). The historically low prices for natural gas in the United States are usually cited as the primary reason for this trend, although the cost of constructing a plant still is often lower than is the case for any other generation technology. Innovation in zero- and low-pollution technologies will require an expansive and integrated approach to established, emerging, and new technologies for electricity production, transmission, distribution, and storage. This chapter addresses several major electricity generation technologies with significant potential for dealing with the pollution problem, including carbon dioxide (CO_2) emissions. These technologies include nuclear power, carbon capture and storage (CCS), and renewables. Each has specific characteristics and poses unique challenges for innovation and eventual deployment. The chapter examines these challenges and potential means for overcoming them.

NUCLEAR POWER

Unique Challenges to Innovation and Deployment for Nuclear Power Technologies

In the United States today, nuclear power provides about 20 percent of total electricity and accounts for almost two-thirds of the nation's low-carbon electricity generation, despite recent increases in the deployment of solar and wind power. In spite of its advantages, however, nuclear power faces formidable obstacles that are limiting its use in the United States.

New technologies might help to overcome these obstacles. Moving nuclear technologies forward, however, will require addressing several critical issues, including technical barriers, capital requirements, the regulatory framework, the market for nuclear plants, and risk management. Given the role nuclear power is already playing as a major source of low-carbon electricity generation and the potential for advanced nuclear technologies to expand this role in the future, the committee believes that serious consideration of improving the environment for nuclear innovation is warranted, and offers recommendations in this section as a means to that end.

Nuclear energy also is facing stiff headwinds elsewhere in the world. Rising costs and public concerns over nuclear safety have led some nations to scale back their nuclear growth plans, while others have retreated from nuclear power entirely. In some countries—most notably China, India, Russia, South Korea, and some Middle Eastern countries—there are ambitious plans for nuclear expansion. Several other countries are embarking on nuclear power programs for the first time, and still others are seriously considering doing so. But when all these national plans are combined with the expected retirement of much of the existing nuclear fleet as those plants reach the end of their life, the nuclear role in global carbon mitigation appears likely to grow only slowly in the coming decades and may even shrink (EIA, 2014a). In parallel with these developments, the center of gravity of the global light water reactor-based nuclear energy industry is continuing to shift away from the United States, as suppliers in Russia, Korea, and China gain competitiveness in international markets previously dominated by American, European, and Japanese vendors. As its global presence in nuclear energy diminishes, the United States is less likely to be able to shape the governance institutions needed for safe nuclear operations worldwide, and will also be less able to strengthen the security arrangements with respect to nuclear power and its fuel cycle that are key to achieving nonproliferation goals. For all these reasons, it is important as well to consider the potential role of nuclear innovation in an international context.

Domestic Outlook for Nuclear Energy

Most of the nation's 100 currently operating nuclear power reactors, now largely amortized after having operated for 30 years or more, are supplying competitive, low-cost electricity to the grid. The nuclear fleet also has been directly responsible for the avoidance of both the thousands of premature fatalities and the adverse environmental impacts that would have occurred had the electricity they supplied been generated by fossil fuel power plants instead. However, most currently operating nuclear plants in the United States are expected to be retired between 2030 and 2050, and some will be shut down well before then. Four new reactors are under construction—the first new nuclear builds in decades—and a fifth is now being completed after a long delay. Aside from these five plants, however, there are no firm plans to build any more nuclear power plants in the United States to replace the existing nuclear fleet with new nuclear capacity, and the nation has no strategy for sustaining, let alone expanding, nuclear energy generation.

Five operating reactors have recently closed or will soon do so. Other as yet unannounced retirements may also occur in the next few years. Single-unit plants with relatively high operating and maintenance costs in locations with low wholesale electricity prices and unregulated power markets are most likely to be affected. Continuing uncertainties over the ultimate cost of addressing safety issues raised by the Fukushima accident, as well as the operational impacts of Fukushima-related issues, also may accelerate the retirement schedule of certain plants over the next few years. In its most recent projection, the Department of Energy (DOE) estimates that another 6 gigawatts (GW) of nuclear capacity will be retired before the end of the decade, leaving 98 GW of nuclear capacity in service in 2020, slightly less than the current capacity (EIA, 2014a).

Beyond 2020, the outlook is less certain. More than 70 percent of currently operating reactors have received approval from the U.S. Nuclear Regulatory Commission (U.S. NRC) to continue operating for an extra 20 years beyond the expiration of their original 40-year operating licenses (EIA, 2014a, pp. IF-35). Some are considering applying for a second 20-year operating license extension.

The principal reasons for both early plant retirements and the lack of new nuclear ordering are economic. With natural gas prices at today's low levels, new nuclear plants and even some existing reactors cannot compete with gas-fired combined-cycle plants. The very high initial capital cost and the uncertainties associated with that cost are additional deterrents to new builds. Other factors that discourage utilities from considering new nuclear plant investments include the uncertain outlook for carbon pricing and for the inclusion of nuclear energy in state and federal low-carbon portfolio standards, as well as the continuing uncertainty over the government's plans to manage the spent fuel from nuclear power plants.

According to DOE's "reference" projection, only an additional 4 GW of nuclear capacity will be added between 2020 and 2040. DOE expects there will be no further retirements of nuclear plants during this period, so the nuclear capacity in service in 2040 will remain little changed from today. The assumptions underlying this projection are that annual operating and maintenance costs for nuclear units will remain flat throughout this period (in recent years they have been increasing at about 4 percent per year), and that all of the plants reaching 60 years of life during this period—representing roughly 50 GW of capacity—will apply for and be granted a second 20-year operating license extension.

These assumptions may well be optimistic. The regulatory framework for a second license extension is the same as that for a first extension, although the technical issues around aging are still being fully addressed (NEI, 2015). Operators still will need to assess individual plant safety and economics in determining whether to apply for a second license renewal. It therefore appears plausible that the existing nuclear fleet will shrink during this period rather than remain at its present size (EIA, 2014a).[1]

Since U.S. electricity consumption is projected to grow by 20 percent by 2040, the Energy Information Administration (EIA) estimates that the nuclear share of output will fall to about 17 percent of the total in the reference case, with fossil energy (principally coal and gas) still accounting for 66 percent of all generation at that time. DOE's projections, though, are for far less decarbonization of the electricity system than would be required to achieve an 80 percent reduction in overall U.S. carbon emissions by 2050, as is called for by the United Nations (UN) agreements. According to a recent report of the Deep Decarbonization Pathways Project, a collaborative global initiative to explore how individual countries can reduce their GHG emissions, achieving the 80 percent mitigation target by 2050 would require almost complete decarbonization of the U.S. electricity sector, together with switching a large share of end uses from direct combustion of fossil fuels to electricity or fuels produced from electricity (Williams et al., 2014). Electricity generation would need to approximately double by 2050, while the carbon intensity of the power sector would have to decline to 3-10 percent of its current level.

The Deep Decarbonization analysis shows how this could be achieved through aggressive additions of solar, wind, nuclear, and CCS capacity in various combinations, together with a rapid acceleration in the rate of energy-efficiency gains (Williams et al., 2014, Figure 7 and Table 7). Of the four scenarios studied in that report—high renewables; high nuclear; high CCS; and

---

[1]The Energy Information Administration's (EIA) (2014a) *Annual Energy Outlook* also includes an "Accelerated Nuclear Retirement" scenario, which assumes that operating and maintenance costs for nuclear power plants grow by 3 percent per year through 2040, and that all nuclear power plants not retired for economic reasons are retired after 60 years of operation. In this scenario, 42 gigawatts-electric (GWe) of nuclear capacity is retired by 2040 (see Jones and Leff, 2014).

a mixed case involving a balanced mix of renewables, nuclear, and natural gas with CCS in the electricity sector—the high nuclear scenario is associated with the lowest incremental cost relative to the business-as-usual case (Williams et al., 2014, p. 24).

International Outlook for Nuclear Energy

In other parts of the world, as noted above, the outlook for nuclear energy is mixed. Rising costs and public concerns over safety have led some nations to scale back their nuclear expansion plans, while others are phasing out their programs completely. But in still other countries there are ambitious plans for nuclear expansion, and some growth in the global nuclear sector overall is anticipated in the coming decades.

Today 444 reactors (388 GW) in 31 countries are providing more than 11 percent of the world's electricity (NEI, 2016). About 77 percent of this capacity is in OECD countries, but most new nuclear capacity is being built in non-OECD countries. Sixty-three reactors (62 GW) are currently under construction (NEI, 2016), the highest total in 25 years (WNA, 2015). Almost two-thirds of the new capacity is in China, Russia, India, and South Korea, with China alone accounting for more than 33 percent of the total. Another 509 reactors (372 GW) are on order or in the planning stages, with 60 percent of this capacity again in China, Russia, India, and South Korea (and with China alone accounting for more than one-third of it). Several newcomer countries are in the early stages of implementing nuclear power programs or are seriously considering doing so, including the United Arab Emirates, Turkey, Bangladesh, Vietnam, Jordan, Lithuania, and Saudi Arabia.

On the other hand, since the Fukushima accident, several countries, including Germany, Italy, Switzerland, and Spain, have decided to phase out their nuclear programs. Japan itself, which before Fukushima had the world's third-largest nuclear power program and was planning additional nuclear plant construction on a large scale, may also decide to phase out its reactors, all of which were shut down following the accident. Several other countries that previously were considering entering the nuclear field have decided not to do so. As in the United States, moreover, a large wave of nuclear plant retirements is likely to occur around the world before 2050. The average age of the world's nuclear power plant fleet is almost 30 years (IAEA, 2016, Figure 7), and the International Energy Agency (IEA) projects that almost 200 reactors will be retired before 2040 (IEA, 2014).

Overall, the outlook is for moderate growth in the global nuclear energy sector. In its "New Policies" scenario, the IEA projects an increase of 60 percent in world nuclear generating capacity to 624 GW by 2040, with the share of nuclear power in global electricity generation increasing slightly to 12 percent at that time (IEA, 2014). The EIA projects an almost 90 percent increase in world nuclear capacity over the same period, and a slightly higher nuclear share of

14 percent in 2040 (EIA, 2013c). As in the United States, however, the net nuclear contribution globally, after taking account of current plans for both new construction and plant retirements, falls far short of what appears likely to be required if the world is to stay within the limit on average temperature rise of 2° C over the preindustrial equilibrium (IEA, 2014).[2]

Prospects and Obstacles for Innovation in Nuclear Power

The gap between current plans and the possible need for significant nuclear power growth in the future thus is wide, both in the United States and internationally. A continuation of current policies and approaches is unlikely to be sufficient to close this gap. Although plants built today are much safer than those from 40 years ago when the first reactors at Fukushima were built, nuclear energy is likely to require additional safety improvements, lower costs, and a better economic case, as well as more stringent and reliable security against the threats of nuclear proliferation and terrorism. Also, to compete effectively with incumbent, high-carbon technologies and fuels as well as new low-carbon alternatives, nuclear power will have to be adapted to the evolving characteristics of electric power grids, new ways of delivering electricity services, and the diverse needs of generators in advanced and developing countries.

Improvements in nuclear management and governance will be essential to achieving safe and secure nuclear operations and addressing public concerns about nuclear safety. But technological innovations in nuclear power plants and the nuclear fuel cycle may also be necessary to realize the potential of nuclear energy. Innovators seeking to commercialize new nuclear technologies in the United States face formidable obstacles, however. Long lead times, the high costs of development and demonstration, and the lack of a clear regulatory framework are all deterrents to private investment in new nuclear technology development. Government support for nuclear innovation, which played a decisive role in the early years of the nuclear energy industry, has been far more limited in recent years and has been criticized for its lack of direction. The current absence of a regulatory framework is especially problematic for private nuclear reactor developers, some of whom are reportedly planning to build their first full-scale reactors overseas because they think doing so in the United States will be impossible under the existing regulatory regime.

Commercializing nuclear-related innovations is an expensive, lengthy, and risky process. Even incremental improvements to existing nuclear technologies can take many years to develop, and if more far-reaching innovations are to be available commercially when age-related attrition of the existing nuclear fleet begins in earnest in the 2030s, large-scale development will need to begin now and be sustained throughout this period. The necessary level of public and

[2]See the IEA (2014) 450 parts per million (ppm) scenario.

private investment will be substantial, and will require an effective partnership between the federal government and the nuclear industry.

Despite these obstacles, a growing number of groups in the U.S. industry, the national laboratories, and universities have launched efforts to develop more advanced nuclear power reactor and fuel-cycle technologies. These technical developments are intended variously to reduce economic costs and financial risks, enhance safety, facilitate nuclear waste management, and lessen the risks of nuclear proliferation. These initiatives range from incremental improvements in current light water reactor technologies to the development of alternative reactor systems with different types of fuel and coolant and different approaches to siting, construction, operation, and waste management, including sodium-cooled, gas-cooled, and molten salt-cooled reactors (see, e.g., Buongiorno et al., 2015; Forsberg et al., 2014; Hejzlar, 2013; Nathan, 2013; NuScale Power, 2013; Rawls et al., 2014). Several other countries also are developing advanced nuclear technologies (see, e.g., Chen, 2012; Kim et al., 2014; IAEA, n.d.; Sun, 2013). The output capacity of some of these designs is in the 1,000 megawatt (MW) range, similar to that of large conventional light water reactors, but others are more compact, in some cases ranging below 50 MW in size. Many are small, modular designs and would be constructed primarily in factories rather than in the field, as now. Most rely to a greater extent than conventional light water reactors on passive mechanisms to ensure safety, and some concepts approach the limit of "walkaway" safety, in which no external intervention by either operators or active engineered systems would be necessary for safe shutdown. Some of these designs are optimized to minimize waste production or to burn nuclear waste from other reactors. Others, by offering the capability for rapid and efficient load following, would be well suited to grids with large amounts of intermittent solar and wind power generation.

Views vary as to the economic feasibility of these new technologies and the lead times required for their commercialization. Their developers believe they have the potential to provide large amounts of low-carbon electricity safely and economically; others are more skeptical. But all would probably agree that there is at present no clear pathway to determining the commercial viability of these advanced systems.

Commercializing a new reactor technology would likely cost billions of dollars, and given the formidable financial risks involved, it is not credible that such an effort could be undertaken in the absence of public risk and cost sharing. DOE is carrying out early-stage research, mainly at its national laboratories, on a fairly broad range of reactor technologies. With one exception, however, it is not currently providing support for the downstream stages of the innovation process, and it has no plans at present to build prototypes of advanced reactors.[3] (The

[3]Until recently, DOE planned to build a prototype high-temperature gas-cooled reactor, the so-called Next Generation Nuclear Plant Project (NGNP), as is called for in the Energy Policy Act of 2005. DOE recently announced its intention not to proceed with Phase 2 design activities for this project (see GAO, 2014).

exception is DOE's Small Modular Reactor [SMR] Licensing Technical Support Program. Under this nearly $500 million program, DOE is seeking to accelerate the commercialization of two small modular light water reactor systems by supporting the certification and licensing of these designs.[4]) Advanced reactor designs also are hampered by the paucity of suitable national facilities that can conduct tests at the high temperatures and high neutron flux densities that are required for many new designs.

The current nuclear plant licensing framework, administered primarily by U.S. NRC, is tailored to light water reactor technology. It has developed over many decades, and it is generally regarded around the world as providing a strong environment for licensing of established light water reactor power plant designs. It is less well suited to the task of licensing more advanced concepts. Even for small modular light water reactor designs, the cost to their private developers of navigating the regulatory process has been estimated at several hundred million dollars, and as noted above, DOE has provided complementary funds to help support such efforts. (Previously, DOE provided support for licensing of the AP600 and AP1000 pressurized light water reactor designs developed by Westinghouse.) For innovators developing non-light water reactor technologies, the regulatory hurdles are greater. For these technologies, there is no clear regulatory pathway at present in the United States, and the cost and time required to develop this pathway will be substantial.

Some experts, including many at U.S. NRC, believe that existing regulatory procedures, standards, and criteria could be applied to the new designs with relatively modest changes. Others argue that attempts to apply the current regulatory framework to reactor systems with fundamentally different design and safety features are unlikely to succeed, and that new frameworks will be needed to treat these differences effectively.

U.S. NRC has recently been considering a different regulatory approach that would place more emphasis on risk- and performance-based standards. In principle, this approach could lead to a so-called technology-neutral licensing framework for advanced reactors (U.S. NRC, 2007, 2012). Some developers, however, eager to move forward with their projects and anxious about the prospect of a lengthy, broad-based rulemaking for a generic licensing approach, may see an advantage in the development of regulatory standards and criteria specifically for their technologies.

A related issue concerns the procedures to be followed for advanced reactor licensing. Currently, U.S. NRC follows a one-step procedure in which the developer must make large investments in the technology before the design can be certified. The presumption underlying this approach is that once a design has been certified, a large number of identical reactors will be built based on the

---

[4]One of the two companies participating in this program has substantially reduced its funding for SMR development and extended the project timeline into the early 2020s (see Nuclear Engineering International, 2014).

original license. While this is a sensible approach for licensing incremental advances in existing light water reactor technology, it is more likely that the designs for non-light water reactor technologies will evolve rapidly in the early stages of technology adoption as new information is generated by construction and operating experience. A one-step design certification framework would appear to be inconsistent with such a trajectory. Moreover, a regulatory approach that would require investors to commit billions of dollars in "all-or-nothing" funding for essentially the entire cycle of development, design, and commercial design optimization without a safety ruling from U.S. NRC would be a strong deterrent to privately financed efforts to commercialize advanced reactors.

An alternative approach, somewhat similar to the way in which the Food and Drug Administration (FDA) licenses new drugs today, would involve a staged licensing process,[5] in which U.S. NRC would conduct a series of interim safety reviews and issue limited licensing decisions bearing on each successive stage of the development cycle, including pilot-scale prototype and full-scale precommercial demonstration projects, gathering additional information at each stage. The early regulatory feedback would reduce the financial risks to private developers by providing clear signals as to whether the new technology was likely to be able to meet final regulatory criteria, and would allow design changes to be made earlier in the development cycle when they would be less expensive in terms of both time and money.

At present, neither the technical nor the procedural aspects of U.S. NRC's approach to advanced reactor regulation are known. Over the years, U.S. NRC staff have identified a number of specific technical and policy issues that would be associated with the licensing of advanced reactors,[6] and in 2012 the agency produced a report requested by Congress addressing its overall strategy for preparing for the licensing of advanced non-light water reactors (U.S. NRC, 2012). But much about U.S. NRC's approach remains uncertain, as the committee observed during its fact-finding activities. This uncertainty has already led some developers to decide to move some of their development activities outside the United States, even though a U.S. NRC license would

[5]The FDA's process follows, roughly, three stages. Stage one involves data from laboratory and other controlled-environment testing, such as animal testing. This stage would be analogous to simulations and laboratory testing for nuclear technologies. Stage two involves clinical trials. This stage would be analogous to test bed testing and demonstration projects. Step three is the actual application to the FDA, when the applicant summarizes the outcomes from stages one and two and formally requests FDA review. The FDA then conducts a preliminary review of the outcomes to determine whether a full review is warranted. If so, the FDA conducts a full review (see http://www.fda.gov/Drugs/DevelopmentApprovalProcess/HowDrugsareDevelopedandApproved/ApprovalApplications/InvestigationalNewDrugINDApplication/default.htm).

[6]For U.S. NRC papers, memoranda, and other documents, see http://www.nrc.gov/reactors/new-reactors/regs-guides-comm/related-documents.html.

likely provide some commercial benefit in international markets (Behr, 2011). Other developers may follow suit. There are concerns that U.S. NRC may have neither the experience nor the resources to undertake this new and unfamiliar task in a timely way. Additional resources and a clear mandate, perhaps provided by Congress, may be needed to ensure timely action to establish a predictable, well-defined licensing process for advanced reactors in the United States.[7]

Other energy regulations may reinforce the disincentives for investment in nuclear innovation. Examples include the absence of a carbon price and the existence in many states of standards designed to promote the adoption of a portfolio of low-carbon technologies from which nuclear technologies have specifically been excluded. Another major obstacle is the structural underinvestment in innovation of all kinds by U.S. utilities (see Chapter 3). In addition, given widespread expectations of a prolonged period of ample natural gas supplies at relatively low prices, most electric power companies currently have little interest in developing and deploying alternative technologies for central station baseload generation. The disincentives for such investments have been reinforced in parts of the country where expanding wind and solar supplies are driving wholesale power prices down to low or even negative levels at times of high wind and solar output. Elsewhere, in states where vertically integrated utilities are still subject to traditional cost-of-service regulation, regulators often are reluctant to allow utilities to pass technology risk on to ratepayers even if there is a realistic prospect for a long-term reduction in costs.

A third obstacle that uniquely deters nuclear innovation in the United States is the continued lack of progress in resolving the spent fuel management issue. The absence of an agreed-upon national policy and plan for interim storage and final disposal of spent fuel is a major impediment to private investment in the development of advanced nuclear power plant technologies.[8]

All the technologies referred to above involve the fissioning of heavy atoms to release energy. The technology of controlled thermonuclear fusion, in which energy is released during the fusing of light atoms, is very different, and no functional fusion power reactor has yet been built despite many decades of R&D and the investment of billions of dollars. In magnetic confinement fusion, one of the two main approaches to achieving fusion, low-density deuterium and tritium fuel is heated to 100 million degrees while remaining contained for long enough by a powerful magnetic field for the fuel to react. In inertial confinement

---

[7]At the time of this writing (2016), bills had been introduced in the U.S. Senate (S.2795) and the House (HR4979) that recognize the need to update the nuclear licensing regulatory framework.

[8]For example, California law prohibits construction of any new nuclear power facilities until the California Energy Commission has determined that the federal government has identified and approved a demonstrated technology for either (1) the construction and operation of nuclear fuel rod reprocessing plants, or (2) the permanent disposal of high-level nuclear waste (1976 Cal. Stats., Ch. 196, § 1).

fusion, intense lasers or particle beams are used to compress and heat up a small frozen pellet of deuterium and tritium fuel (the "target"), yielding a microburst of energy (NRC, 2013a).

Significant advances have been achieved in both magnetic and inertial confinement fusion, but neither approach has yet demonstrated that it can produce more power than must be provided for operation. Progress has been slower than expected, and it remains to be seen whether a cost-effective, power-producing reactor can be developed. Nevertheless, the promise of fusion is so great that continued work is considered worthwhile. Sufficient fusion fuel exists to supply the entire world's energy needs for millions of years. Furthermore, fusion power plants produce no GHGs and, if appropriately designed, little or no long-lived radioactive waste. However, achieving fusion at the scale needed for energy generation at a competitive cost is a formidable challenge, and the large and costly facilities needed to demonstrate its feasibility are so expensive that international collaboration is and will remain necessary.

Findings and Recommendations on Promoting Innovation and Deployment for Nuclear Power Technologies

A number of actions could be undertaken in the United States to create a more amenable environment for innovation in nuclear power technologies. These actions include reforming the regulatory and licensing framework; providing better support for demonstration projects; improving and expanding international cooperation for testing, demonstration, and deployment; enacting legislation to address spent fuel concerns; developing new mechanisms for addressing the funding gap for demonstration and adoption of low-carbon energy technologies; and reorienting the U.S. fusion program.

> **Finding 5-1:** The current U.S. nuclear regulatory system has evolved for the purpose of licensing mature light water reactor power plant technology, but U.S. NRC will likely need to develop new regulatory approaches to address the needs of advanced, non-light water reactors.
>
> **Recommendation 5-1: U.S. NRC, on an accelerated basis, should prepare for a rulemaking on the licensing of advanced nuclear reactors that would establish (1) a risk-informed regulatory pathway for considering advanced non-light water reactor technologies, and (2) a staged licensing process, with clear milestones and increasing levels of review at each stage, from conceptual design to full-scale commercial deployment.**

To implement this recommendation, U.S. NRC might accelerate its efforts to allow for consideration of advanced technologies based on risk- and performance-informed criteria rather than technology-specific prescriptive specifications. A staged licensing process for advanced technologies might provide interim reviews at each stage, from conceptual design, through precommercial pilot and full-scale demonstration facilities, to commercial deployment. Finally, U.S. NRC might consider stationing a small team of highly capable U.S. NRC experts, tasked with developing safety requirements for new kinds of reactors in cooperation with nuclear developers from the private sector, at an advanced research and test facility dedicated to that task. There they would work collaboratively with developers to learn about facility construction and operations and build the expertise needed for commercial licensing.

Finding 5-2: Pilot- or full-scale nuclear reactor demonstration projects are likely to cost hundreds of millions of dollars or more.

Much of this cost is not for the reactor itself but for the associated site costs, power block, and other infrastructure. A flexible, "plug and play" platform for qualified nuclear innovators could significantly reduce the cost of demonstrations. By providing sites for such facilities and by working with interested state governments to enable public financing assistance for demonstration projects, the federal government would encourage the emergence of new development consortia of nuclear innovators, prospective owner-operators, and financiers. As elaborated in Chapter 3, a national test bed would help innovators find partners and resources for effective testing of advanced technologies. Given the scale of capital required and the technological complexity of next-generation nuclear technologies, a dedicated facility capable of supporting private-party initiatives to test and demonstrate innovative nuclear technologies would be of particular benefit. Thus, a dedicated nuclear test bed would be a key component of the Technology Test Bed and Simulation Network proposed in Chapter 3.

The test bed site would be capable of hosting a broad range of advanced nuclear technology test and demonstration facilities. It would offer power and water supplies; postirradiation examination facilities for studying fuels and other materials; fuel transportation and storage facilities; security infrastructure; and comprehensive site characterization services, including environmental and seismic information. The fee structure for site services would be clearly specified in advance. These facilities and services would be available to both domestic and international nuclear development consortia.[9] Responsibility for

[9]At the time of this writing (2016), Idaho National Laboratory (INL) had test facilities, including a test reactor. This reactor, though, is a traditional light water design. INL's facilities are available for free following a peer-review approval of research proposals. Also at the time of this writing, DOE's Nuclear Energy Advisory Committee was nearing

safety oversight of the site would be shared between DOE and U.S. NRC. U.S. NRC would be responsible for facility licensing of prototype and demonstration advanced reactors as required by the staged licensing review process described previously.

Breaking the congressional deadlock on spent fuel management is essential to making progress on this issue. An especially important step to signal progress is to authorize away-from-reactor storage for spent fuel. The federal government should move to implement the recommendations of the Blue Ribbon Commission on America's Nuclear Future, which lay out a comprehensive and practical approach to nuclear waste management.

> **Finding 5-3:** The development of advanced nuclear technologies is very costly. Other countries also are working on such technologies and may provide a more hospitable environment for their development and deployment.
>
> **Recommendation 5-2: While the U.S. government should be cognizant of the importance to its environmental, economic, security, and climate policy goals of maintaining a healthy environment for nuclear innovation domestically, it should also support and encourage expanded international cooperation in testing, demonstration, and commercial deployment of advanced nuclear technologies.**

Some private U.S. nuclear developers will look overseas to carry out or participate in testing, demonstration, and scale-up projects. The federal government should provide support for these efforts by enabling related contributions by U.S. national laboratories, universities, and regulatory organizations. This may also require a review and rationalization of U.S. nuclear export controls consistent with national security goals and policies on nuclear nonproliferation. Given the long lead time, large expense, and high regulatory and market risks of developing and demonstrating advanced nuclear technologies, it is unlikely that private companies will pursue these activities successfully without complementary public investments.

Financing and managing the demonstration and early-adoption stages of the innovation process for large-scale energy technologies, including nuclear, is a continuing challenge. There is a mismatch between interests and capabilities. As detailed in Chapter 3, venture equity investors are structured to finance technology development but not major project assets. Project investors are structured to finance large assets but not to take on the risk of technology scale-

---

completion of a report on an advanced test and/or demonstration reactor that would inform DOE's Office of Nuclear Energy. The final report was expected in April 2016.

up. Buyers of power, such as utilities, are regulated to keep short-term costs down, but the cost of electricity generated by demonstration and "first few" projects will often be higher than that of electricity generated by the incumbent technologies, and this cost gap may persist for some time. The federal government has stepped in to help fill the gap in the past, sometimes with private-sector cost sharing, but many of these attempts have failed, and troubled projects have been common. New public funding mechanisms are needed that would also draw on private capital while avoiding dependence on the annual congressional appropriations process. One possibility is to build on existing state-level public benefit charges instead of creating a new funding mechanism, perhaps with an added incentive from a federally mandated innovation surcharge on either electricity sales or transmission. This type of approach would create a financing pool that could incentivize the formation of regional public/private energy innovation and deployment partnerships. One such approach would entail the establishment of Regional Innovation Demonstration Funds, as described in Chapter 3.

Lastly, the potential benefits of both magnetic and inertial confinement fusion are great, and significant technical advances continue to be made, even though progress has been slower than expected, and commercialization remains a distant prospect. Nonetheless, the tremendous potential benefits of fusion power warrant considering the adoption of a flexible U.S. investment strategy for fusion R&D that would incorporate a sensible balance between domestic and international collaborations as part of an overall program of strong support for the development of cleaner long-term energy options. A recent and laudable effort at expanding possible avenues for fusion development comes from an Advanced Research Projects Agency-Energy (ARPA-E) program designed to "create and demonstrate tools to aid in the development of new, lower-cost pathways to fusion power and to enable more rapid progress in fusion research and development."[10] Still, given the long timeline and large expenditures likely necessary to create a commercially viable system, it is important to give careful consideration to positioning the U.S. fusion R&D program appropriately relative to other long-term energy options, while balancing the multiple competing demands within research programs against the limited resources available to them and also retaining sufficient flexibility to adapt to new discoveries and opportunities.

[10]For an Accelerating Low-Cost Plasma Heating and Assembly (ALPHA) program overview, see http://arpa-e.energy.gov/sites/default/files/documents/files/ALPHA_ProgramOverview.pdf.

CARBON CAPTURE AND STORAGE

The timing and scale of the energy and environmental challenges described throughout this report necessitate the use of a portfolio approach to achieve an increasingly clean energy sector. Given the likely role of fossil fuels in the future electric power generation mix for years to come and the dramatic reductions in GHGs that can be realized through CCS technologies, moving these technologies for both coal and natural gas generators through the development, demonstration, and deployment stages remains critical. This section examines the role CCS could play in the future global power generation sector; the status of CCS power-sector projects around the world; the range of market and nonmarket barriers to CCS; and discrete, implementable actions the federal government can take to support CCS technology development and deployment.

The Role of Carbon Capture and Storage in the Future Global Electricity Portfolio

The development of CCS technologies for fossil fuel-based electricity generating stations through research, successful technology demonstration, and eventual deployment at scale is an important component of an overall strategy for achieving an increasingly clean energy sector. To a large extent, the importance of CCS is driven by the sheer magnitude of available quantities of fossil fuels across the globe and their associated low prices. Prices for fossil fuels are currently very low. In February 2016, the average price of coal per million BTUs (MMBtu) delivered to the electric power sector was $2.17. While average delivered prices of natural gas have fluctuated, the recent price of $2.33 per MMBtu is near historical lows (the average 2008 price, by comparison, was $12.40 per MMBtu) (EIA, 2016e; FERC, n.d.).[11] When production profiles of power generation technologies are considered, the levelized cost of electricity (LCOE) also shows the current cost advantage of fossil fuel-based generation. EIA estimates the LCOE for an advanced carbon-emitting gas plant entering service in 2022 to be more than 70 percent lower on a dollar per megawatt hour (MWh) basis than that for a pollution-free solar photovoltaic (PV) plant, and approximately 43 percent less than the cost for an offshore wind plant (EIA, 2015f), (2016g). (See Chapter 2 and Appendix B for more detail on estimates of the LCOEs for various technologies.)

Thus, EIA, IEA, and private-sector reference forecasts for energy use all project fossil fuel-based energy (i.e., coal, natural gas, and petroleum) to make up approximately 60-70 percent of energy inputs for power generation between

---

[11]EIA Form EIA-923, "Power Plant Operations Report," and predecessor form(s), including Form EIA-423, "Monthly Cost and Quality of Fuels for Electric Plants Report."

2035 and 2040—a figure that is relatively consistent with current levels of use.[12] Additionally, even under more optimistic scenarios that assume successful commitments to and plans for reducing emissions, such as IEA's "New Policies" scenario, fossil fuel generation still provides the majority of all electricity capacity (53 percent) in 2035.[13] Therefore, not only is "legacy" power generation fossil fuel-based, but recent and anticipated capacity additions globally have been and will continue to be dominated by fossil fuels under a wide range of future scenarios. In fact, given the variability and intermittency of most renewable sources of power, fossil fuel plants are likely to continue to be used to compensate for the fluctuations in wind and solar generation (Oates and Jaramillo, 2013; Valentino et al., 2012).

> **Finding 5-4:** Fossil fuels will remain a large and important component of the fuel mix for electricity generation in the United States and around the world for many decades to come.

There is general consensus that, to avoid the most dangerous and costly effects of climate change, the global average surface temperature increase should be limited to a maximum of 2° C over the preindustrial equilibrium (UN, 2015). This increase corresponds to an atmospheric concentration of GHGs of approximately 450 parts per million (ppm) or less, on a CO_2-equivalent basis (IPCC, 2014b). However, because most GHGs, unlike conventional pollutants, remain in the atmosphere for generations, climate stabilization at any given temperature will require global aggregate CO_2 emissions to fall at a rate far below that at which natural processes can remove them from the atmosphere. The long residence time and global dispersion of CO_2 emissions in the atmosphere, in particular, make this reduction exceptionally challenging, and global emissions will need to fall to a small fraction of their current level by the end of the century to limit global average temperature increases to the 2° C level (IPCC, 2007). Because fossil fuels will be such a large and important energy source for many years to come, mitigating CO_2 emissions will require decarbonizing fossil fuel-based electric power generation.

---

[12]EIA's *Annual Energy Outlook 2014* "Reference Case" projection anticipates that 68 percent of total U.S. electricity generation in 2040 will be supplied from coal, natural gas, and petroleum. This is virtually identical to the percentage supplied from coal, natural gas, and petroleum in 2012 (calculated from EIA [2014a, Table A8, p. A-18]). IEA's *World Energy Outlook 2012* "Current Policies" scenario anticipates that fossil fuels will supply 66 percent of global electricity generation in 2035 (calculated from IEA [2012, Table 6.2, p. 182]). BP's *Energy Outlook 2035* projects that fossil fuels will make up more than 60 percent of primary energy inputs to power generation in 2035 (BP, 2014).

[13]This figure (53 percent) was calculated using IEA (2012, Electricity Capacity Table, p. 554).

To date, various independent analyses have underscored the importance of CCS strategies in any successful effort to decarbonize power generation. For example, the Sustainable Development Solutions Network (SDSN) and the Institute for Sustainable Development and International Relations (IDDRI) have been engaged in ongoing analysis of deep decarbonization pathways for 15 nations (Australia, Brazil, Canada, China, France, Germany, India, Indonesia, Japan, Mexico, Russia, South Africa, South Korea, the United Kingdom, and the United States) at different stages of development, remaining cognizant of their differing circumstances and capacities (Williams et al., 2014). More than half of those countries would need to supply a substantial amount of their electricity from fossils fuels using CCS (Williams et al., 2014).

Other experts also have consistently highlighted the critical role of CCS technology in meeting the world's climate goals. The Massachusetts Institute of Technology (MIT) states that (1) per Btu, coal is a low-cost "mainstay of both the developed and developing world, and its use is projected to increase"; (2) "because of coal's high carbon content, increasing use will exacerbate the problem of climate change unless coal plants are deployed with very high efficiency and large scale CCS is implemented"; and (3) "CCS is the critical future technology option for reducing CO_2 emissions" while allowing coal to meet future energy needs (MIT, 2007, p. x).

However, even under a dominant regime of switching from coal to natural gas (e.g., because of price impacts or lower emissions), technologies for drastically reducing emissions, such as CCS, will remain critical given the relatively high carbon content of natural gas compared with alternative fuels (e.g., renewables, nuclear) (C2ES, 2013). On a CO_2 emission rate basis (pounds of CO_2 per MWh), the conventional combustion of natural gas releases approximately 50 percent of the emissions produced by coal (EIA, 2014d). As reported by Biello (2014), Intergovernmental Panel on Climate Change (IPCC) Working Group III Co-Chair Ottmar Edenhofer stated in 2014, "We depend on removing large amounts of CO_2 from the atmosphere in order to bring concentrations well below 450 [parts-per-million] in 2100." Biello goes on to note, "Ultimately, he [Edenhofer] said, keeping global temperature rise to 2 degrees without any CCS would require phasing out fossil fuels entirely within 'the next few decades'" (Biello, 2014).

Finding 5-5: Efficient and cost-effective technologies for capturing and either storing or utilizing CO_2 and other GHGs from power plants will be a necessary and important component of a portfolio of measures for abating GHG emissions.

Status of Carbon Capture and Storage Projects, Market Barriers, and the Need for More R&D

As described in Chapter 2, CCS technologies have not yet reached performance or price levels that would make the widespread capture of CO_2 from power plants commercially competitive, especially absent a price on emissions. In particular, while many component technologies are now or will soon be available, they have yet to be integrated into large-scale commercial projects (IEA, 2013).

Currently, 15 large-scale[14] CCS projects are operational, 7 are under construction, and another 33 are in development or in the early planning stages around the world. These 15 projects represent a wide range of industries, including power plants, natural gas processing, fertilizer production, hydrogen production, and others (Global CCS Institute, 2016). The number of such projects in operation and under construction is double the number in 2000 (Global CCS Institute, 2016). Within the power sector, large-scale CCS projects are just now being realized (MIT, 2016). In 2014, the Boundary Dam Integrated Carbon Capture and Storage Demonstration Project near Estevan, Saskatchewan, in Canada commenced operations to become the world's first large-scale CCS project for power generation, with a CO_2 capture capacity of 1 million tons per annum (Mtpa). In the United States, one plant began generating power in 2016, construction began on another in 2014, one other is still in the planning stages, and another was canceled in 2016. Elsewhere in the world, 16 full-scale power generation CCS projects are in the planning stages as well (MIT, 2016).

Globally, 11 pilot-scale power generation plants with CCS (ranging from 1 MW to 50 MW) have completed demonstration (3 in the United States between 2008 and 2011), with another 8 currently in operation (2 in the United States) and 3 in the earlier planning stages (MIT, 2016). In addition, construction began in 2016 on a natural gas-fired plant that uses oxy-combustion technology to produce a high-pressure and high-quality "pipeline-ready" CO_2 by-product (Net Power, 2014). Dedicated geologic storage has been demonstrated at a handful of these pilot-scale plants, and is also planned for several large-scale plants. However, all current large-scale power plants capture and transport carbon only for enhanced oil recovery (EOR) applications (MIT, 2016). (For more detail on the technology readiness of CCS technologies, see Chapter 2 and Appendix D.)

Although these data show that facilities and projects have demonstrated some of the critical aspects of CCS processes and engineering practice within

---

[14]The Global CCS Institute measures projects in terms of the facility's storage capability. Large-scale projects are defined as those capturing more than 0.8 million tons per annum (Mtpa). In contrast, MIT is measuring only power plants with CCS and differentiates large-scale and pilot projects by their electricity generating capacity, with pilot projects having a capacity less than 50 MW.

various regions and several different sectors, applications in the power sector still are only emerging. The "energy penalty" (the increase in energy input per unit of energy output) associated with power generation plants with CCS is not inconsequential: a pulverized coal plant, an integrated gasification-combined cycle (IGCC) plant, and a combined-cycle natural gas plant use 31 percent, 16 percent, and 17 percent more energy, respectively, than their non-CCS counterparts (Rubin et al., 2007a). Compared with theoretical minimums, as outlined in the technology readiness assessments in Appendix D, current technologies also use three times more energy to capture and compress CO_2 than they would without CCS. Except for EOR applications, neither the thermodynamics nor current markets presently favor CO_2 "utilization." Theoretical calculations, however, do show that there is much to be gained from improving the technology.

Additionally, with respect to cost, a coal-fired IGCC power plant with CCS entering service in 2022 is projected to produce electricity at an average LCOE of approximately $82/MWh (in 2015 dollars), while the average LCOE for an advanced combined-cycle natural gas power plant with CCS is projected to be $87/MWh. Meanwhile, the LCOEs for new conventional coal-fired power plants and advanced combined-cycle natural gas plants would average an estimated $99/MWh and $59/MWh, respectively. Table B-1 in Appendix B outlines estimated costs of various emitting and nonemitting power generation technologies. It is important to note the distinction between those technologies that can dispatch on demand and those that cannot. The true costs of the latter are often higher because of the need for backup generation, storage, or other methods of dealing with their intermittency and variability. Even when various externalities are considered, both coal- and natural gas-fired power generation technologies with CCS technology still exhibit higher LCOEs than those without CCS.

Overall, while power generation plants with CCS cannot be expected to be less expensive than their non-CCS counterparts on a purely capital expenditure/technical basis, there is significant opportunity for research, development, and demonstration in CCS technology to begin leveling the playing field between fossil fuel-based power generation with CCS and alternative zero-carbon technologies such as wind. This opportunity means that while several pilot projects and one large-scale plant with CCS technologies are now operational, additional innovation and technological development are needed to make CCS technologies commercially competitive at the scale required for carbon abatement.

Key Nonmarket Barriers to the Development and Deployment of Carbon Capture and Storage Technologies

Operational concerns about CO_2 capture and transport risks, as well as about the safety and integrity of CO_2 storage in underground structures, are major nonmarket barriers to permanent geologic storage of CO_2.

The operational concerns include perceptions about possible effects on water tables and other issues related to caprock and injection operations. Moreover, these concerns extend beyond those of the public and of environmental nongovernmental organizations (NGOs). In a February 2014 conversation with the National Association of Regulatory Utility Commissioners (NARUC), several commissioners expressed concerns about the lack of certainty with respect to suitable storage sites.[15] DOE estimates that 180-240 gigatonnes of CO_2 storage is available in oil and gas reservoirs and in unmineable coal reservoirs. DOE also estimates that 1,610-20,155 gigatonnes of storage is available in saline formations. While saline formations offer substantial opportunities, they are not as well identified or characterized as oil, gas, and coal reservoirs (NACAP, 2012).

The safety concerns include the possibility of carbon leakage, loss of well integrity, induced seismicity, and potential human health or other environmental impacts. Singleton and colleagues (2009) studied different notions of risk and public risk perceptions around the geologic storage of CO_2. They concluded that while the public eventually will perceive CCS risks as similar to those associated with existing conventional fossil fuel technologies, the perceived risks will be higher in the interim because of the emerging nature of CCS and the paucity of demonstrations proving its safety (Singleton et al., 2009).

Another major obstacle to the development and deployment of CCS is the current lack of a uniform regulatory framework for managing the access to and use of underground pore space, siting and constructing CO_2 pipelines, permitting or licensing storage activities on federal lands, and managing the long-term stewardship of closed injection sites (MIT, 2007). Additionally, several issues surrounding long-term liability and ownership of the performance/behavior of the CO_2 in underground storage remain unresolved. Open questions include how best to address operational liability (liability associated with the actual capture, transport, and injection activities of CCS); climate liability (associated with potential leakage and contribution to thwarting of climate goals); and in situ liability (associated with possible CO_2 migration within the rock formation or induced seismic activity) (de Figueiredi et al., 2005). Combined, these issues heighten the perceived risk faced by firms seeking to employ CCS. Finally, the inability of electric power utilities to routinely recover the costs of

[15]Personal communication, NARUC commissioners, February 11, 2014.

demonstration plants from their governing public utility commissions has created a general disincentive for utilities to act as innovators (MIT, 2016).[16]

The past inability of the government to take the lead on critical CCS issues has not inspired confidence among the power sector and represents another nonmarket barrier to CCS deployment and development. Since 2003, numerous restructuring and resiting efforts, cancellations, and other delays in DOE planned CCS demonstrations (FutureGen and FutureGen2.0) have left the public wary of the technical aspects of CCS and the industry wary of the type of sustained government support that can be anticipated for emerging technologies.

> **Finding 5-6:** The risks involved in transporting and storing CO_2 and the lack of a regulatory regime are key barriers to developing and deploying technically viable and commercially competitive CCS technologies for the power sector at scale.

Actions to Promote the Development and Deployment of Carbon Capture and Storage Technologies

One of the most important actions that can be taken to promote the development and timely widespread deployment of CCS technologies in the power sector is the implementation of enough demonstration projects to prove the technologies' viability and efficiency at scale. CCS experts concur. MIT authors have urged that large-scale projects be undertaken to demonstrate the technical, economic, and environmental performance of an integrated CCS system as soon as possible. They also have suggested that several integrated large-scale demonstrations with appropriate measurement, monitoring, and verification are needed in the United States over the next decade, with government support (MIT, 2007). These demonstrations are important not only for establishing the technology itself but also for gaining the public's trust (both in the effectiveness of the technology for large-scale storage and in government leadership). Implementing a comprehensive regulatory regime for the operation of CCS power generation projects is a second priority for realizing large-scale CCS (MIT, 2007).

Given the centrality of the cost differentials between coal- and natural gas-fired power plants with CCS and other clean power generation technologies, there are several actions the government can take to increase the competitiveness of cleaner fossil fuel power generation.

As discussed in Chapter 3, new approaches and increased funding are needed to strengthen the energy innovation system. Widespread adoption of CCS will require significant reductions in the cost and the energy penalties associated with CCS technologies. Given the importance of commercially

---

[16]See Chapter 6 for additional detail on utility regulation, including cost recovery under "cost-of-service" regulatory systems.

competitive CCS options to any global transition to a low-carbon future, increased research and development (R&D) efforts are needed to reduce the cost of CCS, including both retrofit and advanced combustion technologies. For example, DOE conducts CCS research at the National Energy Technology Laboratory (NETL), but the research is carried out under the Clean Coal Research Program and thus limited to coal.[17] Given the fuel switch from coal to natural gas that is currently under way, it is important that CCS research be conducted on natural gas as well. Moreover, NETL currently lacks a test bed type of resource that could enable large-scale demonstration and testing of CCS technologies for either coal or natural gas.

Given the fiscal pressures on discretionary federal spending, the difficulties of demonstrating commercial projects at scale under federal/DOE guidelines, and other complicating factors, the committee believes that Congress's consideration of this approach as a means of accelerating the development and demonstration of critical CCS technologies is warranted. One mechanism that might be considered for this purpose is the Regional Innovation Demonstration Funds proposed in Chapter 3. Additionally, cooperation between the federal government and regional utilities could help avoid jurisdictional issues that might occur with a federally mandated program. For example, DOE could cooperate with NARUC to design and develop possible funding mechanisms for supporting innovation efforts such as expanded research, development, and demonstration of a range of cleaner energy technologies that could include CCS. Again, this could be an important target of the proposed Regional Innovation Demonstration Funds.

The use of Pioneer Project Credits could provide incentives for innovation in and the deployment of CCS technologies. A limited number of targeted production tax credits could be offered to a pioneer tranche of natural gas-fired power plants employing carbon capture technologies. Such credits could be offered using both a reservation system (to assure project financiers that the credit could be used by the plant when built) and a reverse auction feature (to achieve the greatest possible benefits at the least cost to taxpayers). Prospective projects would compete against one another through a reverse auction, perhaps on the basis of dollars per ton of emissions avoided. This approach would promote continuous innovation while helping to ensure that the projects could be built at the lowest possible subsidy cost. Only those plants that achieved significant progress in planning (such as by securing key permits) would be allowed to reserve credits, and only those that actually operated and successfully reduced emissions would be awarded the production tax credit they had bid for through the reverse auction. Credits that were reserved but unused after a set period of time could be returned to the credit pool and made available to other

---

[17]Given the large stock of existing coal-fired power plants in the United States, China, India, and other countries, it is likely that the stock of coal-fired power plants globally will increase in the coming decade, and DOE's priorities for NETL need to include retrofitting of existing coal plants.

projects. In this manner, first-of-a-kind plants could be financed and built, and technologies could be tested under real-world conditions at scale.

Some of the early carbon capture projects have offset or plan to offset part of the high cost of carbon capture by leveraging the value of the captured CO_2. Since CO_2 is miscible with crude oil, it can be used to increase the amount of oil recovered from older oil reservoirs. Indeed, CO_2 from natural underground sources and natural gas processing facilities is routinely used today to produce significant amounts of oil, particularly in the U.S. Permian Basin. The market value of CO_2 used for oil recovery will vary with the market price of oil and other factors, but it is well below the level needed to serve as the sole justification for the cost of carbon capture from a power plant, at least at today's oil price levels.

Congress could consider tax credits designed to bridge the price differential between the market value of CO_2 used for EOR and the costs of carbon capture. These would best be narrowly and uniquely structured tax incentives for private-sector innovation at every stage, from R&D to construction of fossil fuel-fired power plants equipped with CCS. They could include Pioneer Project Credits for natural gas-fired power generation with CCS, CO_2/Enhanced Oil Recovery Credits, and tax-exempt bond financing. These tax credits could be paid for through the elimination of current subsidies to incumbent, mature technologies. Such a tax credit for carbon capture technology applied to electricity generation and other industrial processes, made available through the same cost-saving innovation-enhancing reverse auction process outlined earlier, would only need to cover the difference between an oil producer's willingness to pay for CO_2 and the cost of its capture and delivery. Because the federal government derives revenue from each additional barrel of oil produced in this manner, some analyses suggest that the tax credit could be revenue-neutral or even revenue-positive over time.

Congress also could consider allowing the incremental capital spending associated with carbon capture (not the entire facility) to be financed using tax-exempt private activity bonds. This is the method often used from 1968 to 1986 to finance pollution control facilities such as flue gas desulfurization equipment and other advanced pollution control equipment at power plants owned and operated by investor-owned utilities. It also is the method often used today to finance privately owned solid waste, sewage, and hazardous waste facilities.

In addition, placing an economy-wide price on CO_2 emissions equivalent to its externalities—for example, through a carbon tax—would, in the long run, be a direct and highly efficient way for Congress to help level the playing field between entrenched fossil fuel-based CO_2-emitting power generation technologies and increasingly clean power generation technologies such as CCS. Given that fossil fuel facilities with CCS will cost more than equivalent facilities that lack the expense of carbon capture, as well as limited economic opportunities to use captured CO_2, carbon pricing or an equivalent regulatory requirement may be a prerequisite for widespread deployment of CCS.

In a previous study (NRC, 2011), a National Research Council committee found that carbon pricing would be an important element of a comprehensive national CO_2 mitigation program. According to that committee, "Most economists and policy analysts have concluded…that putting a price on CO_2 emissions (that is, implementing a 'carbon price') that rises over time is the least costly path to significantly reduce emissions and the most efficient means to provide continuous incentives for innovation and for the long-term investments necessary to develop and deploy new low-carbon technologies and infrastructure" (p. 58). This issue is explored more fully in Chapter 2 (see Recommendation 2-2).

A supportive policy scheme for CCS projects would aid in the continuing demonstration of the technologies themselves and in harnessing public acceptance of long-term geologic storage of CO_2 as the focus for the next phase of CCS development. Congress needs to establish a comprehensive regulatory framework for the transport and safe geologic storage of CO_2. Such a framework would include two key components. First, it would feature a federal opt-in program for CO_2 pipelines. This program would enable the Federal Energy Regulatory Commission (FERC) to supervise a mechanism whereby operators of such pipelines could apply to FERC for federal siting authority. Such authority, if granted, would allow the use of federal eminent domain authority to site a pipeline. In turn, the pipeline operator would be obligated to operate as a common carrier (but without FERC or other regulation of rates). Second, Congress would enact legislation declaring that underground CO_2 storage is in the public interest and eliminate the uncertainties of the Class VI well category under the existing Underground Injection Control regulation. This new legislation would ensure access to pore space, establish arrangements for the management of long-term stewardship and liability for storage sites once they had been closed, and institute GHG accounting programs.[18]

Another possibility would be for Congress and the Environmental Protection Agency (EPA) to establish a framework for the development of long-term performance standards, formulated as agreements with industry in accordance with what is known as the "Dutch Covenant" model. This would be a framework for a collaborative public/private approach rather than an adversarial arrangement.

Recommendation 5-3: Congress should direct the EPA to develop a set of long-term performance standards for the transport and storage of captured CO_2. This effort should include establishing management plans for long-term stewardship and liability for storage sites once they have been closed, as well as GHG accounting programs.

---

[18]See Morgan and McCoy (2012) and the model legislation contained therein.

Electric power utility regulators have expressed concerns over the current lack of certainty about suitable dedicated geologic CO_2 storage sites, hampering development. DOE could address this barrier and immediately begin facilitating the undertaking of large-scale demonstration projects by supporting a formal and comprehensive site survey, led by the U.S. Geological Survey (USGS), to identify and characterize suitable underground storage sites for the United States. Funding should be made available for updating this survey on a regular basis.

The risk of being unable to find a suitable site for CO_2 storage in a timely and economical fashion is considerable. No commercially available services for CO_2 storage currently exist, and while there are firms that conduct site characterizations and will drill test wells to verify a site's suitability for storage, those activities have high costs in both time and money, with no guarantee of a satisfactory outcome for CO_2 storage.

The ZeroGen project in Australia illustrates the challenges that can arise in identifying appropriate storage sites. That project spent 3 years and AUD$100 million on exploration, drilling, and testing, eventually leading to the conclusion that the area initially identified as promising could not sustain the injection rates required for the proposed 390 MW IGCC project. One of the key lessons learned from this project was that a "…large amount of expensive data gathering should be expected and while success rates *might* be higher than in the oil and gas exploration sector, failure rates and costs and delays are likely to be significant" (Garnett et al., 2014, p. 218).

Recommendation 5-4: USGS should identify and characterize CO_2 storage sites.

RENEWABLE FUEL POWER-GENERATING TECHNOLOGIES[19]

The Role of Renewables in an Increasingly Clean Energy Future

Nearly every model and forecast of an increasingly clean energy system includes an expanding role for renewable electricity generation. Wind and solar tend to dominate the renewable portion of these forecasts, but other renewable sources, such as hydro (small, large, low-head), biomass, geothermal, and offshore, also have potential to contribute to a clean energy portfolio. Without taking account of price and the design of the nation's electricity system, the size of U.S. renewable resources is adequate to meet the country's long-term electricity needs. As discussed in this section, however, advances in economic, technology, and market structures will be needed if renewables are to be utilized

[19]This section draws on Lester and Hart (2012, Chapter 2).

effectively over the mid- to long term for a major portion of the country's electricity system.

Chapter 3 presents issues, findings, and recommendations related to effective pathways for innovation, scale-up, and deployment of increasingly clean power technologies. It argues for the need for a complex transition to newer technologies, both large-scale and distributed; new electricity grid models; new roles for the demand side of the energy market; a robust perspective on gaps that need to be addressed in the U.S. energy innovation system; regulatory reform from the local to the federal level; and differentiated roles whereby the federal government, states, regions, and the private sector would partner or take the lead. Across this diverse energy landscape, renewables are likely to play an increasing role in the clean energy future of the United States and the world. Renewable resources are by definition sustainable, with stable, predictable economics for a given project and small or no consumable fuel costs, and are not subject to the volatility that characterizes the prices of conventional fuels. While some renewables (biomass in particular) produce emissions, most produce zero use-phase emissions and have modest environmental and health operating impacts. These characteristics attract investment and attention around the world.

Current Challenges for Renewable Energy

Renewables have seen impressive cost declines in recent years, but the electricity they produce still generally costs more than most electricity generated from fossil fuels, particularly natural gas, in the United States. While the data are variable, Bloomberg New Energy Finance (BNEF) (2014a, p. 15) reported that "over a five-year period to the first quarter of 2014, the worldwide average levelized cost of electricity has declined by 53 percent for crystalline silicon PV systems, and 15 percent for onshore wind turbines." Further, BNEF reported in October 2015 that the "levelised cost of electricity for H2 2015 shows onshore wind to be fully competitive against gas and coal in some parts of the world, while solar is closing the gap" (BNEF, 2015). "Over the same years, the cost per MWh of coal- and gas-fired generation has increased in many countries, with the notable exception of the US where gas prices remain much lower than elsewhere" (Best Energy Investment, n.d.). Lawrence Berkeley National Laboratory notes that prices for installed solar PV declined by 12-15 percent from 2012 to 2013, and U.S. distributed solar prices fell an additional 10-20 percent in 2014, with declines continuing into 2015, continuing a 6-year trend (Bolinger and Seel, 2015). The total cost reduction over the period 2009-2013 was close to 50 percent. Wind energy prices have fluctuated over the last decade, with increases from 2004 to 2009 being driven largely by rising capital costs for materials, primarily steel, but also including iron, copper, aluminum, and fiberglass (Lantz et al., 2012). Subsequent years have seen a reversal in that trend, with a capacity-weighted average installed project cost of approximately

$1,630 per kilowatt (kW), down from the cost of more than $2,200 per kW observed in 2009-2010 (Wiser and Bolinger, 2014).

In addition to price considerations, many renewable technologies are variable and cannot be dispatched in the traditional manner. High penetrations of variable generation may need to be balanced by flexible supplies or more responsive demand, including smart metering and distributed storage. The unpredictability of some renewable generators could pose management challenges for today's electricity grid with substantial increases in deployment.[20] For these reasons, modernizing the electricity regulatory system and grid management models (as discussed in Chapter 6) is important if the benefits of these technologies are to be fully captured.

Findings on Renewables

The challenges that need to be addressed if the deployment of renewables is to increase are structural in nature. As the National Research Council (2010a, p. 322) concluded:

> The primary current barriers are the cost-competitiveness of the existing technologies relative to most other sources of electricity (with no costs assigned to carbon emissions or other currently unpriced externalities), the lack of sufficient transmission capacity to move electricity generated from renewable resources to distant demand centers, and the lack of sustained policies.

According to the National Research Council (2010a, p. 4), establishing renewables as a major source of energy in the future will require innovation not just in renewable technologies but also in grid technologies. "Achieving a predominant (i.e., >50 percent) level of renewable electricity penetration will require new scientific advances (e.g., in solar photovoltaics, other renewable electricity technologies, and storage technologies) and dramatic changes in how we generate, transmit, and use electricity."

National Renewable Energy Laboratory (NREL) (2012) models a series of scenarios to analyze the grid-integration implications of generating 30-90 percent of U.S. electricity from renewable sources. The report concludes that it could be technologically feasible for renewable energy resources to "supply

---

[20]The day-ahead predictability of wind is subject to fairly serious errors. At the 95 percent confidence interval required by the Electric Reliability Council of Texas (ERCOT), for example, errors are a bit more than 30 percent (see Mauch et al., 2013). And despite its inherent predictability, solar forecasting is not a mature science. For example, much of the variation in solar power is due to cumulus clouds that cause very fast and deep variations that have thus far proven resistant to forecasting techniques (see Curtright and Apt, 2008).

80 percent of total U.S. electricity generation in 2050 while balancing supply and demand at the hourly level," with contributions coming from "all regions of the United States...consistent with their local renewable resource base" (p. 3). The report includes current existing nuclear and IGCC units but does not allow for new additions of either nuclear or IGCC units or fossil fuel technologies with CCS. In addition, the report includes only renewable technologies commercially available as of 2010. The authors also assume that grid technologies would improve system operations to "enhance flexibility in both electricity generation and end-use demand" (p. 5), and that both transmission infrastructure and transmission capacity access would expand. However, the authors caution that because solar PV and wind have little dispatchability, high "levels of deployment of these generation types can therefore introduce new challenges to the task of ensuring reliable grid operation" (p. 12). They also note that higher incremental costs would be a significant barrier to the deployment of renewable technologies at high levels and that improving performance and lowering costs would have the greatest impact on overcoming that barrier (NREL, 2012).

According to a different analysis published 2014, improving the penetration of renewables to 20-30 percent of electricity generation would be feasible if there were changes in the management and regulation of the power system (Apt and Jaramillo, 2014). The authors identify variability in power output as a key technical barrier that leads to a number of operational and regulatory challenges, and suggest a number of solutions for overcoming this barrier. Those solutions include better prediction of variability and strategies for reducing it; changes in the operation of power plants, reserves, transmission systems, and storage; improved planning of renewable capacity expansion; and implementation of new regulatory paradigms, rate structures, and standards.

> **Finding 5-7:** The variability of renewable generation does not prevent the future expansion of renewable technologies, although high penetration of renewables likely will require investments in innovation to improve grid technologies, storage, and regulatory paradigms, among other changes.

Still, a broad range of needs and challenges for renewables will need to be addressed if these energy sources are to gain lasting economic and environmental value. These include

- innovation (both improvements in renewables and breakthroughs in storage);
- continuing reinvestment in renewable and grid technologies to support growth and drive consistent learning;
- consistent markets that encourage competition and cost-effective investments in economies of scale (e.g., Lueken and Apt [2014], argue that the existing regulatory and market structures are inadequate

to encourage the required amount of storage for renewable integration);

- supportive, transparent, and flexible regulatory regimes that can adapt and evolve alongside a similarly evolving energy system; and
- investment in re-architected grids that are more reliable and efficient and can be designed to effectively manage, store, and use energy with significant renewable penetration and leverage distributed resources and smart demand.

These findings are aligned with the focus of the present report on innovation, the policy and financing consistency needed to support the development of competitive markets, pricing of environmental externalities, and the modernization of the electricity system necessary to fully embrace the value of renewables.

Renewables and Economic Growth Opportunities

Ongoing government support for innovation and encouragement of private-sector investment in renewable technologies could help the United States be a technology leader in the development and deployment of these technologies. U.S. expertise in innovation, entrepreneurship, manufacturing, and business is creating opportunities for a major U.S. role in the rapidly growing markets for renewables. Innovation is needed in systems, components, manufacturing, and integration. For example, despite increasing competition from China, which played a central role in the recent decline of the German solar industry, the outlook for solar PV production in the United States remains promising.

Recent announcements and discussions of plans for new solar PV manufacturing plants in New York State suggest a growing economic opportunity in renewable manufacturing. In September 2014, for example, SolarCity broke ground on a massive solar panel manufacturing facility in Buffalo, New York. The facility, when, and if, fully completed in early 2017 "is supposed to make a gigawatt worth of solar panels a year, in the one million square foot facility. SolarCity…[has] agreed to work with the state to spend $5 billion over the course of 10 years to build out and operate the factory, creating local jobs" (Fehrenbacher, 2014). In another potential manufacturing project, the world's number two solar thin-film manufacturer, Japan's Solar Frontier (with a major ownership share by Shell as a subsidiary of Showa Shell Sekiyu), is currently exploring the development of a factory in upstate New York (Ayre, 2014).

It is important for manufacturing investments in the United States to focus a portion of their product output on U.S. markets. Global corporations increasingly see renewable generation technologies addressing global markets, with opportunities to site multiple manufacturing facilities close to major market adoption regions of the world and with opportunities to combine economies of scale with faster innovation cycles closer to customers and market channels. Leveling the playing field for global corporations considering investment opportunities for renewables in the United States and in other global market centers would require removing barriers to U.S. competitiveness while encouraging private-sector investment (NRC, 2012). The federal and state governments need to carefully consider tax, trade, and other policies that would encourage renewable manufacturing investments in the United States.

U.S. Markets for Renewables

Renewable resources vary considerably across the United States. Combined with regional electricity markets, state-specific policies, regulatory and market structures, and several thousand utility jurisdictions, this variation means a diversity of markets for renewables. One consequence of this diversity is important opportunities to learn and share the lessons from the most robust markets. Renewables are approaching competitiveness in some regions of the United States and exhibiting cost decline trends, with increasing competition, market economies of scale, improving technology, and supply chain and value chain efficiencies all helping to drive market improvements. Regions with the most cost-effective renewable resources and market development efforts offer recent examples of this approaching commercial competitiveness. In Colorado, for example, Xcel Energy in 2013 issued a broad "all source" solicitation so it could consider the most competitive proposals for wind, solar, and other resources, including natural gas. Xcel selected a diverse portfolio that would add 317 MW of natural gas, combined with 450 MW of wind and 170 MW of solar. Xcel describes how the "strong competition between resources and even between different types of resources yields a number of low cost resource combinations that could meet Xcel Energy's needs" (Xcel Energy, 2013). This is but one example of how improvements in renewable technology and industry experience, combined with market pull and competitive solicitation mechanisms, are contributing to progress toward the competitiveness of renewables in regions with strong renewable resources.

While states have a range of pricing and procurement policies, incentives, standards, and models, many parts of the United States encourage competition for wind projects to win power purchase contracts and enable low-cost financing for construction. It is important to continue these trends in the near and mid terms while avoiding problems with inflexible policies such as feed-in tariffs and similar structures that lock in higher-than-necessary prices and delay market competition, maturation, and innovation.

Recommendation 5-5: As renewable technologies approach becoming economically competitive, states should seek to expand competitive solicitation processes for the most cost-

effective renewable generation projects and consider the long-term power purchase agreements (PPAs) necessary to enable low-cost capital for project financing.

The diversity of U.S. market policies on renewables with such market structures as renewable energy credits (RECs) has generally avoided policies that lock in higher costs. At the same time, the lack of consistent polices for market scale and the patchwork nature of renewable portfolio standards (RPS) as enacted by the states have hampered innovation and private-sector investment in renewables. In addition to varying prices for electricity, higher capital costs and slower market maturation than is necessary have resulted from inconsistent standards across states, differing pricing models, differing market mechanisms for receiving off-take and interconnect agreements, uncertain siting practices, and an inability to site and finance related transmission investments.

Although they have declined, installed PV prices in the United States remain twice as high as those in Germany and substantially higher than those in the United Kingdom, Italy, and France. A Lawrence Berkeley National Laboratory study attributes this disparity largely to differences in "soft costs," which may be driven partly by differing levels of deployment scale (Barbose et al., 2014). Reductions in soft costs remain a major part of DOE's SunShot initiative.

Finding 5-8: Consistent siting, streamlined permitting, clear and responsive interconnection processes and costs, training in installation best practices, and reductions in other soft costs can have a significant impact on lowering the cost of solar and other distributed generation renewable technologies.

Recommendation 5-6: DOE and national laboratory programs should provide technical support to states, cities, regulators, and utilities for identifying and adopting best practices—such as common procurement methods, soft cost reduction approaches, PPA contracts, structures for subsidies and renewable energy certificates, and common renewables definitions (taking into account regional resources)—that could align regional policies to enable more consistent and efficient markets that would support the adoption of renewables.

Renewable Portfolio Standards

State RPSs are an option commonly used to force utilities to increase their utilization of renewables, and they have created significant but still inefficient market pull. RPSs require that either utilities or, in jurisdictions with retail competition, retail electricity suppliers use renewable energy or obtain RECs[21] for a minimum amount of their electricity sales, or that utilities procure a minimum amount of renewable generating capacity in their portfolio of electricity resources. An RPS sets a schedule of renewable energy or capacity to be obtained by specific years. Requirements generally increase over time. Many RPSs include a set-aside or carve-out that requires a minimum percentage or amount of the overall standard to be met using a specific technology, typically solar energy. An alternative or clean energy standard is comparable to an RPS, but permits some portion of the requirement to be met through investments in energy efficiency or the use of various nonrenewable alternative energy resources. For purposes of simplicity, the discussion here refers to state standards that include renewable energy requirements as RPSs.

RPSs have been important for the development of renewable energy resources. Currently, 29 states and the District of Columbia, accounting for 64 percent of U.S. electricity sales, have RPSs.[22] Approximately 46 GW of new renewable generating capacity had been developed in these 29 states by the end of 2012, equaling two-thirds of all nonhydro renewable electricity generation capacity additions in the United States since 1998 (Heeter et al., 2014). An additional 8 states have adopted voluntary renewable energy goals (Heeter et al., 2014).[23] Together these 36 states account for nearly three-quarters of U.S. electricity sales (EIA, 2013a).

---

[21]RECs are often used to provide a uniform system for tracking the purchase and use of renewable energy. Such tracking ensures that the financial incentives created by a portfolio standard flow to the owners of covered renewable resources. Tracking services are provided by independent system operators (ISOs)/regional transmission organizations (RTOs) (e.g., ERCOT and PJM-Environmental Information Services [EIS]), states (e.g., Michigan Renewable Certification System, North Carolina Renewable Energy Tracking System, New York State Energy Research and Development Authority, and Nevada Tracks Renewable Energy Credits), and independent services (e.g., Midwestern Renewable Energy Tracking System, New England Power Pool Generation Information System, North American Renewables Registry, and Western Renewable Energy Generation Information System).

[22]Arizona, California, Colorado, Connecticut, Delaware, District of Columbia, Hawaii, Illinois, Iowa, Kansas, Maine, Maryland, Massachusetts, Michigan, Minnesota, Missouri, Montana, Nevada, New Hampshire, New Jersey, New Mexico, New York, North Carolina, Ohio, Oregon, Pennsylvania, Rhode Island, Texas, Washington, and Wisconsin as of March 2013 (see www.dsireusa.org).

[23]Indiana, North Dakota, Oklahoma, South Dakota, Utah, Vermont, Virginia, and West Virginia.

States have differing definitions of qualifying resources. Of the states with standards or goals, 10 allow solar thermal resources to qualify, 9 include energy efficiency, 6 allow combined heat and power, and 4 count certain nonrenewable resources (Heeter and Bird, 2012). Of the 29 states with standards, most provide some additional incentive that benefits solar energy resources. Fourteen states and the District of Columbia have a set-aside that must be met with solar resources;[24] 2 have a set-aside for distributed generation;[25] and 3 provide triple or double credit for solar electric, distributed generation, or nonwind resources.[26]

Early RPS results have been achieved with relatively small impacts on retail electricity prices. In part, this small retail price increase is attributable to the directly measurable costs of RPSs. It may also be due in part to the federal production tax credit.[27] REC purchases in competitive retail markets and utility-reported purchase costs in other cases have been small relative to total utility revenue. Most RPS policies include a cost containment mechanism: either a cap on total compliance costs as a percentage of average retail rates or an alternative compliance payment. These mechanisms typically limit compliance costs to 1-4 percent of average retail rates in the case of overall caps or 6-9 percent of average retail rates in the case of an alternative compliance payment and no overall rate cap (Heeter et al., 2014). A recent survey of state-level costs found that while there are substantial variations among states and from year to year, "Focusing on the most recent historical year available [2010 to 2012], estimated incremental RPS compliance costs were less than 2 percent of average retail rates for the large majority of states" (Heeter et al., 2014, p. v). However, incremental renewable energy costs—the additional cost per MWh for renewable energy in excess of the cost per MWh of nonrenewable generation—can be significant. The survey found that in restructured states, average renewable energy or REC purchases added $2-48 per MWh of renewable energy obtained. In other states, utilities reported that general RPS obligations, excluding the typically more expensive set-asides for solar or distributed generation, had incremental costs ranging from negative $4 to an additional $44

---

[24]Delaware, District of Columbia, Illinois, Maryland, Massachusetts, Missouri, Nevada, New Hampshire, New Jersey, New Mexico, New York, North Carolina, Ohio, Oregon, and Pennsylvania as of March 2013 (www.dsireusa.org).

[25]Arizona and Colorado (www.dsireusa.org).

[26]Michigan, Texas, and Washington (www.dsireusa.org).

[27]A prior National Research Council study found some complementarity between state-level RPSs and the federal production tax credit (PTC). That study commissioned modeling that produced estimates suggesting that the combined impact of both policies on new builds of wind power is only slightly greater than the impact of either policy alone. This finding implies that much of the wind power could be built with just the RPS (or the PTC), and thus a (possibly substantial) portion of the costs of compliance may have been paid for by the PTC. That study also found, however, that energy and economic models were not well suited to modeling the impacts of tax policies (NRC, 2013c).

per MWh of renewable energy. However, there is no standard methodology for reporting RPS costs. The costs in the survey may omit certain integration and system operating costs, and historical costs may not be representative of future costs as RPS requirements increase. Few states have conducted benefit-cost analyses of their RPS requirements (Heeter et al., 2014).

A modest RPS can either increase or decrease market prices. RPSs subsidize renewable resources and bring additional resources with low variable operating costs into energy markets. If there is a sufficient upward slope in the supply curve for generation, displacing higher-cost resources (and potentially lowering demand for the fuel serving those generation resources) can depress prices in independent system operator (ISO)/regional transmission organization (RTO) energy markets. In organized electricity markets with capacity pricing, however, the reduction in energy market prices may be offset by higher administratively determined capacity prices. Additionally, RPSs effectively impose a tax on retail suppliers that must pay for more expensive renewable resources or RECs. This has the effect of increasing prices. How these potentially offsetting effects play out given a low RPS requirement depends on the supply curves for renewable and nonrenewable generation and market structures. However, the impact on market prices becomes less ambiguous when RPS requirements increase. According to Fischer (2010, p. 117), "both the analytical and numerical modeling suggest that rate reductions are only likely at lower RPS shares. At higher RPS shares, in contrast, the implicit tax quickly dominates and electricity prices increase rapidly."

The development of state-level RPSs is, in part, a response to the lack of a more comprehensive national policy for reducing carbon emissions. Local development of renewable resources provides an immediate, visible representation of state action to address climate change, even if its impact on global GHG emissions is minimal. An analysis by Resources for the Future suggests that state RPSs reduced U.S. CO_2 emissions by 4 percent in 2010. The authors concluded that, "given that by 2010 the RPS have been in effect for only a few years in many states, this is a fairly significant impact. The gap between the two cases [with and without state RPSs] is likely to continue to widen as RPS are fully implemented across the nation" (Sekar and Sohngen, 2014, p. 10).

Consideration is being given to the continuation or expansion of RPS requirements to achieve higher levels of renewable energy deployment.[28]

[28]RPS policies may be subject to reconsideration in some states. Ohio recently enacted a 2-year suspension of its RPS requirement, which in the absence of further legislative action would defer full compliance from 2025 to 2027. (Ohio also suspended energy-efficiency standards; its action on energy efficiency raises issues that deserve careful study in terms of the suspension's potential to increase customer costs and slow economic growth.) Proposals have been made to repeal RPSs in 18 of the 29 states that have adopted such standards. See http://web20.nixonpeabody.com/energyblog/Lists/Posts/Post.aspx?ID=81&Title=Tough+Times+For+Renewable+Portfolio+Standards (accessed June 20, 2014).

Identifying the factors that may be important to the adoption of RPSs can involve a review of the arguments advanced and/or a study of the circumstances surrounding their adoption.[29] RPS policies also may reflect efforts to limit other environmental externalities, reduce the cost of renewable technologies through enhanced learning by doing, enhance the reliability of distribution with distributed generation, promote the security of energy supplies, preserve water resources, and/or create local jobs in an emerging clean energy industry and potentially capture first-mover advantages (Fischer and Preonas, 2010).

Although RPSs have reduced CO_2 emissions, RPSs are not the most cost-effective means of doing so. As indicated in a previous National Research Council (2010c) report, pricing GHG emissions provides the critical foundation for cost-effective reductions in GHG emissions and the basis for innovation and a sustainable market for renewable energy resources. Most studies of alternative renewable energy policies agree that a carbon tax or cap-and-trade system would reduce GHG emissions more cost-effectively than RPSs (see, e.g., Fischer and Newell, 2008; Fischer et al., 2013; Palmer and Burtraw, 2005; Palmer et al., 2010; Tuladhar et al., 2014). For example, Palmer and colleagues (2010) compare the cost of achieving a reduction in CO_2 emissions using a national RPS—a 25 percent renewable standard by 2025—with that of a cap-and-trade policy achieving the same emissions reduction. They estimate that the RPS would have an average cost of $14 per ton of CO_2 reduced, while the same reduction could be achieved in a cap-and-trade program for $4 per ton, or less than one-third the cost. This finding is not surprising. In requiring the use of renewable technologies, RPSs fail to recognize other actions, such as the substitution of gas- for coal-fired generation, that might reduce emissions more cost-effectively. RPSs treat all renewable generation equivalently, as if all renewable generation sources displaced comparable nonrenewable sources and had an equivalent net emissions impact. They lower the cost of selected clean energy technologies, but do not incorporate the social costs of carbon and other environmental externalities into the price of polluting resources, potentially distorting price signals for both consumers and other market participants (Borenstein, 2012; Nordhaus, 2013). Market-based environmental regulation that appropriately prices both GHGs and other environmental externalities will tend to produce and provide incentives for emission reductions more cost-effectively relative to comparable technology requirements.

Technology deployment incentives can create innovation benefits by supporting learning by doing.[30] However, the appropriate incentive for promoting learning by doing is likely to be lower than the deployment incentives currently available from RPSs, tax credits, and other renewable energy programs.

---

[29]For one assessment of factors correlating with the adoption of RPSs, see Lyon and Yin (2009).
[30]Learning by doing is discussed in Chapter 2.

General RPSs are not tailored to improving the reliability of distribution, promoting the security of energy supplies, preserving water resources, and/or creating jobs. There are many ways to improve the reliability of distribution. Greater reliance on renewable resources might help Western Europe reduce imports of natural gas or help California limit cooling water requirements. However, greater reliance on renewables also could increase dependence on imported rare earth elements. And while renewable generation will create some jobs, potentially offsetting job losses can be associated with increased power costs. Policies specifically designed to achieve these secondary benefits could be more cost-effective in meeting stated objectives. A cap-and-trade program could efficiently reduce and price GHG emissions. If a binding emissions trading system were implemented, RPSs would not necessarily produce additional emission reductions, but would likely increase overall compliance costs (Fischer and Preonas, 2010). In circumstances where a binding emissions trading system reasonably reflects pollution costs and creates an effective market signal for emission reductions, it would be beneficial for states or regions to evaluate the impacts of replacing RPSs with a cap-and-trade system and more modest, targeted incentives to produce learning-by-doing benefits. Savings from reduced renewable energy deployment incentives could be redirected to earlier stages of the innovation system, where greater public resources are needed to support the development and demonstration of new low-carbon technologies (see Chapter 3). Ultimately, when possible, states would benefit from seeking to ensure that their policies evolve to establish best practices for competition featuring market signals and other such mechanisms for participants.

Removing barriers to participation by government agencies and departments as active customers in state and regional markets for renewable energy would also improve opportunities for renewable generation. Steps to this end would include enabling federal facilities, such as military bases, to sign PPAs for renewable power. The Department of Defense in particular offers opportunities for renewables because of the size of its energy demand and budget, its global footprint, and its mandate to use renewables for 25 percent of its total energy needs by 2025. Executive Order 13693 goes a long way toward encouraging the federal government's role as a leading consumer of increasingly clean energy by setting renewable and alternative energy and energy-efficiency targets for federal facilities.

Finding 5-9: Few states have conducted benefit-cost analyses of their RPS requirements and evaluated options for evolving their RPSs, learning from other states, and taking innovation and evolving models into account.

Recommendation 5-7: State and regional authorities should regularly arrange for independent evaluations of the effectiveness and cost of their policies for encouraging

deployment, competition, cost declines, reductions in financing costs, and other aspects of renewable energy technologies. They also should be encouraged to adopt evolving best practices for competition, including market signals and mechanisms, in their state renewable policies and programs.

Incentives, Subsidies, and Diverse Technology Market Growth

Subsidies are critically important to emerging technologies, particularly those with life-cycle benefits that are not priced. As discussed above, despite significant cost reductions in renewables in recent years, renewables in most areas of the United States still are not cost-competitive and continue to require subsidies as an important component of establishing a path to their eventual competitiveness. These subsidy levels have been declining for maturing renewables, with onshore wind having developed into the most cost-effective renewable. As of 2013, solar "cash incentives (rebates and performance-based incentives) have fallen by 85-95 percent from their historical peak in 2001-2002, and incentive reductions from 2012-2013 equal 40-50 percent of the drop in installed prices" (Barbose et al., 2014, p. 2). Federal subsidies, such as the investment tax credit for solar and the production tax credit for wind, that are targeted at deployment have demonstrated their effectiveness. That being the case, subsidies are best not designed to be permanent, but to have sunset provisions and to phase out over time on a technology-by-technology basis, with market tests that consider cost and progress toward unsubsidized competitiveness.

This type of approach to supporting technology development and early deployment does not constitute picking winners and losers. Incentives are applied across a class of technologies, not to a single product or company, with pricing mechanisms that decline over time, but there also are market tests for appropriate phase-out under actual market development conditions. These approaches encourage private-sector investment in innovation, infrastructure, and scale-up while avoiding the stop-and-start impacts of short-term extensions of tax credits.

Another principle for federal and state renewable policies involves appropriate regulatory and financing structures for projects with differing scales:

- Small, distributed generation projects (such as rooftop solar) reduce but are unlikely to economically eliminate owners' purchases of electricity in retail markets. Appropriate pricing of interconnection with the power company and the grid and pricing for backup services, such as the price of utility-provided power when the distributed generation project is not producing enough power for the customer, are important to maximize the value of distributed generation projects.

- Community-scale renewables provide virtual generation and energy management to multiple customers, and over time are expected to be combined with efficiency, demand response, microgrids, combined heat and power, storage, and other distributed energy approaches and resources. Community-scale projects need pricing and rate models that encourage innovation and competition, recognize the value of this distributed virtual system, and also incorporate the evolving role of the utility in integrating and managing the electricity system as an "Internet" network of networks.
- For utility-scale renewable generation, long-term PPAs have proven effective in incentivizing a number of projects around the country. The long-term PPAs have enabled low-cost project financing, since most renewable projects have high capital costs and little or no variable fuel costs.

Across all technologies and scales, it is important to emphasize that renewable deployment needs to take place in an increasingly competitive market, and to continue to reward learning and economies of scale, as well as projects with the best economics. Effective federal, state, and local policies need to be consistent with growing market signals that look out at least 5 years to encourage investment in innovation and development that will continue to bring down costs.

Financing

Reducing the cost of capital for renewable and clean energy projects is an important component of leveling the playing field. While this issue is discussed in greater detail in Chapter 7, it is important to acknowledge here the importance of federal and state policies to reducing the financing costs for zero-emission technologies.

Federal tax changes that would extend real estate investment trust (REIT) and master limited partnership (MLP) financing structures to zero-emission technologies could also help address financing issues related to renewable projects. In addition, many states are developing or considering so-called "green banks" to leverage low-cost state borrowing in combined public/private financing structures for lower-cost renewable and energy-efficiency deployment projects. The federal government might consider encouraging banks to participate in these green bank programs through approaches similar to the Community Reinvestment Act, which has enabled the deployment of increased private capital for local projects.

Utility Adoption Issues/Barriers

Chapter 6 reviews challenges associated with the regulatory structure of the electric power industry. Utilities play critical roles in the deployment of renewables, but current business models for utilities tend to be insufficient for adequately incentivizing the adoption of renewables. State and federal incentives and regulations need to encourage utilities to change their priorities and decision processes, but without being unduly burdensome.

6

Modernizing the Electric Power System to Support the Development and Deployment of Increasingly Clean Technologies

Developing and deploying cost-effective increasingly clean energy technologies will require an electric power sector with systems, regulation, and infrastructure that encourage and accommodate those technologies. Developing such a power sector will, in turn, require technological changes to the power system and fundamental changes in the regulation and operation of electric power utilities. Power systems—the electric power transmission and delivery grids—will need to become capable of integrating new technologies and in greater quantities. To achieve this goal, regulators will need to implement regulations that give utilities incentives to become fully engaged in innovation and the demonstration of new technologies, with rules that permit reasonable and nondiscriminatory access to the transmission and delivery systems.

Since the restructuring activities that began in several states in the 1990s, the electric power industry has been under pressure to change in a number of ways. While restructuring efforts mostly stopped by the early 2000s, several states enacted policies to encourage higher adoption rates of specific increasingly clean power generation technologies, principally for electricity from renewable sources. The growth of renewable and distributed energy resources, the expansion of energy-efficiency programs, slowly growing or declining utility sales, low natural gas prices, and the need to invest in the grid to maintain its reliability and security have prompted consideration of the significant changes in utility technical, business, and regulatory models needed to facilitate the truly wide-scale adoption of increasingly clean power technologies. However, the industry is in the early stages of evaluation of these changes. Investors will not fund the development of increasingly clean technologies without a realistic opportunity to capture market share and earn economic profits as these options become cost-effective. Current utility technical, business, and regulatory models present barriers to the development of new technologies and the entry of new firms, especially in the case of distributed and variable generation technologies. This chapter describes such barriers and the opportunities to develop new regulatory frameworks and business models that could both improve industry performance and create opportunities for increasingly clean energy technologies while enhancing the customer experience and delivering value for both customers and investors. The transmission and delivery systems are both complex, and present many challenges and opportunities; the emphasis here is more on distribution and distributed and variable resources, with some coverage of transmission.[1]

This chapter begins with a brief review of the challenges and opportunities currently faced by the U.S. electric power industry. It then describes the current electric power system—its structure and its regulatory framework. Finally, the chapter lays out the features of a modern power system that would support the development and deployment of increasingly clean energy and energy-efficiency technologies.

CHALLENGES AND OPPORTUNITIES FOR THE ELECTRIC POWER INDUSTRY

Challenges

The electricity industry is facing significant new expectations and requirements to replace aging infrastructure, mitigate the effects of storms and other disruptive events, secure the electric system and critical infrastructure that depends on electric power against cyber and physical attacks, and maintain system stability. At the same time, the industry is dealing with retiring coal and some nuclear generation and integrating variable large-scale renewable and distributed resources. Moreover, current utility business models often rely on volumetric increases in sales to provide funds for new investments. With slowly growing or declining sales, many utilities lack the revenue growth used historically to fund new investments. This trend could leave the nation with an outdated power system and prove costly to consumers.

---

[1]The full suite of issues surrounding transmission (and delivery) could take up an entire report on its own. The committee notes that at the time of this writing (summer 2016), another National Academies of Sciences, Engineering, and Medicine study was under way that was charged with examining how the transmission and delivery systems can evolve to become more reliable and resilient, including "greater reliance on distributed power generation." Readers are encouraged to consult the report of that study once it becomes available.

Aging Infrastructure

Much of the existing U.S. electric power infrastructure was built more than 40 years ago and is in need of replacement and modernization. The American Society of Civil Engineers (ASCE) estimated in 2011 that maintaining this infrastructure would require $673 billion in new investment by 2020. To put this figure in perspective, the total market capitalization of U.S. investor-owned utilities equaled $504 billion as of December 31, 2013 (EEI, 2014). ASCE forecasts significant economic consequences if the electric sector fails to close the investment gap:

> As costs to households and businesses associated with service interruptions rise, GDP will fall by a total of $496 billion by 2020. The U.S. economy will end up with an average of 529,000 fewer jobs than it would otherwise have by 2020....In addition, personal income in the U.S. will fall by a total of $656 billion from expected levels by 2020. (ASCE, 2011, p. 20)

In its 2013 report on the state of America's infrastructure, ASCE notes a recent decline in investment in electricity distribution systems. It reports that aging equipment has resulted in an increasing number of power disruptions and that "significant power outages have increased from 76 in 2007 to 307 in 2011" (ASCE, 2013, p. 61; see also EPRI, 2013). More than 90 percent of customer service interruptions can be the result of distribution outages.

Increasing Reliability Problems and Outages

Electric utilities experienced an increasing number of weather-related outages from 1992 to 2012 (Executive Office of the President, 2013).[2] Such outages cost the U.S. economy between $25 billion and $70 billion per year (Campbell, 2012; Hines et al., 2009). These costs could rise with increasing reliance on information and communication systems as well as digital devices and control technologies that depend on access to reliable sources of electricity. The costs also could rise with the evident increase in the frequency of severe weather events. A power outage can impact the economy of an entire region, as illustrated by Superstorm Sandy, which caused outages for more than 8 million customers in 21 states and an estimated $65 billion in damages. These outages left fuel pumps at gas stations unable to function and curtailed operation of the Colonial Pipeline, which brings refined petroleum products from the Gulf of Mexico (DOE, 2013). And the high cost of power outages to consumers is

---

[2]Outages from non-weather-related events do not appear to exhibit the same trend (see Hines et al., 2009).

apparent from the growing number of customers that have installed their own backup means of power generation. It has been estimated that more than 12 million commercial and industrial customers have installed more than 200 gigawatts (GW) of backup generating capacity (Gilmore and Lave, 2007), while the penetration of residential distributed generation is growing by more than 20 percent per year, with an estimated 3 percent of residential customers having installed backup generators (Generac, 2014).

Security Concerns

Electric utilities also have been placed on the front lines in defending the power system from cyber-security and physical attacks. The Department of Homeland Security's Industrial Control Systems Cyber Emergency Response Team (ICS-CERT) reported responding to 198 cyber incidents in fiscal year 2012 across all critical infrastructure sectors. Forty-one percent of these incidents involved the energy sector, particularly electricity (DHS, 2013). Most experts agree that the risk of a significant attack on the power system is significant, and its consequences could be large (see, e.g., Dlouhy, 2013). The 2003 Northeast blackout was an example of "the failure of a software program"—in this case "not linked to malicious activity"—that left operators unaware of system conditions and "significantly contributed" to a region-wide power outage. The operators' lack of awareness and resulting failure to return the system to a reliable state helped create conditions in which tree contacts with transmission lines would ultimately trigger the outage (U.S.-Canada Power System Outage Task Force, 2004, p. 131).[3] The blackout impacted more than 50 million people and cost the U.S. economy an estimated $6 billion.[4] A large-scale cyber attack or combined cyber and physical attack could potentially have even greater costs, triggering sustained power outages over large portions of the electric grid and prolonged disruptions in communications, food and water supplies, and health care delivery. The investment requirements associated with tracking and mitigating security risks are substantial and will increase as understanding of these risks continues to evolve.

Significant Growth in Distributed and Variable Generation Capacity

Utilities are seeing significant growth in customer-sited, distributed generation. Combined heat and power (CHP) has reached significant scale in the

---

[3]While the information system failures in this case were not the result of a malicious attack, the task force nonetheless found "potential opportunities for cyber system compromise of Energy Management Systems (EMS) and their supporting information technology (IT) infrastructure," and supported new cyber and physical security standards.
[4]The Department of Energy estimated costs of $6 billion—close to the $6.4 billion midrange estimate prepared by Anderson Economic Group. For a summary of estimates of the blackout's costs, see Electric Consumers Resource Council (ELCON) (2004).

United States—more than 80 GW in 3,700 industrial and commercial facilities (DOE and EPA, 2012). While the rate of development of these facilities declined after 2005 when the Federal Energy Regulatory Commission (FERC) lifted the requirement of the Public Utilities Regulatory Policy Act that utilities in competitive markets purchase power from these facilities, CHP installations are once again increasing. An additional 870 megawatts (MW) of CHP generation was added in 2012, with a further increase estimated to have occurred in 2013 (BNEF, 2014b; Chittum and Sullivan, 2012). In an August 2012 executive order, the White House targeted a further increase of 40 GW by 2020 (White House, 2012). Rooftop solar photovoltaic (PV) systems added 1,458 alternating-current MW in 2014 and 2,158 alternating-current MW in 2015 (EIA, 2016i).[5] Distributed generation imposes new investment, control, and protection requirements on distribution systems that historically were designed for one-way power flow to customers.

In addition to distributed generation, total capacity additions of solar (which includes rooftop solar PV, utility PV, and solar thermal) and wind added nearly 10,000 alternating-current MW in 2014, representing 27 percent and 26 percent of U.S. electricity generation capacity additions in that year (EIA, 2016i). These additions to capacity are driven in part by public policy and utility rate designs that recover a portion of fixed costs through kilowatt hour (kWh) charges (EIA, 2014i). The rebound in wind capacity in 2014 was likely driven by the extensions and modifications of the federal production tax credit (PTC) (EIA, 2016i). In 2003, renewable resources other than hydroelectric represented less than 2 percent of U.S. power generation capacity; by 2012, these sources—primarily through the growth of wind and solar—represented more than 7 percent of generation capacity.

Variable resources create significant integration and investment challenges. For example, the California Independent System Operator (ISO) forecasts that by 2020, the combination of a reduction in solar output and an increase in demand in the evening could produce as much as a 13 GW increase in net load on its system over a 3-hour period (California ISO, 2013). Such rapid changes in net demand would have to be offset by significant additions to flexible generating capacity, storage, or responsive demand. Additionally, a future power grid must be built to accommodate continuous changes in wind and solar generation as wind speeds change and clouds pass overhead.

Financial Challenges for Utilities

Electric utilities have increased their total capital expenditures since 2007. However, that increase has had a negative impact on their cash flow. In 2013,

---

[5]The Energy Information Administration (EIA) collects electric capacity data in alternating-current megawatts. Solar PV generators produce electricity in direct-current megawatts. Generally, PV systems are associated with an AC-to-DC ratio of between 80 and 90 percent (EIA, 2014c).

capital expenditures and dividends of investor-owned electric companies (IOUs) exceeded net cash from operations by $23.5 billion. Over the period 2007 to 2013, IOUs posted a net cash deficit of more than $155 billion (EEI, 2014). Although utilities have been able to finance this deficit during a period of relatively favorable interest rates, whether the industry will be able to sustain the required pace of investment is unclear. As of the first quarter of 2016, 36 percent of electric utilities had a Standard and Poor's credit rating of BBB or lower (EEI, 2016).

As discussed in earlier chapters, electric utilities also are facing slowly growing, flat, or declining sales. Total U.S. electricity sales declined by 0.1 percent in 2013 and have fallen in 5 of the last 6 years. The Energy Information Administration (EIA) and many utility load forecasters are predicting less than 1 percent growth in electricity consumption (EIA, 2014j), continuing a long-term decline in the growth of electricity demand that began in the 1950s (Barbose et al., 2013; EIA, 2012; Faruqui, 2013). Many factors are likely to continue limiting growth in electricity sales, including expanded energy-efficiency programs; revisions to state and federal efficiency codes and standards; the falling cost and associated growth in customer-sited PV generation; growth in gas-fired CHP generation; and fuel switching from electricity to natural gas for certain heating, cooling, and industrial process applications.

Most utility costs for electricity distribution are fixed and represent investments in poles, wires, and other equipment or systems. In the conventional regulatory model, fixed costs often are recovered through volumetric rates charged on a per kWh basis. Slowly growing or declining sales remove a key source of revenue for new investment, while new investments tend to produce increases in customer rates. These higher rates can further depress sales, eventually creating a self-reinforcing cycle that could over time undermine the conventional utility business model (Kind, 2013).

Opportunities

While the challenges facing the electric power industry are substantial, there are also significant opportunities for improvement.

Improved Reliability through Distributed Resources

Distributed resources, if integrated under appropriate interconnection standards, in microgrids, or in automated distribution systems, offer the potential to improve grid reliability and resilience for customers that place a high value on uninterrupted service. These distributed resources may include increasingly clean energy technologies such as CHP, PV, and efficient fuel cells. Another possibility is the use of electric vehicles for load balancing and distributed storage. This concept, sometimes dubbed vehicle-to-grid, is viewed as being potentially useful for storing excess electric power at night, when wind generation tends to peak. The concept has been studied under circumscribed and controlled model conditions. However, limited data have been collected on the value of uninterrupted service to different customers, nor has there yet been a full-scale demonstration of vehicle-to-grid (Centolella and McGranahan, 2013; Sullivan et al., 2009; Tomić and Kempton, 2007).

Opportunities to Improve System Efficiencies

Generation capacity factors and the average utilization of transmission and distribution assets are often below 50 percent (EIA, 2013a, Tables 3.1A and 4.7A)—far below the levels typical of other capital-intensive industries, in which asset utilization is commonly 75 percent or greater (Board of Governors of the Federal Reserve System, 2012). Better load factors could improve asset utilization and reduce investment requirements and costs. A great deal of the demand for electricity is associated with the thermal inertia of buildings or devices that have flexibility with respect to the timing of electricity usage. Smart thermostats and other intelligent devices can improve load factors while preserving the quality of service experienced by consumers. By optimizing equipment operation and reducing energy usage when homes or buildings are unoccupied, such technologies also can reduce overall energy consumption.

Opportunities to Create and Capture Additional Value through Increased Research and Development

Electric utilities spend 0.2 percent of their revenues on research and development (R&D) (Battelle Memorial Institute, 2011, p. 21; Lester and Hart, 2012)—less than one-tenth the average rate for all sectors of the U.S. economy and much lower than the rate in the most productive sectors (AEIC, 2010; Anadon et al., 2011, p. 222; NSF, 2010). Utilities could make a more significant contribution to the development and demonstration of advanced increasingly clean and energy-efficiency technologies, but realizing these improvements will often require new business and regulatory models. As a result, suppliers to the industry are unlikely to provide sufficient support for the development of advanced technologies without utility leadership. Fortunately, policy makers and industry stakeholders are beginning to actively consider changes in utility business and regulatory models.

DESCRIPTION OF THE CURRENT ELECTRIC POWER SYSTEM

This section describes the current structure of the electric power sector and the existing utility regulatory framework.

Current Structure of the Power Sector

The U.S. power sector encompasses diverse and multilayered systems of ownership, operation, and governance. Municipal and cooperative systems account for the vast majority of the nation's more than 3,000 electric utilities and 26 percent of electricity sales and revenues. Investor-owned electric utilities account for 60 percent of industry revenues, and independent power producers for the remaining 14 percent of revenues.[6]

Regulatory Authorities and Structure of the Industry

The power system in the continental United States comprises essentially three mostly distinct large interconnections: the Eastern Interconnection, the Western Interconnection, and the Electric Reliability Corporation of Texas (ERCOT). Each operates as an integrated machine that instantaneously matches generation and use and directs the flow of power.

Regulation of the system is balkanized and complicated. In general, each state, each territory, and the District of Columbia regulate many aspects of the electric power system within their jurisdictional boundaries. Typically, the regulation is codified by the state legislature and carried out by a public utility commission. These regulatory commissions regulate other utilities, such as telecommunications and water, as well.

State public utility commissions regulate the retail rates, distribution reliability, and service of investor-owned electric utilities. For nearly the first century of the industry, states took the stance that electric power was a natural monopoly like other utilities, the assumption being that competition is unstable, and avoiding duplication by competitive firms actually lowers prices. Absent an effective competitive market to create efficient prices, states have accepted that they must regulate these natural monopolies in a way that produces the same result that would occur if effective competition did exist. Indeed, "the single most widely accepted rule for the governance of regulated industries is to regulate them in such a way as to produce the same results as would be produced by effective competition, if it were feasible" (Kahn, 1970, p. 17).

Through the 1990s, several states took steps to restructure their electricity markets so that electricity supply became "unbundled" from transmission and distribution.[7] In those states that now have retail competition, state public utility commission authority may be limited to distribution, the acquisition of power for default service, and certain types of rules applied to retail suppliers. In other states, utilities tend to be vertically integrated, and state commissions may

---

[6]Some power marketers are owned by holding companies that also own utilities.
[7]For a succinct description of "restructuring," see Regulatory Assistance Project (RAP) (2011).

regulate the planning and construction or acquisition of generation facilities.[8] Eighteen states and the District of Columbia permit retail competition for electricity for some or all consumers. These states account for more than half of electricity sales (in megawatt hours [MWh]). In those states with unlimited retail competition, a majority of industrial and commercial customers purchase power from competitive suppliers. In Texas, Illinois, and Ohio, more than 50 percent of residential electricity consumers also purchase power from competitive electricity suppliers. State commissions generally do not regulate cooperatively and municipally owned utilities, although the state may regulate the siting of major facilities and have jurisdiction over energy emergencies.

FERC regulates the transmission and wholesale sale of power in interstate commerce. It approves reliability standards for the bulk power system developed by the industry through the North American Electric Reliability Corporation (NERC) and NERC's eight regional reliability organizations.[9] However, FERC cannot, on its own, propose and adopt new reliability standards. It also cannot address the reliability of distribution systems, which accounts for more than 90 percent of customer outages.

In large regions and the state of California, ISOs and regional transmission organizations (RTOs) are responsible for the planning and operation of the bulk power transmission grid and the commitment and dispatch of central station generating units.[10] FERC regulates ISOs and RTOs with the exception of ERCOT, an interconnection entirely within the state of Texas that is regulated by the Texas Public Utilities Commission. While the California ISO is regulated by FERC, the governor of California currently appoints its board.

Each ISO and RTO operates a system of markets including day-ahead and real-time energy markets that coordinate the operation of central station generation and efficient utilization of transmission assets. Additionally, PacifiCorp, a utility operating in six western states, and the California ISO have agreed to create an energy imbalance market to coordinate generation dispatch in real time. Other utilities in the West are considering participating in this market. With the exception of Georgia and Oregon, which permit limited competition for large commercial and industrial customers, utilities in the District of Columbia and each of the states that allows retail competition also participate in regional ISO or RTO wholesale markets.

---

[8]Electricity generators and utilities also are subject to extensive regulation by federal and state environmental protection agencies.

[9]The eight regional reliability organizations are Florida Reliability Coordinating Council; Midwest Reliability Organization; Northeast Power Coordinating Council; ReliabilityFirst Corporation; SERC Reliability Corporation; Southwest Power Pool, RE; Texas Reliability Entity; and Western Electricity Coordinating Council.

[10]California ISO, ERCOT, ISO New England, Midcontinent ISO, New York ISO, PJM Interconnection, and Southwest Power Pool. Canadian RTOs/ISOs include Alberta Electric System Operator and Independent Electricity System Operator.

Federal agencies and power marketing administrations, including the Tennessee Valley Authority, Bonneville Power Administration, Southeastern Power Administration, Southwestern Power Administration, and Western Area Power Administration, operate significant generation facilities in some regions and account for 6.5 percent of the nation's total generating capacity (APPA, 2014).

In addition to investor-owned and publicly owned utilities, about 12 percent of electricity customers are served by rural electric cooperatives (NRECA, 2016). Cooperatives are private, nonprofit businesses that are owned by the customers and are incorporated under the laws of the state in which they operate. They are governed by a board of directors that is elected from the membership (NRECA, 2016).

Infrastructure of the Present Power System

Today's power grid was built to deliver electricity produced by a few large power plants. Electricity is produced at central power stations at high voltages, and gradually stepped down to lower voltages as it flows through the transmission and distribution network until it is delivered to the user at a voltage that is considered safe for residential and commercial use. Figure 6-1 depicts the structure of today's power system (EPRI, 2014). Most consumers are billed based on the quantity (kWh) of electricity they use over a fixed period of time (e.g., a month), information that is collected either by utility employees (meter readers) who physically visit individual meters to note usage, or in some areas by meters that can send an electronic record of use directly to the distribution company for billing purposes. In each case, however, the flow of electricity and the flow of information are unidirectional, in opposite directions: electricity flows from the utility to the end-user, while information about usage flows from the consumer back to the utility. Planning for electricity generation capacity in this conventional power system typically is centered on a few key stakeholders (e.g., ISOs/RTOs, utilities owning generation assets, state public utility commissions and power siting boards), and focuses on larger generation facilities and transmission lines. This approach has achieved virtually universal access to electricity and an average annual reliability of 99.97 percent in the United States (IEEE, 2011). While this level of reliability was accepted in an industrial economy, it is lower than that achieved in many other developed countries (compare CEER [2012] and Eto and LaCommare [2008]) and may not be optimal for many customers in today's digitally based economy. Unfortunately, the traditional power grid will not support the level of distributed energy technologies that will occur based on current trends and demands for increasingly clean and more efficient, reliable, and resilient electric power.

Stakeholders with diverse and often conflicting interests are active participants in ISO/RTO committees and parties to FERC and state regulatory

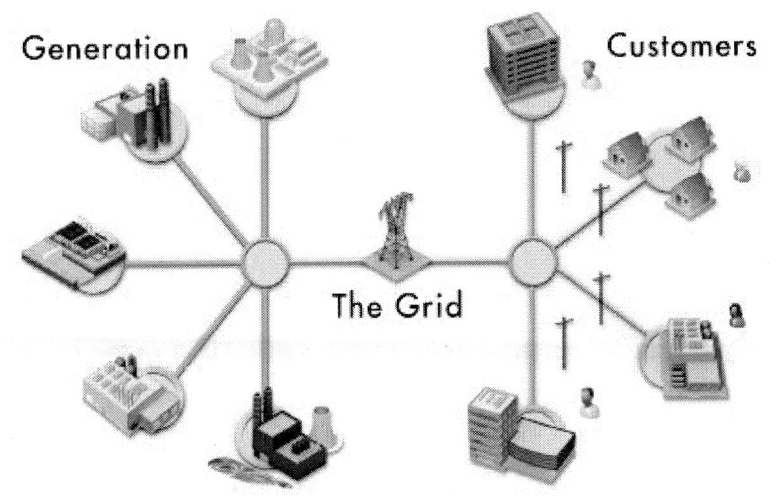

FIGURE 6-1 Today's power system, characterized by central generation, transmission, and distribution of electricity to end-use consumers.
SOURCE: EPRI, 2014.

proceedings. The complex organizational and governance structure of the power sector has facilitated experimentation and presents a challenge for the development of national energy policy.

The Current Utility Regulatory Framework

Electric utilities are mature organizations, often with conservative cultures. Well-aligned incentives and the engagement of policy makers, regulators, and external stakeholders with utilities are likely to be important to enabling utilities to embrace innovation and fully support the adoption of cost-effective increasingly clean energy and energy-efficiency technologies that can deliver net value to customers. This subsection describes the traditional cost-of-service model for regulation of electricity distribution and vertically integrated electric utilities and its limitations. The next section outlines alternatives that might better align utilities with the deployment of advanced increasingly clean energy and energy-efficiency technologies where such technologies would benefit society and utility customers.

As noted above, utility regulation has historically been intended to replicate the pressures of competitive markets for services even though the utilities' services are provided on a monopoly basis. The regulation of electricity distribution has focused on minimizing utility costs and avoiding the undue exercise of monopoly power. Regulators have been charged with ensuring that utilities provide adequate service and do not charge unreasonable or discriminatory prices. However, this is only a part of the function of regulation.

The objectives of regulation also include supporting investments that deliver net value to customers, ensuring the quality and reliability of service that customers value, and encouraging innovation to create dynamic efficiency gains. Given new challenges and customer expectations, regulators and policy makers have begun to question how best to realize both sets of objectives.

State public utility commissions have used a cost-of-service approach to set the rates charged by electricity distribution and vertically integrated utilities.[11] This process establishes the total of all costs prudently incurred to provide service, then sets rates necessary to enable the utility to recover the costs incurred during the year under review and realize a return on invested capital.

Given current conditions, the cost-of-service model has significant limitations (Malkin and Centolella, 2014):

- **Quasi-judicial proceedings**—Rates typically are set through time-consuming, quasi-judicial proceedings in which the utility files a lengthy application and testimony detailing the cost basis for a requested increase in rates. In some cases, the parties reach an agreement stipulating to a result that is recommended to the commission. Such agreements can provide parties greater flexibility, but if one or more parties do not agree, the case can revert back to a litigated process. Litigation can work well when the relevant questions can be answered on the basis of historical facts; however, it may not provide an ideal basis for making the types of risk and value judgments that utilities and regulators increasingly face.
- **Status quo fallacy**—Utilities frequently are asked to justify any significant changes from practices previously accepted by the regulator. To do so, they often must demonstrate that a new practice will lower their costs. However, this focus on incremental utility costs assumes that the utility will continue to provide the same fixed set of services. In reality, distribution utilities are increasingly expected—and in many cases required—to perform new functions. A conventional utility cost analysis may present a barrier to investments that expand future options, lower societal and environmental costs, and diminish incentives for innovation that could provide long-term benefits.
- **Misaligned incentives**—In the current environment of increasing costs and slow growth, cost-of-service regulation often fails to provide appropriate incentives for investment, efficiency, and innovation. With cost-of-service regulation, there is a lag between the time a utility makes a capital expenditure and the time it begins to recover its costs following a subsequent rate case. This lag has a

[11]For detailed discussions of cost-of-service regulation, see Bonbright et al. (1961), Kahn (1971), Phillips (1988), and RAP (2011).

negative impact on cash flow and can impair a utility's ability to earn its authorized return, which can in turn cause the utility to defer discretionary investments that would otherwise benefit customers. However, simply shortening this lag time can reduce the incentive for efficient operations. An assumption of cost-of-service regulation is that the interval between rate cases will create an efficiency incentive because the utility retains any firm-wide cost savings realized during that period. But if the utility has to file frequent rate cases, it has little opportunity to benefit from such cost savings. Cost reductions will be passed on rapidly to customers, as the utility's expenditures in one year become the basis for its allowed revenue in the next.

- **Barriers to innovation**—Innovation may introduce regulatory risk for a utility. If a new system fails to perform as expected, the utility may see its costs disallowed. Although firms in competitive markets have an opportunity to earn higher profits when innovation delivers value to their customers, utilities are seldom rewarded for assuming the risks of innovating. While firms in competitive markets can rapidly innovate, learn, and, if necessary, redirect their efforts, a regulated utility may need to cycle through a lengthy regulatory review process and justify changes from previously approved practices. The time between when the utility identifies a valuable commercial innovation and the innovation's full implementation can extend to as long as a decade.

Finding 6-1: To expedite innovative solutions, it will be necessary to redesign business models and regulatory incentives currently designed for a centrally controlled system so they are built on a customer-driven model with multiple solutions.

A MODERN POWER SYSTEM THAT WOULD SUPPORT THE DEVELOPMENT AND DEPLOYMENT OF INCREASINGLY CLEAN ENERGY AND ENERGY-EFFICIENCY TECHNOLOGIES

This section describes the features of a modern power system that would support the development and deployment of increasingly clean power and energy-efficiency technologies—its technical features, a supportive regulatory approach and specific regulatory policies, new utility business models, and workforce development.

The committee's recommendations in these areas are directed at both federal and state policy makers and state utility regulators. While much of the focus of this study was on national policies, the committee recognized that state utility regulation plays a central role in creating the conditions necessary for the development and deployment of cost-effective increasingly clean energy and energy-efficiency technologies. In addressing its recommendations to the states, the committee recognizes that different regions have different industry structures and opportunities that may require tailoring its recommendations to local conditions.

A modern electric power system that supports and encourages the development and deployment of increasingly clean power and energy-efficiency technologies will have certain essential features that can be identified now and may require others that will become evident over time. Perhaps the most essential feature is the further refinement and implementation of a regulatory framework and business models that align incentives for power generators, system operators, and utilities of all types with key objectives of reducing or eliminating pollution and other unpriced environmental harms, ensuring system reliability, safeguarding physical and virtual assets from malicious or accidental harm, improving and upgrading grid infrastructure, and protecting consumers from unfair pricing or other harms. A system that can produce these outcomes is one that (AEE, 2014)[12]

- encourages innovation in power generation technology, transmission and delivery infrastructure, and service models;
- empowers customers by giving them tools and options for managing their electricity costs;
- improves the design, operation, and coordination of power markets;
- moderates future customer bill increases relative to what otherwise would be experienced; and
- creates sustainable business models for firms in the power sector.

Special attention to the last point is warranted because, simply put, the job of creating and running a modern electric power system that encourages and produces these outcomes must be financially attractive for firms and their investors. Power-sector business models, however, are built largely in response to regulatory environments. Legislatures must create and regulators must implement a regulatory system of markets and incentives that support and encourage these investments.

> **Finding 6-2:** Regulatory and business models that encourage firms to invest in developing and deploying increasingly clean power and energy-efficiency technologies are critically important.

---

[12]This paper was produced as part of working group effort involving a number of electric power utilities and other firms. See the paper for a complete list the utilities and other organizations involved.

Technical Features of a Modern Power System

The challenges and drivers described above reveal an electric power industry that is starting to make transformative changes in energy production and use. Supporting these new patterns of electricity production and use will require a power grid that is physically and institutionally different from the grid of today.

A modern grid would support multiple actors at more (e.g., distributed) points of generation and/or consumption, and respond quickly and efficiently to variability in loads. Its main feature is that the distribution network is integrated with other components of the grid through active management and operation (IEEE, 2011). Figure 6-2 depicts a modern, integrated electricity grid as envisioned by the Electric Power Research Institute (EPRI) (IEEE, 2011), whose structure is very different from that of the current system as illustrated earlier in Figure 6-1. The distinguishing characteristic of this integrated grid is the multidirectional flow of both electricity and information (data) between energy supply and energy use and the suite of advanced smart-grid technologies that would enable the efficient management of these flows.

The ability of customers to route excess electricity back to the grid for use by another customer elsewhere in the system has system-wide advantages. First, customers can benefit from decreasing their net power consumption or actively participating in power markets. Second, distributed energy technologies can enhance the overall reliability experienced by customers, provide distribution voltage support and improve voltage quality, and reduce system losses. The power system's resiliency can also be enhanced, as portions of the grid with appropriate control technologies can continue to function during system outages through the use of islanding techniques (Passey et al., 2011). Third, environmental impacts can be reduced because new (possibly higher-polluting) generation and capacity additions can be avoided, and resources that are no longer cost-effective to maintain can be replaced. Overall, enabling this multidirectional flow of electricity allows value to be gained from the use of distributed energy technologies, especially given storage capability.

The multidirectional flow of information and the integration of advanced information and communication technologies with the operation of the power system are key features of a modern integrated power system. The system enables customers to participate and provides system operators with detailed real-time data that can be used to optimize system operations. Advanced metering infrastructure (AMI) is one of the enabling technologies. It measures and records electricity usage at short intervals and can provide the data to both the customer and the utility. The most advanced AMI has built-in two-way communication capability for real-time data. Communicating thermostats and other smart energy-using devices in customers' homes and businesses, together with access to information on anticipated electricity prices, can optimize the

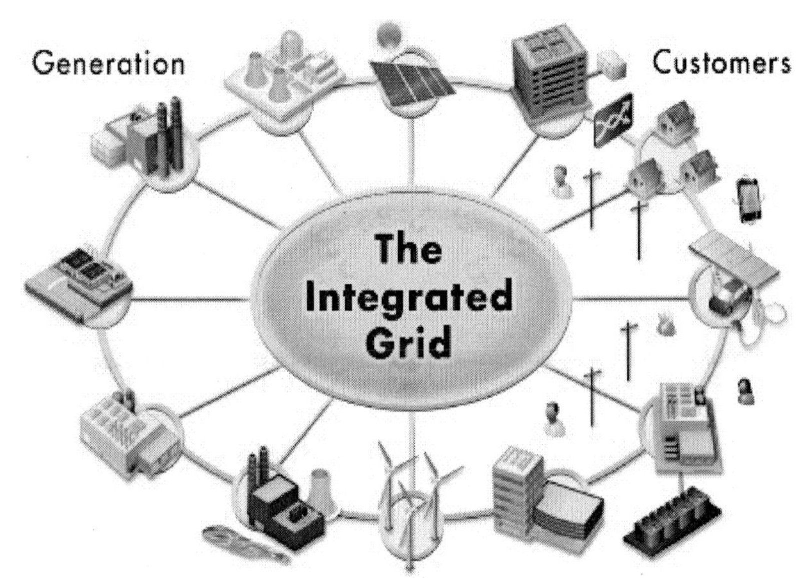

FIGURE 6-2 Concept of an integrated grid with multiple customer sites for distributed energy resources, networked with other points of generation as a distributed energy network.
SOURCE: EPRI, 2014.

timing of energy demand. Utilities also can forecast changes in customer demand and use this information in optimizing system reliability and efficiency.

A Supportive Regulatory Approach

This subsection describes an evolution in utility regulation that could support the development of a modern, integrated power system such as that described above. Establishing a well-functioning modern grid that provides increasingly clean and more efficient, reliable, and resilient electric power will require a supportive regulatory and business environment. A wide variety of policies have already begun to stimulate and drive U.S. investment in smart-grid technologies, such as that resulting from the American Recovery and Reinvestment Act of 2009, which provided more than $3.4 billion in stimulus funding for smart-grid technology development and demonstration (plus $615 million for smart-grid storage). With the expiration of that act's funding, new regulatory models have begun to emerge to support needed upgrades and technology investments.

The Importance of Regulatory Frameworks

Utilities base their business models on the regulatory frameworks within which they operate. Various alternative regulatory models could incentivize and reward the development and deployment of increasingly clean power and energy-efficiency technologies and the necessary supporting systems and infrastructure.

Over time, regulators have taken steps to adapt to changing conditions, including experimentation with alternative regulatory models.[13] Some of these alternative models provide greater support for new investment. These models may involve prior regulatory review of utility plans to align them with regulatory objectives, and also may be conditioned on utility commitments to making specific improvements. Such alternative models include the following:

- **Annual rate cases with a forecast test year**—In some jurisdictions, the utilities forecast their investment expenditures based on prior planning reviews. By using these forecast values in annual rate proceedings, the utilities and their regulators can better match costs and revenues to the prospective level of rates. However, frequent regulatory involvement can make this approach administratively burdensome. Examples in which this approach has supported investment include the Public Service Commission of Wisconsin with its biennial Strategic Energy Assessment and annual rate cases, and the Iowa Utilities Board's preconstruction approvals of new generation.
- **Capital expenditure trackers**—A tracker is a separate rate-adjustment mechanism that allows for the recovery of specific costs outside of the conventional rate case process. Historically, tracker mechanisms were reserved for significant and volatile costs, such as fuel, which are largely beyond the utility's control. More recently, several states have permitted accelerated recovery of specific capital expenditures outside of a cost-of-service rate case. For example, Pennsylvania's Distribution System Improvement Charge allows accelerated recovery of costs associated with approved long-term infrastructure plans.
- **Formula rates**—In this approach, a specific formula for setting rates is established in advance by statute or a prior public utility commission order. The utility then files its cost data, and the information used to determine its allowed rate of return in a standard format. While the formula sets the types of cost that may be recovered, costs may be subject to review based on whether the

[13]For further discussion of alternative models, see McDermott (2012), Pacific Economic Group (2013), and Malkin and Centolella (2014).

expenditures were prudently incurred. Examples of formula rates include FERC's transmission rates and the Illinois Energy Infrastructure Modernization Act.

These approaches can support investment, but they can involve a high level of regulatory oversight. They also offer limited incentives for the utility to reduce its costs and share any cost savings with consumers. For example, capital cost trackers have been criticized for diminishing efficiency incentives and for allowing rate increases for the cost of new capital additions without consideration of countervailing cost reductions. Similarly, some commentators have criticized formula rates on the grounds that they fail to encourage cost-efficiency and productivity improvements (Costello, 2009).

Other alternative models are designed to provide strong incentives for reducing costs. These models include the following:

- **Multiyear revenue and price caps**—Under this model, changes in utility revenues or rates can be indexed to inflation and adjusted for a targeted rate of productivity improvements and any extraordinary events. Alternatively, the regulator may set annual step changes or freeze revenues or rates for the duration of the rate plan. These multiyear rate plans can promote cost reduction by enabling the utility to share in any cost savings and absorb cost increases during the years covered by the plan. In the absence of strong reliability standards or incentives, however, they have been associated with a reduction in spending on operations and maintenance and an increase in the average duration of customer outages. In addition, unless the multiyear plan is tied to a reasonable utility business plan for new investment and changes in its operations, the revenue or rate cap may not match the rate levels needed for required capital investments.
- **Sliding-scale rate plans**—In a few states, including Alabama, Louisiana, and Mississippi, regulators determine a target return for the utility and set rates based on cost and revenue forecasts to achieve the return target, subject to a predetermined ceiling on rate increases. The regulator also sets a range of authorized earned returns. The utility's actual earnings are later reviewed, and if the earned returns are within the authorized range, the utility may retain or must absorb all or a share of any variance between its target and actual earnings. The opportunity to retain earnings within the authorized range provides an incentive for the utility to be efficient. If actual earnings exceed the authorized range, however, the utility may be required to return the excess earnings to customers. Sliding-scale plans also can incorporate performance incentives based on reliability, customer satisfaction, or other metrics. The sliding-scale approach may be considered a light-

handed form of regulation and has not attracted significant support from policy makers in other regions.

An Emerging Regulatory Model in the United Kingdom

Regulators, utilities, and other stakeholders are actively seeking to define regulatory models that support needed investments, incentivize cost savings, and encourage innovation. U.S. regulators have taken note of a rate-setting framework being implemented by the utility regulators in the United Kingdom, Office of Gas and Electric Markets (Ofgem). New York State Department of Public Service (2014) commented favorably on this model in its report in New York's widely followed "Reforming the Energy Vision" (REV) proceeding. Ofgem is implementing an approach for the regulation of network companies called RIIO (Revenue set to deliver strong Incentives, Innovation and Outputs).[14] RIIO is an incentive-based framework intended to mimic the effects of competitive markets by linking revenue to output metrics, innovation, and cost savings. It encourages transmission and distribution utilities to focus on delivering net long-term value to customers. RIIO's major components include the following:

- **Revenues set based on a review of the utility's business plan**—The review of the utility's business plan includes benchmarking of planned operating expenses and an engineering assessment of capital expenditures.
- **Cost savings shared with customers**—RIIO includes an earnings-sharing mechanism with large sharing factors. To the extent that a utility's actual earnings exceed its authorized return, 50 percent to 60 percent is refunded to customers, while if costs are higher than anticipated and earnings fall below the authorized level, the utility may have to absorb up to 50 percent of the loss in earnings. The precise sharing percentages can vary among utilities based on the regulator's assessment of a utility's cost projections.
- **Clearly defined results-based metrics and output incentives**—Incentives can be bidirectional, either increasing or decreasing earnings. The regulator may adjust output metrics and incentives during the rate plan, with adjustments being applied to the remaining years of the plan. Ofgem has proposed or adopted performance incentives related to the following:
  - The frequency and duration of outages—Incentives are based on studies of the value placed by different customers on uninterrupted service.

[14]For additional information on RIIO, see Ofgem (2010b, 2013) and Jenkins (2011).

  - Customer satisfaction—Incentives may include an up to 1 percent up or down adjustment in revenue based on customer surveys and an additional incentive of up to 0.5 percent of revenue based on an independent panel's assessment of the utility's stakeholder engagement practices.
  - Environmental impacts—Incentives may be based on reductions in line losses; the visual impact of power lines (undergrounding); and reductions in greenhouse gas emissions, including leakage of sulfur hexafluoride (SF_6), a potent greenhouse gas used in insulating transformers and other electrical equipment.
  - Social obligations—Incentives address issues of fuel poverty and assistance to vulnerable customers in accessing available services.
  - Timing and efficiency in connecting customers—New customers purchase electric service from competitive suppliers. Incentives are based on utilities' performance in connecting customers.
  - Meeting worker and public safety standards.
- **Application of the revenue cap to total expenditures**—At the start of the rate plan, the regulator fixes the percentage of revenue that will be recovered in each rate year, with the residual being capitalized. Once this ratio has been established at the beginning of the plan, it does not change based on the nature of the utility's actual expenditures. The utility has the flexibility to take advantage of learning and modify its spending plans to meet its output objectives as efficiently as possible. An annual rate adjustment aligns revenue with authorized levels.
- **Innovation programs**—Ofgem is funding innovation programs for the piloting of large projects, small projects, and the rollout of proven solutions. These programs enable third parties to partner with the utility to deliver cost savings, carbon reductions, or other environmental benefits. An expert panel disburses multiple rounds of funding.[15]
- **Limited revenue reopeners**—While Ofgem's general approach is to require utilities to manage business risks, it may define circumstances in which rate plans may be reopened to address changes in underlying economic assumptions or unknowns such as new cyber-security requirements.
- **End-of-period adjustments**—Ofgem tracks asset health and may implement an additional positive or negative incentive at the end of the rate plan to ensure that assets have been appropriately maintained,

---

[15]Ofgem publishes an annual report on projects funded through its Network Innovation Competitions. As of this writing, the list of projects funded in 2015 was the most recently available (https://www.ofgem.gov.uk/sites/default/files/docs/innovation_competitions_brochure_webready_0.pdf).

replaced, or upgraded. Ofgem also may allow recovery near the end of the rate plan for investments designed to produce benefits during the next rate plan. Ofgem may allow utilities to carry forward into the next rate plan a share of cost savings realized near the end of the current plan.

RIIO is an example of a regulatory authority attempting to balance incentives for cost savings with performance incentives based on specific output metrics. In many respects, Ofgem was dealing with concerns comparable to those facing U.S. regulators. The U.K. power industry faces aging infrastructure, a changing power generation mix with increased reliance on variable renewable generation, and limited revenue growth. In developing its reform program, Ofgem sought to engage consumers in defining desired results. It also recognized that accelerating innovation could play a key role in making power and energy affordable, as well as meeting the nation's climate objectives. Ofgem's electric power innovation programs are currently providing funding of more than $60 million for projects designed to test advanced technologies for facilitating the integration of renewable generation, cutting distribution losses, reducing generation requirements through distribution voltage optimization, and improving the flexibility and operation of transmission and distribution networks (Ofgem, 2014). The United Kingdom is in the early stages of implementing RIIO, with the first plans now in place.

There are important differences between the regulatory environment in the United Kingdom and that in the United States. RIIO builds on 20 years of U.K. experience with price cap regulation. Both the regulator and utilities had accumulated skills and tools to help them develop a long-term performance-based rate mechanism. Moreover, the regulatory process in the United Kingdom is more consultative than that in the United States and lacks a comparable history of contentious rate case litigation. For example, the regulator in the United Kingdom is able to offer a utility a menu of different incentive contracts designed to incentivize the utility to disclose accurately its expected cost for meeting performance metrics.[16]

Taking differences in their regulatory environments into account, several U.S. regulators are considering how to adapt the RIIO framework to their own circumstances with some core results-based concepts, including

- revenues based on forward-looking business or grid modernization plans;

---

[16]This practice is known in the United Kingdom as an information quality incentive and more generally as a menu of contracts approach to setting rates. For background and a description of how the approach is implemented, see Cossent and Gómez (2013) and Ofgem (2010a, p. 66).

- multiyear revenue caps that provide an incentive for the utility to pursue efficiency improvements and retain a share of the resulting cost savings or bear a share of the resulting cost overruns;
- caps on total expenditures that give utilities the flexibility to shift spending between operating and capital expenditures to meet requirements efficiently as new information becomes available;
- earnings-sharing mechanisms that allow customers to benefit from cost savings or bear a share of costs incurred during multiyear plans;
- output-based, bidirectional performance incentives for reliability, energy efficiency, customer satisfaction, and other performance metrics; and
- funding set aside specifically for research, development, and other innovation projects.

Advancing Consideration of Alternative and Emerging Regulatory Models

There has been or is ongoing consideration of alternative and emerging regulatory and utility business models in many states, such as California,[17] Hawaii,[18] Illinois,[19] Maryland,[20] Massachusetts,[21] Minnesota,[22] New York,[23]

---

[17]See California Public Utilities Commission, *In the Matter of Order Instituting Rulemaking Regarding Policies, Procedures and Rules for Development of Distribution Resource Plans Pursuant to Public Utilities Code Section 769*, Order Instituting Rulemaking, Public Utilities Commission of California Rulemaking 14-08-013 (August 20, 2014).

[18]See Public Utilities Commission of Hawaii, *In the Matter of Public Utilities Commission regarding Integrated Resource Planning,* Docket No. 2012-0036, Decision and Order No. 32052, Exhibit A: Commission's Inclinations on the Future of Hawaii's Electric Utilities (April 28, 2014).

[19]Illinois enacted a formula rate statute—the Energy Infrastructure Modernization Act—to support grid modernization. This formula rate framework is scheduled to sunset in 2017. Additionally, the Illinois Commerce Commission is currently considering how best to provide competitive suppliers access to customer information while protecting customer privacy.

[20]See Maryland Public Service Commission Staff, Report on Performance Based Ratemaking Principles and Methods for Maryland Electricity Distribution Utilities, *In the Matter of the Electric Service Interruptions in the State of Maryland Due to the June 29, 2012 Derecho Storm*, case no. 9298 (July 1, 2014).

[21]See Massachusetts Department of Public Utilities (2013).

[22]The Great Plains Institute recently partnered with Xcel Energy, Minnesota Power, the Center for Energy and the Environment, George Washington University Law School, and other stakeholders to review regulatory models in Minnesota in what is called the e21 Initiative (see http://www.betterenergy.org/projects/e21).

[23]See New York State Department of Public Service (2014).

and Texas.[24] Given the power industry's current challenges, state regulators and policy makers would do well to investigate and consider such regulatory models that align utility incentives with achieving long-term cost savings, providing net value to customers, promoting public policy objectives, and encouraging innovation.

Many state regulatory commissions have limited staff and will require tools and training beyond what they currently have if they are to develop and effectively implement alternative models (see Fox-Penner, 2014). Several commissioners consulted during the course of this study emphasized that they would welcome and greatly value assistance from the Department of Energy (DOE)[25] to help train and educate commissioners and staff. In particular, they suggested creating a national program that would provide additional resources and training, and perhaps serve as both a coordinator and repository for best practices and lessons learned as many states move forward with regulatory reforms.

> **Finding 6-3:** Many state regulatory commissions require additional analytical tools, training, and other resources to develop and implement effectively regulatory models that support and encourage the development of increasingly clean energy and energy-efficiency technologies.
>
> **Recommendation 6-1: DOE should develop information, tools, and programs that would facilitate state regulatory commissions' consideration and implementation of regulatory models tailored to meeting current challenges. These resources would be a natural extension of the Electricity Policy Technical Assistance Program already operated by the Office of Electricity Delivery and Energy Reliability.**[26]

Specific Supportive Regulatory Policies

This subsection reviews utility regulatory policies designed to advance the cost-effective deployment of advanced increasingly clean power and energy-efficiency technologies.

---

[24]The Texas market is among the most open to retail competition and has fostered the development of a wide variety of retail supply options (see Compete Coalition, 2014).
[25]Or possibly other national organizations, such as the National Association of Regulatory Utility Commissioners.
[26]For an assessment of specific information and tools that DOE could consider developing, see DOE EAC (2014a).

Automation of Customer Preferences in Energy-Using Devices

A low-pollution energy future will require more efficient integration of variable low-carbon resources into power system operations. Today, integration requirements can limit the use of renewable generation or require the continuing operation of additional fossil fuel-fired units primarily for the purpose of offsetting changes in the output of variable resources.

Additionally, a significant portion of the energy used in buildings is wasted. One study estimates that 39 percent of residential energy consumption is wasted, with the majority of that waste due to heating and cooling of unoccupied spaces or overheating or overcooling of the home to achieve comfortable temperatures in some parts of the home (Meyers et al., 2010). Given the reach and declining cost of communications with distributed devices and advances in data analytics, automated systems may be able to significantly reduce such waste and lower carbon emissions. Unlike prior generations of programmable thermostats, a modern smart thermostat can

- sense when a home or portion of a building is unoccupied;
- identify when consumers with smartphones are arriving back in their home neighborhood;
- automatically fine tune operational schedules to address seasonal changes;
- reduce run times of air conditioner compressors on less humid days when the compressors are not needed;
- balance heat pump operations to provide desired comfort and reduce the use of less efficient auxiliary heating elements;
- prompt customers to change their furnace filters when accumulated dirt has reduced the filters' efficiency; and
- provide reports on the efficiency of energy use and recommendations for energy savings.[27]

One study estimates that the installation of smart thermostats could reduce the energy used by residential air conditioners in southern California by more than 11 percent (Nest Labs, Inc., 2014).

Many people rely on automated customer choice technologies to perform a variety of functions in their lives. An example is booking flights. Consumers can enter the date on which they want to fly, the hours they prefer to travel, and the number of connections they are willing to make. An application then sorts through thousands of flights and suggests the least expensive options consistent with consumers' preferences. In the same way, one can enter a preferred temperature and program a smart thermostat to give it 1-2° of temperature

---

[27]See, for example, https://nest.com/thermostat/saving-energy/#we-didnt-think-thermo stats-mattered-either.

flexibility. Today's communicating thermostats can access forecasts of local temperatures and humidity, sense whether anyone is at home and determine when the house generally is unoccupied, and learn the building's characteristics and the efficiency of its cooling and heating systems. Using precooling and smart operating strategies, such thermostats have reduced peak use of residential air conditioners by 50 percent in 100° F-plus Texas temperatures (Nest Labs, Inc., 2013) and cut demand in a Nevada utility program by more than 3 kW per household.[28] In the Nevada program, annual electricity and natural gas usage in participating homes was cut by 3.6 percent and 6.4 percent, respectively.[29]

The impact of such automation could be great. One California study estimated that the thermal inertia of residual air conditioning, given no more than 1° C of temperature flexibility, water heaters with up to 4° C of flexibility, and refrigerators with up to 2° C of flexibility, could permit smart devices to shift 20 GW or more of the state's residential demand during more than 2,000 hours of the year and provide at least 8-11 gigawatt hours (GWh) of energy storage throughout the year. This estimate suggests that during much of the year, smart devices have the potential to shift a majority of California's residential electricity demand to different time intervals. For California residential customers, shifting electricity demand to lower-cost intervals could reduce their estimated energy cost for air conditioning (at wholesale market prices) by about 10 percent and their energy costs for water heating and refrigeration by up to 40 percent or more (Mathieu, 2012; Mathieu et al., 2012).[30] A 2011 National Energy Technology Laboratory study concluded that enabling system operators to send signals to smart energy-using devices could reduce peak demand by more than 25 percent and produce billions of dollars per year in economic, reliability, and environmental benefits (Goellner et al., 2011). Most uses of electricity, including heating and cooling buildings, heating water, and refrigeration, have thermal inertia or, in the case of most pumping loads, batch processes, and charging of electric vehicles and other devices, flexibility in the timing of power use. Moreover, smart devices could respond continuously to help system operators offset ramping of variable resources or, if carbon and other environmental impacts were appropriately priced, to shift consumption to periods when resources with an optimal combination of lower costs and environmental impacts would be used.

---

[28]Application of Nevada Power Company d/b/a NV Energy for Approval of its 2014 Annual Demand Side Management Update Report as it relates to the Action Plan of its 2013-2032 Triennial Integrated Resource Plan, Volume 5—Technical Appendix, http://pucweb1.state.nv.us/PDF/AxImages/DOCKETS_2010_THRU_PRESENT/2014-7/39345.pdf (hereafter Nevada Power Application). See also Kerber (n.d.).

[29]Nevada Power Application.

[30]Estimates were based on estimated device saturations in 2020. As a point of comparison, the contemporaneously prepared forecast for 2020 of residential noncoincident peak demand was 29,105 GW and of residential average hourly electricity consumption was 11,959 GWh (California Energy Commission Staff, 2012).

The barriers to a future in which smart devices can reduce the costs and environmental impacts of energy use by implementing the preferences of ordinary consumers are not primarily technological or economic; rather, they are largely regulatory and policy-related. Consistent with previous recommendations of the National Research Council (NRC, 2010c), pricing carbon and other environmental externalities would help ensure that the changes in consumption patterns over the long run associated with increasingly responsive demand would reduce both total societal costs and those costs currently reflected in electricity prices.

Important Roles for Regulators to Enable Automation

FERC and state utility regulators could take several steps to enable greater use of automated technologies to optimize demand participation in power markets.

First, in organized power markets, wholesale settlements are in many instances based on distribution utility load shapes, not on the actual load patterns of each retail supplier's customers. Where settlements are based on customers' actual demand profiles, energy service companies (ESCOs) can have a competitive advantage by packaging energy with demand-optimizing technologies. For example, a competitive retail supplier could provide a smart thermostat and offer a lower rate to customers with less peak coincident load shapes.[31] Basing wholesale settlements on each ESCO's actual load shape could be accomplished with, but does not require, AMI, as many AMI meters can record interval data.[32] Alternatively, usage from conventional meters could be allocated to time intervals based on sensors at a sample of each supplier's customers. Given an emerging role for smart devices, accessing such data could be as essential to effective competition in power markets as access to transmission was seen to be 20 years ago. Offering customers financing or demand-side management incentives for adopting smart devices also could accelerate their adoption. In a regulated environment where rates are not directly tied to each customer's contribution to system and distribution circuit peak demands, a significant portion of the economic savings produced by changing demand patterns may be enjoyed by nonparticipating customers.

---

[31]Technologies that automate customer energy choices do not require the use of dynamic retail prices. However, such prices may produce additional economic-efficiency benefits. Devices that automate customer preferences could make it easier for customers to take advantage of two-part and dynamic pricing to control their energy bills (Centolella, 2012).

[32]AMI meters include a two-way communication capability. Advanced meter reading is an older technology that provides an automated way to collect meter data.

> **Finding 6-4:** Customer adoption of smart devices may be important to provide the information needed to operate an efficient competitive power market.

> **Finding 6-5:** Basing wholesale settlements on the actual load shapes of the customers of each ESCO can provide incentives for customers to adopt smart devices.

Second, ISOs and regional transmission organizations (RTOs) typically settle with load participants in organized wholesale power markets on an hourly basis. Incentives could be enhanced and responsive demand could play a greater role if load, like generation, were settled on a 5- or 15-minute interval basis. Customer-specific and more granular wholesale settlements could encourage utilities and competitive retail suppliers to work with their customers to automate and manage more efficiently the timing of flexible demand, offering lower prices to customers who use automation to create efficient changes in their usage profile.

> **Finding 6-6:** The settlement of load in wholesale markets on a 5- or 15-minute interval basis instead of hourly would enable and provide an incentive for a much greater role for automated demand in maintaining reliability, balancing variable resources, and reducing peaks in demand.

Third, although most ISOs and RTOs develop short-term price forecasts, only the New York ISO and ERCOT publish such information. Information based on these indicative "look-ahead" forecasts could be used to position demand for anticipated system conditions and would be highly beneficial if made available to devices all the time, everywhere they are available, in a standard format, as inexpensively as possible.[33] The Federal Power Act directs FERC to "facilitate price transparency" and to "provide for the dissemination,

---

[33]The ISO and RTO "look-ahead" price forecasts are not settlement prices. As with any forecast, there will be differences between the indicative forecast and after-the-fact prices in the real-time market. Nonetheless, making available information based on "look-ahead" forecasts could enhance demand participation. First, an intelligent device could now consider both the day-ahead price and a "look-ahead" forecast that incorporated information about operating-day conditions and reliability events. Second, unlike an hourly day-ahead price, the "look-ahead" information could provide more granular interval data, enabling short-term demand participation. Third, once "look-ahead" forecasts were being made available, ISOs and RTOs would have an incentive and the opportunity to improve the publication of information about forecast prices. Providers of the data analytics underlying responsive devices would have a comparable incentive to use the information to help customers take power when it was least expensive. State and federal collaboration on appropriate privacy protections and data sharing with system operators also could help enhance forecast accuracy.

on a timely basis of information about [wholesale] prices...to...the public." FERC is authorized, if necessary, to "establish an electronic information system" for this purpose (16 USC §824t). If wholesale price forecasts were coupled in a systematic and predictable way to retail prices, system operators could provide smart devices with market price forecasts that would enable those devices to improve their performance in response to information from the system operators regarding anticipated market conditions.

Fourth, system operators have the capability to incorporate response curves that reflect statistically predictable relationships between prices and demand into forecasts used for both operations and planning purposes.[34]

Recommendation 6-2: System operators should consider utilizing their capability to build response curves that reflect predictable price-demand relationships to enable flexible demand that responds to short-term prices, and incorporate those curves into the forecasts they use for operations and planning purposes.

Fifth, state regulators could spur more demand for smart devices that are connected to the home or building, such as a smart thermostat or commercial building energy management system, by allowing these devices to be financed through on-bill repayment programs. These programs would permit the financing of energy management and energy-efficiency devices to be linked to the premises and be transferred from one owner or tenant to the next, which could prove an effective way for customers to finance the devices.

In contrast with current demand-response programs, the response of intelligent devices need not depend on a payment to a curtailment service provider[35] or the calculation of a baseline. This feature reduces administrative costs and inconvenience to customers (Bresler et al., 2013), avoids dissimilar treatment of otherwise comparable customers (Borenstein, 2014), and minimizes opportunities for abuse.[36] According to the Board of Managers for PJM, one of the nation's largest RTOs, PJM's long-term vision is that "Price Responsive Demand, which allows more customers to respond directly to market prices and to voluntarily reduce their consumption when wholesale prices rise, is the

---

[34]The recommendation for the straightforward recognition of these relationships also appears in Centolella and Ott (2009). Such recognition would avoid imposing on millions of price-responsive customers who receive no payment in the wholesale markets burdens and penalties comparable to those applied to supply-side resources.

[35]As a result, this approach can be fully consistent with the Circuit Court's decision in *Electric Power Supply Association v. F.E.R.C*, Case No. 11-1486, U.S. App. LEXIS 9585 (D.C. Cir. May 23, 2014), rehearing pending.

[36]*Enerwise Global Technologies, Inc.*, Order Approving Settlement, 143 FERC ¶ 61,218 (June 7, 2013); *Rumford Paper Co.*, Order Approving Settlement 142 FERC ¶ 61,218 (2013).

ultimate solution to demand participation" (PJM, 2009). Technology is now offering a means of democratizing demand participation and significantly improving system efficiency.

Volt/Volt Ampere Reactive (VAR) Optimization (VVO)

In a conventional distribution system, voltage is increased at the substation and may be boosted at intermediate points to levels consistently above minimum requirements to ensure that as power usage changes over time and voltages drop through the length of the distribution circuit, minimum voltage levels are consistently maintained for customers at the end of each line. VVO programs can reduce electricity generation requirements on many distribution circuits by 2-5 percent using modern solid-state power electronics on distribution circuits and smart inverters in distributed resources or, in some cases, a range of control technologies in conjunction with load tap changers (LTCs), regulators, and capacitor banks (DOE, 2012a; EPRI, 2012). They do so by managing voltage in real time, leveling and reducing unnecessary voltage levels across the circuit, and thereby reducing both losses and the apparent power delivered to customers' meters. This can occur without reducing the overall power quality needed by customer devices. VVO appears to represent a significant and often cost-effective means of improving energy efficiency and reducing emissions. The same advances in power electronics also can help integrate distributed and variable resources, reduce peak demand, ensure consistent voltage levels for end-use devices, and improve power quality on the grid.

Despite recent advances, however, VVO programs have not yet been widely adopted. Some of this delay is due to the continuing evolution of the technology. However, there also are several nontechnical barriers, including the following:

- Regulators may not recognize VVO as an energy-efficiency program since it occurs on the utility side of the meter. VVO will reduce the energy usage recorded at the meter. For a utility that is recovering fixed costs through volumetric charges and does not have a revenue decoupling mechanism, failing to adjust rates for the lower metered energy usage associated with VVO means a loss in earnings.
- Measurement and evaluation techniques, metrics, and associated standards have been slow to develop. Universally accepted approaches for measuring and verifying results on an ongoing basis as loads and circuit characteristics change over time do not yet exist.
- The impacts of VVO technologies will vary from circuit to circuit. Moreover, universally accepted planning tools for identifying those locations in distribution systems that could benefit the most from the use of different approaches to VVO do not yet exist.

Finding 6-7: Volt/VAR optimization has the potential to enable significant decreases in the amount of power generation required to support transmission and delivery and to increase system quality and reliability, but faces several nontechnical barriers.

To take full advantage of VVO, state regulators could investigate and consider cost-effective distribution utility VVO programs. DOE could play a key role in facilitating the cost-effective deployment of VVO technologies by supporting the development of planning, benefit-cost analysis, and measurement and verification tools and standards and promoting the sharing of experience and best practices. Additionally, DOE and the electricity industry could consider establishing a cooperative program to promote the understanding of these technologies, their potential benefits, and consideration of options for removing regulatory disincentives among regulators and industry stakeholders (DOE EAC, 2014a).

Recommendation 6-3: DOE should support distribution utility VVO programs and facilitate the cost-effective deployment of VVO technologies by supporting the development of planning, benefit-cost analysis, and measurement and verification tools and standards and promoting the sharing of experience and best practices.

Dedicated Innovation Budgets and Roles for Utilities

Utilities have historically devoted a very small percentage of their revenue—about 0.2 percent—to R&D. Some jurisdictions are addressing this funding gap by setting aside dedicated funds for R&D and fostering innovation. California energy consumers support energy R&D through both the Electric Program Investment Charge, a public goods charge that funds research programs managed by the California Energy Commission (2016), and a unique joint venture between the state's electricity distribution utilities and Lawrence Livermore National Laboratory—California Electric Systems for the 21st Century (CES-21).[37] Under the state's Energy Infrastructure Modernization Act, Illinois electricity distribution companies have an innovation accelerator and venture fund—the Energy Foundry—that supports innovative energy technology companies.[38] The Massachusetts commission recently decided to establish dedicated funding for utility R&D as part of its Grid Modernization program

[37]For more information on CES-21, see California Office of Ratepayer Advocate website at http://www.ora.ca.gov/general.aspx?id=1864.
[38]For more information, visit the Energy Foundry website at http://www.energy foundry.com.

(Massachusetts Department of Public Utilities, 2014). And New York utility customers pay a systems benefit charge to fund the New York State Energy Research and Development Authority, which supports energy research, development, and innovation programs.[39]

Ofgem's RIIO framework offers another example of such an approach. Under RIIO, funds are set aside for a package of innovation stimulus policies comprising the Network Innovation Competition (NIC), the Network Innovation Allowance (NIA), and the Innovation Roll-out Mechanism (IRM). The NIC is a U.K.-wide competitive funding opportunity that is open to any network utilities. Utilities must demonstrate that they have a process in place to facilitate collaboration with other (e.g., non-network) companies, and funds are awarded by an independent panel based on the extent to which proposed projects

- accelerate the development of a low-carbon energy sector and/or deliver environmental benefits and have the potential to deliver net financial benefits;
- provide value for the money for network electricity/gas customers;
- Create knowledge that can be shared across energy networks in Great Britain (GB) or create opportunities for rollout for a significant proportion of GB networks; and
- are innovative (i.e., not business as usual) and have an unproven business case, but the innovation risk warrants a limited trial research, development, or demonstration project to demonstrate its effectiveness.

Utilities also need to demonstrate that the incentives within price control regulation are not sufficient to justify the project.

The NIA is a set-aside allowance that each utility receives to fund small-scale innovative projects. A set-aside innovation budget can enable utilities to grow their innovation capabilities and support projects, primarily at the pilot or small demonstration scale, without the risk of such funds being diverted for ongoing operations.

Finally, the IRM enables utilities to apply for additional funding within the price control period for the rollout of initiatives that have demonstrable, cost-effective low-carbon and environmental benefits.

Recommendation 6-4: State regulators and policy makers should implement policies designed to support innovation. For example, they could evaluate approaches in which utility or energy customer funds are set aside to support state and regional innovation programs.

---

[39]For more information, see the Authority's website at http://www.nyserda.ny.gov/About.aspx.

Chapter 3 includes discussion of and recommendations for increased utility involvement in innovation, particularly at the demonstration stage and as partners in regional innovation networks.

Energy Efficiency and Energy Management Financing: On-Bill Repayment

On-bill repayment is another potential tool for financing energy efficiency and energy management. For detail, see Chapter 4.

New Utility Business Models

Over the last two decades, the power system in much of the country has been fundamentally changed by open-access transmission,[40] the development of ISO and RTO markets,[41] market-based pricing of wholesale generation,[42] demand response in wholesale markets,[43] regional transmission planning,[44] and competitive retail supply. Nonutility generators now provide about 40 percent of the nation's electricity (EIA, 2016d, Tables 3.1.A and 3.3.A). By 2012, ISO and RTO demand-response programs were playing a major role in organized markets, with the potential to provide up to 10.7 percent of capacity requirements in ISO New England, 7.3 percent in the Midcontinent ISO, 7 percent in PJM, 5.8 percent in the New York ISO, and 5.2 percent in the California ISO (FERC, 2013). In jurisdictions that permit retail competition for power supply, more than 17 million households and a substantial majority of businesses have shopped for power from competitive retail electricity suppliers (Compete Coalition, 2014). And in Texas, arguably the most open market in the country, all consumers are served by competitive retail electricity suppliers, and these suppliers are offering more than 300 different packages of pricing and services to help customers manage their energy bills (Compete Coalition, 2014).

In many primarily southern and western states, changes have been less dramatic. Utilities remain vertically integrated, with bundled retail rates for generation, transmission, and distribution services being regulated by state

[40] *Promoting Wholesale Competition through Open Access Non-discriminatory Transmission Services by Public Utilities; Recovery of Stranded Costs by Public Utilities and Transmitting Utilities, Order 888,* 61 FERC 61,080 (April 24, 1996); *Preventing Undue Discrimination in Transmission Services, Order 890,* 72 Fed. Reg. 12,226 (March 15, 2007).

[41] *Regional Transmission Organizations, Order 2000,* 81 FERC 61,285 (December 20, 1999).

[42] *Market-Based Rates for Wholesale Sales of Electric Energy, Capacity and Ancillary Services by Public Utilities, Order 697,* 119 FERC 61,295 (June 21 2007).

[43] *Wholesale Competition in Regions with Organized Markets, Order 719,* 73 Fed. Reg. 64,100 (October 28, 2008).

[44] *Transmission Planning and Cost Allocation by Transmission Owning and Operating Public Utilities, Order 1000,* 136 FERC 61,051 (July 21, 2011).

commissions. However, utilities in these jurisdictions have nonetheless participated in industry developments. Utilities in a number of these jurisdictions have participated in competitive procurements for generation (Tierney and Schatzki, 2008). Some vertically integrated utilities have invested in grid modernization and advanced metering and have been leaders in offering time-varying and dynamic rates (FERC, 2013; IEE, 2013).[45] New nuclear generating facilities[46] and demonstrations of carbon capture and storage (CCS) technologies[47] are being developed by vertically integrated utilities with the ability to recover generation costs in state-regulated rates. However, projects such as the Vogtle and V. C. Summer nuclear units and the Kemper County CCS facility have experienced schedule delays and cost increases. These delays and cost increases may reflect both (1) risks inherent in these projects and (2) the limited ability of regulation to replicate the incentives created by competitive markets and an opportunity to shift cost and schedule risks to ratepayers. Mechanisms such as the Regional Innovation Demonstration Funds proposed in Chapter 3 that incorporate competition for ratepayer funding would preserve incentives for cost-efficiency and enable demonstrations of potentially transformative nuclear and CCS technologies to be undertaken in competitive generation markets.[48]

Emerging Opportunities for New Business Models

The business model for electricity distribution until recently has remained relatively stable among most utilities. However, it is in distribution as well as in retail energy services that new utility business models are emerging. The need to consider new business models is driven in part by the challenges and opportunities previously discussed: replacement of aging infrastructure,

---

[45]For grid modernization, advanced metering, and time-varying and dynamic pricing programs supported by the American Recovery and Reinvestment Act of 2009, see https://smartgrid.gov/recovery_act/project_information.

[46]The Tennessee Valley Authority (Watts Bar 2) and vertically integrated utilities in two traditionally regulated states (Vogtle 3 and 4 [Georgia, lead utility: Southern Company] and V. C. Summer 2 and 3 [South Carolina, lead utility: South Carolina Electric and Gas]) are the principal owners of the new nuclear units under construction in the United States. For information on proposed U.S. nuclear power plants, see http://www.world-nuclear.org/info/Country-Profiles/Countries-T-Z/USA--Nuclear-Power.

[47]Southern Company has proceeded with its Kemper County Carbon Capture and Sequestration project despite cost overruns. However, American Electric Power halted its Phase 2 CCS project at its Mountaineer plant in West Virginia "because they did not believe state regulators would let the company recover its costs" (Wald and Broder, 2011, p. A1). For additional information on CCS projects, see http://sequestration.mit.edu/tools/projects/us_ccs_background.html.

[48]Two carbon capture for enhanced oil recovery projects—Summit Power's Texas Clean Energy Project and NRG's Energy Parish Project—received support from DOE's Clean Coal Power Initiative and are being pursued in a competitive power market.

expectations for greater reliability and resilience, cyber and physical security requirements, integration of variable and distributed resources, slowly growing or declining sales and limited revenue growth, and the opportunities created by new distributed energy technologies.

As discussed earlier, historically, power flowed from central station generation, through the transmission grid, and in one direction from the substation linking transmission and distribution to the customer (see Figure 6-1 earlier in this chapter). Distribution systems were designed based on the assumption that power moved only in this one direction. Distribution investments were sized to meet the peak demands of the customers expected to connect to each circuit. The fixed costs of the distribution system could be recovered through volumetric rates because there was little risk that customers would produce significant energy with customer-sited generation. Indeed, 44 states and the District of Columbia adopted net metering policies that effectively pay small customers retail rates for power delivered into the grid, initially with limited utility opposition.[49] While net metering has become a highly contested issue, it was not viewed as a significant threat to the utility business model when these policies were first adopted (DOE, 2015a).[50]

As illustrated earlier in Figure 6-2, new distributed energy technologies are challenging this model of distribution operations. First, the falling cost of information and communication technology is making it cost-effective to manage demand by improving home and building operations and automating customer preferences for savings and comfort. As a result, there are now millions of end uses for power that could respond in real time to anticipated changes in prices or grid conditions. Utilities will have opportunities to connect to what is being called "the internet of things" to influence the timing of energy demands across their distribution systems in potentially very significant ways. Second, PV and other distributed generation technologies in some regions are becoming cost-competitive with retail rates. While this does not imply that such technologies cost less than providing the same energy services from conventional generation, cost parity with retail rates may make them attractive to customers and lead to their adoption, perhaps at an accelerating rate.[51] Utilities face potentially significant challenges in integrating these distributed technologies with the planning and real-time operation of their distribution systems.

---

[49]For a description of state net metering policies, see http://www.dsireusa.org/resources/detailed-summary-maps/net-metering-policies-2.

[50]For a utility industry perspective, see http://www.eei.org/issuesandpolicy/generation/NetMetering/Pages/default.aspx.

[51]As more customer-sited generation occurs, the throughput in distribution systems will tend to decline. As a result, additional rate increases may well be required to maintain the existing utility infrastructure, which could in turn provide a greater incentive for customers to self-generate.

In the last 5 years, moreover, many utilities have planned and begun implementing programs to modernize their distribution systems. These programs are built on the integration of information and communication technologies into power system planning and operations. Grid modernization or "smart-grid" initiatives have the potential to provide significant cost savings and improvements in reliability and customer value. EPRI estimated in 2011 that national deployments of smart-grid technologies could produce net economic benefits over the 20-year period through 2030 of $1.3-2.0 trillion. To deliver these benefits, utilities would need to invest roughly $17-24 billion per year, with an average benefit-to-cost ratio of between 2.8 to 1 and 6.0 to 1. Similarly, the Smart Grid Consumer Collaborative, representing a broad cross-section of industry stakeholders, found that smart-grid investments would produce a benefit-to-cost ratio of between 1.5 to 1 and 2.6 to 1 and net present value benefits of between $247 and $713 per utility customer (EPRI, 2011; SmartGrid Consumer Collaborative, 2013; see also Schneider et al., 2012). And the early results from such efforts confirm the availability of significant benefits (see, e.g., DOE, 2012a,b,c; EPRI, 2012; Faruqui and Palmer, 2012; Schneider et al., 2012).[52] To ensure that grid modernization works effectively, secure interoperable standards are necessary. Efforts to develop standards and protocols to ensure that different systems and devices can communicate and operate with each other are being undertaken by the National Institute of Standards and Technology (NIST, 2014).

Two Potential Business Models to Address Challenges and Needs in Distribution

The above developments are leading to consideration of two emerging parallel and potentially complementary business models for distribution utilities and/or other market participants: distribution system operator (DSO) and customer energy service provider (CESP) (see Fox-Penner, 2010; Rocky Mountain Institute, 2013). These models may be able to address several challenges that distribution networks will encounter with increasing levels of distributed and variable generation assets. Both models, however, face challenges to their full development and implementation.

First, the efficient integration of distributed energy technologies, distribution automation, VVO, and other characteristics of a smarter power grid will require a more active DSO. Historically, distribution operations could largely ride off of the operation of the transmission system. A system that includes intelligent distribution and distributed energy technologies will require detailed and transparent planning and real-time operational management and coordination. The operation of a system in which distributed technologies can impact both distribution and transmission system operations may require

[52] Additional project evaluation reports can be found at http://smartgrid.gov.

- a federated control architecture connecting transmission and distribution operations;
- integrated modeling and state estimation to give both transmission and distribution operators real-time awareness of power flows across transmission and distribution systems;
- a flexible, advanced information architecture to manage a major expansion in operational data and integrate an evolving set of information systems and applications while maintaining cyber security;
- the ability to commit and dispatch or forecast and coordinate the operation of large numbers of distributed technologies, dynamically manage the topology of mesh or microgrid-based distribution networks, optimize voltage, and simultaneously maintain phase balance across the distribution system; and
- distribution-level market structures that can coordinate settlement of transactions.

These requirements, in part, parallel the types of systems that had to be developed for the operation of RTOs and ISOs. However, efficient operation of a distributed system may have to accommodate a larger number of control points and manage greater complexity. The development of such systems will take time and require a coordinated R&D effort. DOE has taken only partial steps to address such requirements through the Green Electricity Network Integration (GENI) program in the Advanced Research Projects Agency-Energy (ARPA-E), formation of the Office of Electricity and Energy Reliability's Grid Tech team, and support for standards development and the Smart Grid Interoperability Panel. While the DSO role is likely to develop over a number of years, the development of needed operational systems in time to match the pace of cost-effective deployment of distributed energy technologies could prove challenging. With market participants connecting with the distribution system, clearly defined interconnection and interoperability standards will be needed, and distribution planning will need to become more transparent.

> **Finding 6-8:** Clearly defined interconnection and interoperability standards and more transparent distribution planning will be essential for connecting increasing numbers of market participants to the distribution system.

Enhanced, accessible distribution planning tools may be needed to support the development and regulatory approval of distribution plans. DOE has supported the development of distribution planning models, including GridLAB-D™, an advanced distribution system simulation and analysis tool that provides information to users who design and operate distribution systems.

However, GridLAB-D is not widely used by regulatory commissions or other industry stakeholders. Regulators in parts of Europe and Latin America have addressed such gaps by developing reference network models (RNMs). An RNM is a planning tool that, using heuristics and contingency analysis, forecasts the distribution investments reasonably needed to integrate new resources, achieve desired reliability targets, and meet forecast demand in an approximately optimal fashion. RNMs may differ in scope from conventional distribution planning models in automatically generating expansion candidates from a library of standardized equipment rather than relying on a distribution planner to propose candidate investments, and in validating the feasibility of planning decisions both electrically and in terms of physical considerations when integrated with a geographic information system (Domingo et al., 2011). By identifying a reasonable plan that meets distribution planning objectives, an RNM can help regulators examine the impacts of distributed energy resources and evaluate proposed utility distribution investments (see Cossent et al., 2011; Jamasb and Pollitt, 2008; Larsson, 2005).

In a distributed energy system, these operational and planning functions (or some aspects thereof) will be natural monopoly roles. Policy makers will have to weigh considerations related to existing utility capabilities, economies of scope, and the need for transparency and independence when determining whether the DSO/CESP function should be assumed by distribution utilities, ISOs or RTOs, or new independent entities. To the extent that these functions reside within existing utilities, the focus of the distribution utility would change from being primarily a wires function to one that incorporates much greater reliance on information and communication technology, operational models, and data analytics. Definition and development of the roles and functions of DSOs and CESPs is now beginning to be explored in a number of fora.[53]

New distributed energy technologies also will create opportunities for utilities and/or competitive suppliers to offer customers a broader range of energy services. Competitive retail energy suppliers could transition from providing commodity electricity service to managing customer energy bills. Some suppliers already offer packages that include smart thermostats that optimize energy usage and automate customer preferences. Suppliers might support enhanced service quality and reliability with the deployment and operation of distributed generation and storage, such as backup generators or PV and battery storage. And there are firms that currently offer microgrid development services to commercial, institutional, and industrial customers.

A wide range of well-funded firms—including providers of home security services (e.g., ADT), telecommunications companies (e.g., Verizon), cable providers (e.g., Comcast), big box retailers (e.g., Lowes, Home Depot),

---

[53]California Public Utilities Commission Order Instituting Rulemaking Regarding Policies, Procedures and Rules for Development of Distribution Resource Plans Pursuant to Public Utilities Code Section 769. *Order Instituting Rulemaking, Public Utilities Commission of California Rulemaking 14-08-013*, 2014 (August 20, 2014).

manufacturers of consumer electronics (e.g., Samsung, LG) and controls (e.g., Honeywell), and tech giants (e.g., Google, Apple)—are already competing for a share of the market for home energy management services. These firms, as well as start-ups in the market, can be expected to participate in the market for consumer energy services.

Utilities also may become customer-focused energy service providers. Given their existing capabilities and customer relationships, utilities may be able to accelerate the availability of "adjacent energy services." The utility could provide a portal that would give customers access to such services from third-party suppliers. Alternatively, such services might be available directly from the utility or from a utility affiliate. In vertically integrated markets, the utility might offer enhanced reliability services and demand management through its energy-efficiency or demand-response programs. Some jurisdictions might follow a model, comparable to that used during deregulation of telephone services, allowing utilities to offer adjacent services on a competitive basis, subject to light-handed regulation, with a portion of the revenue offsetting the cost of providing regulated distribution services. Other jurisdictions might require corporate separation of potentially competitive services from regulated distribution functions.

Effective DSOs have the potential to provide support for large-scale deployment of cost-competitive increasingly clean distributed power assets. They could do so in a way that would reduce energy costs while providing greater reliability and value to customers. For DSOs to be effective, however, would require timely development of a number of capabilities, including

- an effective control architecture and systems for the federated management of transmission, distribution (including dynamic distribution topologies), and deployment at scale of distributed energy technologies;
- integrated operational modeling and systems providing real-time operator awareness of multidirectional power flows across transmission and distribution systems;
- a flexible information architecture and related interoperability standards to support the expanded availability of power system operational data and the evolution of operating systems and applications;
- operational models, systems, and applications to support the integration and management of more intelligent and dynamic distribution systems and distributed energy technologies;
- advanced cyber-security systems for a power grid that to an increasing extent relies on information and communication technology;
- distribution-level market structures that can coordinate settlement of multidirectional transactions; and

- enhanced distribution planning models that also are transparent and user-friendly to facilitate regulatory and stakeholder review of distribution planning decisions.

Finding 6-9: The creation of effective DSOs and CESPs will require the timely development of key capabilities.

For these systems to be effective, they will need to be developed based on widely accepted interoperability standards. They also will need to be made ready to defend against and respond to attacks and accidents. Poorly designed systems and those developed without appropriate security will be particularly vulnerable to cyber attack.

Recommendation 6-5: DOE should undertake a multiyear R&D program to ensure the timely development of the capabilities needed for effective DSOs or CESPs through policy analysis; dialogue; and the sharing of experience and best practices among regulators, utilities, and other stakeholders to advance understanding of the emerging business models. DOE should strongly consider prioritizing the development of robust, well-designed systems that incorporate appropriate security measures to guard against and respond to cyber attacks.

Workforce Development

Utilities currently face a significant, multifaceted workforce challenge—one that can undermine the potential for positive transformation in the electric power sector if not properly resolved. As of 2008, projections showed 50 percent of the electric utility workforce being retirement-eligible within 10 years, representing more than 200,000 skilled employees (Hardcastle, 2008). This attrition would add to an already existing reduction in the utility workforce over the past two decades. The workforce declined precipitously (by approximately 50 percent) in the 1990s and 2000s as the sector restructured and many utilities began participating in competitive markets. Mergers, acquisitions, and cutting of every nonessential cost through minimal hiring were mainstream in the industry. Over time, these practices drove employee numbers downward and created the gap of workers in their 30s and 40s and concentration of workers in their 50s and 60s now characterizing the sector (Hardcastle, 2008; Lave et al., 2007).

To maintain power system functionality, it is imperative that these employees be replaced as they retire. Yet while replacing a workforce of this magnitude represents an already significant human resource challenge, this challenge is exacerbated by the fact that new workers will need both to continue operating legacy systems and to meet new requirements (Lave et al., 2007). The changing nature of the electricity sector—as detailed throughout this chapter and elsewhere in the report—requires a trained and motivated workforce with a very different profile from that of the past. The future utility workforce will be responsible for introducing such technologies as those needed for smart-grid operations, and thus will require employees with greater "niche" skills to support the implementation, maintenance, and operation of systems with primarily digital components. Advanced technologies will require employees comfortable with analytical and mathematical methods, possessing spatial awareness, computer proficiency, and problem-solving skills (Lave et al., 2007). New training programs also will be required as outdated guidance documents and technical manuals (e.g., solar interconnection manuals) become updated. The future utility employee will be responsible not only for learning the new standards and procedures associated with these updates, but also for responding quickly and efficiently to the dynamic technological and regulatory environment that will mark the modern electric power system. This level of flexibility is a key differentiator between a modern utility workforce and the more traditional workforce of today. Recruiting individuals with this initial capability and continually training them once they enter the workforce is itself an inherent challenge.

> **Finding 6-10:** The electric power industry faces a challenging shortage of skilled, appropriately trained workers.

> **Finding 6-11:** The necessary skill base for the electric power workforce has changed and continues to evolve. The future workforce will need to be trained in new technologies, such as smart-grid devices, and to implement and maintain new systems, such as advanced distribution networks engineered for two-way power flows and high levels of distributed generation assets.

The main factors contributing to the challenge of recruiting qualified individuals to plan and manage modern electricity systems are themselves interrelated. The curricula of U.S. educational institutions do not emphasize electric power systems and related electrical engineering and computer science foundations. As one example, 13 U.S. universities currently make up the Power Systems Engineering Research Center—a National Science Foundation industry-university cooperative research center that sees itself as "empowering minds to engineer the future electric energy system" (PSERC, 2016). Each of these universities houses programs in electrical and computer engineering, and students obtain disciplinary degrees (e.g., control systems, operations research, economics) for which the coursework may *include* power system-related courses. Of these 13, however, only 2 list specialized programs of study in electric power systems (PSERC, 2016). In part, the relative absence of such programs is a direct result of the withdrawal of electric utilities in the wake of deregulation and consolidation from what were previously plentiful utility-university partnerships. Prior to deregulation and restructuring, it was common for electric utilities to engage in long-term relationships with local universities. Utilities would provide tuition scholarships, fund internships, provide general "support" funds for programs of study in electric power systems, and even establish designated power systems research centers within local universities. These programs often provided opportunities for students to master the textbook fundamentals and simultaneously engage in real-world training (Russell, 2010).

Unfortunately, the absence of such programs of study also is due to a lack of demand from students, and this represents the second major factor contributing to the overall lack of qualified applicants. The electric utility industry historically has not garnered perceptions of professional status and "achievement," and thus has not attracted individuals interested in mathematics, engineering, and computer science, who have gravitated toward other emerging and more "stimulating" industries, such as aerospace and chemicals manufacturing. The electric utility industry also has developed a reputation for not paying as well as others, offering instead a safe, steady, but "dull" form of employment (Lave et al., 2007). Overall, the combination of these demand-side and supply-side problems in U.S. education in power systems is an important factor in the utility workforce challenge.

It is imperative, then, to recreate a vision of the electric power industry as one that is attractive, stimulating, and worth celebrating for the vital role it plays in people's lives and in driving the nation's prosperity. Doing so will require encouraging new dedicated degree programs in power systems engineering and electronics at the college and postgraduate levels, as well as chaired faculty and other filled tenure-track positions committed to teaching and research in this area (Russell, 2010). Training also could benefit from beginning in high school and grade school curricula—to emphasize, and generate an interest in and comfort with, mathematics, computer science, and analytical problem-solving skills in young people. In addition to classroom learning modules, training could extend to such activities as power plant tours and even brief work-study arrangements (Lave et al., 2007). Finally, compensating entry-level graduates with competitive salaries would help bolster the image of the electric utility industry as one that is serious and values requisite skills.

DOE could provide support for energy and power workforce development activities. This could include support for industry-educator partnerships for training a skilled, technical workforce. Additionally, DOE could, consistent with the recommendations of its Electricity Advisory Committee, coordinate workforce development activities with those of other federal agencies and the private sector, evaluate the impacts of its American Recovery and Reinvestment Act workforce training grants, and make training curricula and content developed through those grants available through a central repository (DOE EAC, 2012, 2014b).

Planning and implementing new (or modified) educational programs across the country will take some time. In the meantime, utilities can undertake—and governments can support—several initiatives that can help bridge the immediate gap in the skilled workforce. Knowledge retention programs will be essential as retirement-eligible staff with technical and institutional expertise exit and inexperienced new hires enter the workplace. Increasing an employee's compensation for participating in additional voluntary new-hire mentor programs, as well as other incentive programs, such as allowing phased retirements whereby employees can reduce their hours gradually over a few years' time, could help ensure that utility and industry-specific knowledge is preserved (DOE EAC, 2014b).

> **Finding 6-12:** Industry-educator partnerships are the most effective way to train a skilled, technical workforce, and can bridge the immediate gap in the skilled electric utility workforce. Governments can support such initiatives to make them more effective.

Utilities and customers also could benefit from expedited deployment of many advanced grid technologies. Outage management systems provide one example. Such systems can save costs associated with incorrect outage reports by verifying power outages at customer facilities. PECO estimates that it avoided 7,500 crew dispatches in 2005 because it was able to see that those customer-reported outages were inaccurate (Pritchard and Evans, 2009).[54] More recently, in 2012 Oncor implemented an integrated system of advanced meters and an outage management system along a 3.2 million meter-long network, capable of remotely sensing outages and restoring power to customers before the outages are physically sensed at the customer site. In the first 6 months, this new system helped Oncor avoid hundreds of power outages (Wolf, 2012). Intelligent, integrated systems such as these not only reduce costs by increasing operational efficiency, but also deliver real value to customers in the form of increased reliability and heightened confidence in the utility's abilities.

---

[54]PECO, the former Philadelphia Electric Company, is a distribution subsidiary of Exelon Corporation.

7

Policies Supporting Increasingly Clean Electric Power Technologies

The federal and state governments have supported the discovery, development, and maturation of new energy sources and technologies since America's earliest days. Coal-, petroleum-, natural gas-, nuclear-, and renewable fuel-based electricity production each benefited in the earliest stages of development from targeted government policies intended to support and develop the industry. Most of these industries have fully matured, and as of this writing, each continues to benefit from targeted policies at both the federal and state levels.

Energy policies have focused on supply, usually aimed at increasing or maintaining production levels and decreasing or stabilizing prices (Adams, 2010). Common policies have included direct subsidies, exemption from or reduction of taxation, import controls, funding for research and development (R&D), indemnification, and the creation of government agencies intended to provide research and other direct support to an industry.

This chapter first provides a brief synopsis of historical supports made available to each major energy source in its nascent stages of creation and development, followed by a basic account of the level of current supports. The chapter then investigates various approaches to lowering the cost of capital and reducing risk to private capital for financing the deployment of increasingly clean electric power technologies.[1] One key approach is to ensure that markets fully account for all costs, including pricing of externalities such as the costs associated with pollution, including oxides of sulfur (SO_x), oxides of nitrogen (NO_x), and greenhouse gases (GHGs). But as discussed earlier in this report, pollution pricing would not address institutional barriers and other market imperfections. Thus, another important approach is to enable financing mechanisms that lower or avoid up-front cost barriers by allowing implementation costs to be financed over time by project savings. The final

[1]For policies related to deployment of energy-efficient equipment, see Chapter 4.

section explores ways of addressing barriers that remain at the deployment stage of increasingly clean electric power technologies.

The committee notes that quantitative measures of historical and current subsidies and the social cost for all environmental damage created by using different fuels and technologies in generating electric power are not currently available. More research is needed in this field, given the long life span of power plants.

HISTORY OF GOVERNMENT SUPPORT FOR NEW ELECTRICITY SOURCES

Coal

The federal government's involvement in the U.S. coal industry has a long history. In 1789, the then-new federal government imposed a modest tariff on imported coal, the goal being to protect nascent American industries from British imports. This tariff grew until, by the War of 1812, the import tariff amounted to 15 percent of the price of British coal (Adams, 2006). The government reduced the tariff by half following the war, but maintained it at levels sufficient to guarantee that domestic sources would dominate the new market until, in the 1870s, the United States became a net exporter of coal.

In 1879, Congress established the United States Geological Survey (USGS). While the USGS mission was to be scientific and military in nature, it was specifically charged with charting anthracite in Pennsylvania and coalfields generally. In effect, through USGS, the federal government subsidized coal exploration, and the accumulation of vital industry data became a national-level activity (Adams, 2006).

In the 20th century, the development of competition-restricting collusion between coal producers and railroads, in addition to major labor disputes within the coal industry, led to increased involvement in the industry on the part of the federal government. In 1902, for example, President Theodore Roosevelt set a precedent for federal intervention in the coal industry when a labor strike threatened energy supplies for the entire East Coast. The decision in a 1908 Department of Justice lawsuit against railway companies accused of price manipulation upheld the International Commerce Commission's (ICC) ban on the ownership of mines by railroad companies, and in 1910, the U.S. Bureau of Mines was established, overseeing the creation of federal safety standards for mineworkers. As Adams (2006, p. 77) notes in his study on the political history of U.S. energy systems, "In all these cases, federal intervention in the nation's coal trade preserved the nineteenth-century focus on high levels of production."

Individual states had more extensive involvement with the early development of the coal industry. Beginning in the early 19th century, state governments enacted laws and policies designed to promote the large-scale exploitation of domestic coal reserves and to keep the price of coal stable and accessible. In Pennsylvania in the 1830s, for example, the state legislature passed a measure exempting anthracite coal from taxation. In addition, by refusing to grant exclusive transportation rights to any one company, Pennsylvania lawmakers fostered competition among coal companies and thus kept the price of coal relatively low (Adams, 2010).

Petroleum and Gas

The petroleum industry similarly relied on the support of the federal government in its early development. From early in the 20th century, the federal government used the tax code to encourage the broadest possible exploitation of petroleum and gas reserves. In 1916, the tax code introduced the expensing of intangible drilling costs (IDCs) and dry-hole costs. In 1926, the Percentage Depletion Provision was incorporated into the tax code to allow the deduction of a fixed percentage of gross receipts rather than a deduction based on the actual value of the recovered resources (Pfund and Healey, 2011). The percentage depletion allowance, which still is available to selected taxpayers, is an alternative to cost depletion and is currently available only for domestic production by independent companies up to a maximum of 1,000 barrels per day (or 6 million cubic feet [MMcf] per day for natural gas), and cannot exceed half the net income from the property. The depletion rate is set at 15 percent gross production revenues. The most striking aspect of the percentage cost depletion allowance is that it can exceed the cost of the original investment over the life of the property, in effect providing a complete government subsidy for costs associated with the property's purchase and maintenance (NRC, 2013c).

Through the mid-1980s, subsidies provided to the oil and gas industries constituted the largest federal energy tax provisions in terms of revenue loss. Between 1918 and 2009, these subsidies amounted to an historical annual average of $4.86 billion (Pfund and Healey, 2011).

Nuclear

The nuclear power sector is unique in that it is historically a product of federal-level policy making. Most federal subsidies for the nuclear industry have taken the form of support for R&D. Over the period 1948 to 2014, Department of Energy (DOE) R&D funding for the nuclear sector amounted to $97.44 billion (in 2013 dollars), nearly twice the amount provided for fossil fuel development, and the next-largest allocation of federal energy-related R&D funding after petroleum and gas (Sissine, 2014). From its earliest beginnings, moreover, this funding focused specifically on a concerted effort to develop a new electricity industry, an effort that slackened only with the decline of interest in nuclear production as a result of safety and financial concerns. By way of illustration, of the total amount spent since 1950 on exploring reactor concepts and potential civilian and military applications of nuclear energy, some $42 billion (nearly 60 percent) was spent before 1975 (MISI, 2011).

Another significant policy in support of nuclear power operations is the Price-Anderson Act (PAA). The PAA has a dual purpose: to "protect the public and...encourage the development of the atomic energy industry" (Heal and Kunreuther, 2010; Rothwell, 2001, quoting the PAA).[2] The PAA provides the nuclear power industry with blanket indemnity for tort liability, the first layer of protection being $200 million in private insurance provided through the American Nuclear Insurers, and the second being a set amount to be provided per reactor by nuclear plant owners following an accident at any nuclear power reactor. The value of this coverage totaled $9.3 billion in 2001, an amount that is certainly higher today because of both inflation and increased buy-in costs levied on the nuclear industry. During the early years of the U.S. nuclear power industry, producers argued that indemnity such as that provided by the PAA was necessary to enable them to shoulder other costs, such as those related to construction (Heal and Kunreuther, 2010; Rothwell, 2001).

In addition, the federal government is responsible for the regulation and safe management and disposal of spent nuclear fuel. Specifically, the Nuclear Regulatory Commission's (U.S. NRC) Office of Nuclear Material Safety and Safeguards (NMSS) develops and implements U.S. NRC policy in this area (U.S. NRC, n.d.). Congress tasked DOE, aided by the new Nuclear Waste Fund, with the collection and storage of spent nuclear fuel, and mandated that DOE create a permanent storage site by no later than January 1998 (Garvey, 2009). While the creation of the permanent storage site at Yucca Mountain has been held up for many years, this mandate remains.

Hydropower

Large-scale hydropower in the United States owes its early development in the 1930s to extensive federal programs. A detailed accounting of those government expenditures is difficult because this development occurred in a complex policy environment. This is the case in part because federal dam-building projects undertaken by the Bureau of Reclamation and the Army Corps of Engineers in the 1930s and 1940s also had such goals as flood control and navigation (Pfund and Healey, 2011). Additionally, large-scale hydroelectric facilities function as wholly owned subsidiaries of the federal government, and so do not need to earn private rates of return. Thus, an argument can be made that they have served as an industry support dating back to the establishment of the Tennessee Valley Authority in 1933 (Tennessee Valley Authority, n.d.), the Bonneville Power Administration in 1937 (Bonneville Power Administration, n.d.), and other federally owned and operated electric power producers (Pfund and Healey, 2011).

---

[2]Price-Anderson Act 42 USC 2012i.

Renewables

Wind, solar, and geothermal electricity production represents a recent sector in the U.S. electricity market. Government support for this sector at the federal level takes the form of tax subsidies—specifically the production tax credit (PTC) and the investment tax credit (ITC).

The PTC was first enacted in 1992 and has been renewed or extended a number of times. It provides a rate of 1.5¢/kilowatt hour (kWh) over 10 years, adjusting with inflation so that as of January 2015, it provided 2.3¢/kWh for the first 10 years of electricity production generated from qualifying wind, geothermal, and biomass sources, or a credit of 1.2¢/kWh for other qualifying renewable sources (Heal and Kunreauther, 2010; Pfund and Healey, 2011). The ITC for alternative energy sources provides a nonrefundable tax credit for building solar, wind, geothermal, fuel cell, and microturbine energy generation facilities. The tax credit is given the year the facility enters service. The ITC first appeared in the Energy Tax Act of 1978 (PL 95-618) (Pfund and Healey, 2011). These credits lower the cost of electricity generated from renewable resources, encouraging their substitution for fossil fuels, and thereby tend to reduce GHG emissions (NRC, 2013c, p. 3).

In recent years, the renewable energy PTC and ITC have at times been allowed to lapse, with subsequent, short-term renewal in tax-extender packages. While eligibility was modified over this term to allow PTC subsidies to apply to a broader range of project starts, these on-and-off subsidies have created significant market uncertainty and have led to layoffs throughout the wind turbine, tower, and component supply chain. At the end of 2012, for example, the PTC was extended for 1 year through 2013. It was then allowed to lapse, and was subsequently extended in December 2014 retroactively for calendar 2014, lapsing again at the beginning of 2015. The latest renewal of the credit was enacted in December 2015 and applied retroactively to January 1, 2015. The most recent extension, in December 2015, included a phase-out schedule that differs for solar and wind. The phase-out for wind begins for construction initiated in 2017, with full phase-out at the end of 2019. The PTC for other eligible renewable energy technologies was extended only for construction initiated through the end of 2016 (DOE, n.d.-c).

Currently, eligible solar facilities qualify for an ITC equal to 30 percent of expenditures for construction commencing in 2016, phasing down to 10 percent in 2023 and beyond. Geothermal facilities qualify for an ITC equal to 10 percent of expenditures for construction initiated in 2016 and beyond, while large wind facilities qualify for an ITC that is gradually phased out until 2020. The ITC for other technologies expires at the end of 2016. Technologies eligible for the PTC can opt for the ITC instead if construction commenced prior to January 1, 2015; for construction initiated after that date, only wind facilities remain eligible to claim the ITC in lieu of the PTC (DSIRE, 2015).

A number of renewable energy technologies also receive tax benefits under the Modified Accelerated Cost-Recovery System (MACRS), which allows the owner to write off the value of some capital assets at a rate that exceeds their estimated useful life. Doing so reduces taxable income in earlier years by allowing a larger depreciation expense than is actually represented by how much of an asset's usefulness is consumed in those early years.[3]

As a whole, subsidies for renewable energy technologies have increased over the past 10 years. Records of the last few years show a spike in investment in renewable sources due to the American Recovery and Reinvestment Act (ARRA), to the point where they exceeded investment in fossil fuel-powered production. However, the subsidy patterns that have defined the last few decades have, in the wake of the ARRA's expiration, most likely returned to favoring fossil fuel (EIA, 2015c; ELI, 2009; Heal and Kunreauther, 2010; Pfund and Healey, 2011).

Renewable energy also is supported by state-level policies.[4] One key policy promoting the deployment of renewable energy is the renewable portfolio standard (RPS) (see Chapter 5). An RPS requires that electric power suppliers[5] use renewable energy or obtain renewable energy credits (RECs) above a minimum threshold amount of their electricity sales, or that utilities procure above a minimum amount of renewable generating capacity in their portfolio of electricity resources. RPSs set a schedule for renewable energy or capacity to be obtained by specific years. Requirements generally increase over time.

As of June 2016, 29 states[6] and the District of Columbia had an RPS in force. These jurisdictions account for 63 percent of U.S. electricity sales (EIA, 2015d). Some sources attribute the development of approximately 46 gigawatts (GW) of new renewable generating capacity from 1998 through 2012 to state RPS requirements (Heeter et al., 2014, p. 1). This amounts to roughly two-thirds of all nonhydroelectric renewable electricity generating capacity additions in the United States since 1998. An additional 8 states have adopted voluntary renewable energy goals.[7] Together these 36 states and the District of Columbia account for more than three-quarters of U.S. electricity sales (EIA, 2015d).

---

[3]The depreciation schedule is based on the type of renewable technology. See Internal Revenue Service (IRS) Publication 946, IRS Form 4562: Depreciation and Amortization for further information.

[4]A comprehensive list of salient state laws and regulations in effect as of the end of October 2013 is available in the Energy Information Administration's (EIA) *Annual Energy Outlook 2014* (EIA, 2014a, pp. LR-4-LR-9).

[5]Either utilities or, in jurisdictions with retail competition, retail electricity providers.

[6]Arizona, California, Colorado, Connecticut, Delaware, Hawaii, Illinois, Iowa, Maine, Maryland, Massachusetts, Michigan, Minnesota, Missouri, Montana, Nevada, New Hampshire, New Jersey, New Mexico, New York, North Carolina, Ohio, Oregon, Pennsylvania, Rhode Island, Texas, Vermont, Washington, and Wisconsin (DSIRE, 2016).

[7]Indiana, Kansas, North Dakota, Oklahoma, South Carolina, South Dakota, Utah, and Virginia (DSIRE, 2016).

> **Finding 7-1:** Short-term tax credit extensions lead to market uncertainty, increase investment risk for technology manufacturers and project developers, and contribute to an uneven playing field relative to market segments with long-term policies or economically mature and competitive technologies.

Subsidies for Research and Development

Government subsidies active at the demonstration phase of new technology development are difficult to track, so little research in this area is available. Because of its high profile, carbon capture and storage (CCS) does offer one possible window on this information. Specifically, the ARRA provided $1.52 billion of support for the exploration and implementation of CCS projects (ELI, 2013, p. 9). However, this was a short-lived program, allowed only under the now-expired ARRA.

From 1978 to 2010, federal funding for energy-related R&D amounted to $121 billion, $45 billion of which is accounted for by nuclear power. Meanwhile, $26 billion has gone to the coal industry, $26 billion to end use, $20 billion to renewable energy, and $4 billion to oil and natural gas (EIA, 2011). During the 10-year period 2005 to 2014, nuclear power remained the primary benefactor of DOE-directed R&D funding, receiving 27.4 percent, while fossil energy received 23.5 percent, renewables 18.5 percent, and end use/efficiency 15.8 percent (Sissine, 2014, p. 7).

Potential for Subsidies to Persist

Analysis of the history of U.S. subsidies for energy technologies, including electric power, suggests that a range of fossil fuel subsidies continue to support technologies and industries even after they have achieved maturity and a notional ability to function independently in an open market. Coal in particular benefits from a wide range of government supports despite being a well-established cornerstone of the U.S. energy economy. USGS continues to provide research services for the industry through its National Coal Resources Data System (NCRDS). In addition, current tax benefits include excess of percentage over cost depletion (Internal Revenue Code [IRC] Section 613), which allows taxpayers to deduct 10 percent of gross income from coal production (ELI, 2013). Other benefits include exploration and development expensing (IRC Section 617) and amortization of coal pollution control (IRC Section 169) (ELI, 2013). There are also a number of policies that some observers consider implicit subsidies to coal, such as exemptions from environmental regulation and rules regarding the assessed value of coal mining leases on public lands for royalty payments (GAO, 2013).

Petroleum and gas receive similar benefits. Intangible drilling costs are supported through oil and gas excess of percentage over cost depletion (IRC Section 613), which allows independent producers to deduct 15 percent of gross income earned from qualifying deposits. This deduction can exceed the cost of the asset developed and thus serves to subsidize its development (ELI, 2009; NRC, 2013c). Also providing tax relief to oil and gas producers are the exception from passive loss limitations for oil and gas (IRC Section 468(c)(3)) and oil and gas development expensing (IRC Section 617).

Finding 7-2: Subsidies can serve important public policy functions when they help to establish industries. When an industry is mature, it is ideally placed for stable competition.

Recommendation 7-1: Policies intended to support increasingly clean electric power deployment should be structured both to be as technology-neutral as possible—that is, performance- or outcome-oriented without regard to specific technology—and to include sunset provisions so that they expire either after a specified length of time or once a certain level of performance has been achieved. Proper use of sunset provisions could ensure that subsidies will not outlive their usefulness in effecting important public policy to assist in the initial development of critical industries. Such sunset provisions are best based wholly on performance criteria, thus eliminating unforeseen favoring of one technology over another.

LOWERING THE COSTS AND RISKS OF FINANCING THE DEPLOYMENT OF INCREASINGLY CLEAN ENERGY TECHNOLOGIES

As discussed above, each of the dominant electricity generation sectors, defined by fuel type, has historically been actively supported by government measures in its economic and market development. Beginning in the second half of the 20th century, and in particular following the oil shock and downturn in nuclear development in the 1970s, this pattern of government support took on a new shape. The current landscape is defined in part by government support mechanisms originally intended for nascent sectors of the economy that have since matured.

Master Limited Partnerships and YieldCos

Subsidies can be both direct and indirect, and policies can take many forms. One policy available for fossil fuels that is essentially unavailable to newer, increasingly clean power sources is the ability to create a so-called master limited partnership (MLP) that is traded on a public exchange. Currently, the use of MLPs is restricted to entities that generate income from qualifying natural resource activities or from transportation or storage of many fuels, including ethanol and biodiesel. The income in an MLP is treated as a "pass-through" for federal income taxes and is not subject to taxes at the entity level. Qualifying income includes earnings from transportation, processing, storage, and marketing of natural gas, crude oil, and related products. MLPs are less costly to the government than many other tax preferences since elimination of the corporate tax at the entity level is to some extent recaptured from the investor, who pays an income tax on all pretax income, not just the after-tax income received in the form of dividends. The Joint Committee on Taxation estimates that for the period fiscal year (FY) 2014-2018, the total cost of energy-related MLPs to the government will be $5.8 billion (Joint Committee on Taxation, 2014). The market value of these MLPs is approximately $500 billion. Legislation allowing renewable energy ventures to use MLPs was introduced in the 113th Congress (S. 795, H.R. 1696) but has not been enacted into law.

Wall Street has created another financing vehicle, known as YieldCos, to lower the cost of capital for increasingly clean energy projects until legislation permits companies to use MLPs. A company or sponsor that has a strong development record in building and operating sources of increasingly clean energy, such as solar farms, and has a pipeline of future development projects can establish a subsidiary. This subsidiary can acquire the parent's operating assets that have long-term power purchase agreements with creditworthy entities, providing the subsidiary with a stable, predictable cash flow from which dividends can be paid. Shares in the subsidiary are sold to the public. The public shares pay dividends to the shareholders at attractive yields that are substantially lower than the return generally required by the equity investors in renewable energy projects. The lower yield acceptable to the YieldCo investor is justified because proven operating projects have eliminated some risks (such as cost overruns, construction risk, and operating risk). The YieldCo is able to view the operating results of the project portfolio before investing.

YieldCos have become attractive investments in cases in which the sponsor has a good development and operating record, which ensures that the dividend is secure. In the current economic environment, incremental yield shares of stock that pay yields well above the 10-year treasury rate are viewed as attractive. Furthermore, the dividends from a YieldCo qualify for lower tax rates relative to ordinary interest income, increasing their value to taxable investors. Still, there is concern that factors such as insufficient new investment opportunities, changes in tax law, difficulty renewing power purchase agreements, or rising interest rates could decrease the value of YieldCos and thus their attractiveness to investors. Nonetheless, they warrant continued analysis given their potential to continue to lower the cost of capital for investments in increasingly clean electric power technologies.

Real Estate Investment Trusts

Real estate investment trusts (REITs) are another investment vehicle that could lower the costs of and increase access to capital for financing increasingly clean energy technologies. REITs resemble MLPs in three ways. First, REITs can be bought and sold on public exchanges, providing a conduit for capital from a large and diversified class of investors. Second, REITs provide investors with a stream of income generated from specific activities; they must pay at least 90 percent of their taxable income to shareholders through a dividend.[8] Third, Congress created REITs as an exemption from the corporate income tax rules. Two additional key requirements for an REIT are that at least 75 percent of its total assets must be in real property or interests in real property,[9] and at least 75 percent of its gross income must come from real-property activities or mortgages on real property.

The National Association of Real Estate Investment Trusts (NAREIT) estimates that there are approximately 1,100 REITs in the United States (NAREIT, n.d.-a). NAREIT also estimates that the number of publicly traded REITs has grown significantly over time, from 34 in 1971, with a market capitalization of roughly $1.5 billion, to 202 in 2013, valued at approximately $670 billion (NAREIT, n.d.-b). REITs therefore could provide a significant source of capital for financing the deployment of increasingly clean energy technologies, as long as those technologies meet the definition of real property or interests in real property. The authorizing statute defines interests in real property as including the ownership, co-ownership, or leasing of improvements on land, but does not define improvements. Instead, that determination is left to the Treasury Department, usually through private letter rulings (PLRs) from the Internal Revenue Service (IRS).

In June 2014, the IRS issued a PLR stating that photovoltaic (PV) modules are not an improvement because they are not inherently permanent and thus do not qualify to be part of an REIT, whereas the mounts and exit wires do qualify. PLRs, though, "are limited to their particular facts and may not be relied upon by taxpayers other than the taxpayer that received the ruling" (IRS, 2014); thus they may not be applicable to REITs other than the one that requested the ruling. Moreover, if PV modules were to be considered an improvement, they could also be at risk of no longer being eligible for other support mechanisms through

---

[8]26 U.S.C. §856(c)(2)(A). REITs formed after January 1, 1980, must pay at least 95 percent of their income to shareholders as a dividend.

[9]Or cash and cash items, or government bonds. Because investors expect income, though, REITs generally do not hold cash investments. 26 U.S.C. §856(c)(4)(A).

the tax code, such as the PTC or ITC. This situation has created a great deal of uncertainty regarding the eligibility of technologies such as solar and wind for REIT status. By contrast, IRS rulings have made clear that pipelines, including natural gas pipeline systems, are eligible for REIT status, exclusive of meters and compressors (IRS, 2014).

> **Finding 7-3:** MLPs and similar tax and financing mechanisms, such as REITs, have a positive impact on lowering the cost of capital for energy projects.

> **Recommendation 7-2: The federal government should consider leveling the playing field by making proven financing mechanisms available to increasingly clean energy projects. One means to this end would be convening a roundtable of experts on increasingly clean energy financing to provide recommendations on new approaches for using federal financing programs to leverage and sustain capital investment in increasingly clean energy projects at all levels of the economy.**

Enabling Financing Mechanisms

One critical challenge to financing increasingly clean electric power technologies is the often high up-front capital costs, even though the technologies may provide lower operating costs. Several mechanisms—such as on-bill repayment, energy service performance contracting, and property assessed clean energy (PACE) financing—could address this barrier to cost-effective projects. These mechanisms enable third-party financing so that electricity customers can finance projects that lower their energy use or shift them to greater use of distributed generation, and allow for the savings realized through those projects to pay for the projects over time. Some entity, however, still must provide the initial capital that will be repaid over time.

One such entity is known as a "green bank." Green banks are in the planning or early deployment stage in a number of states. Essentially, a green bank blends public and private capital to fund the up-front cost of increasingly clean energy improvements. The intent is to spread the risk for either investor and to scale the market for projects. These entities can be housed within an existing state agency with administrative, rulemaking, and underwriting authority. Examples are found in New York and Connecticut. Green banks provide capital for development and implementation through a range of financing mechanisms, which include revolving loan funds, loan loss reserve pools, and commercial PACE (C-PACE) financing.

C-PACE is a financing mechanism used by local governments that allows commercial, industrial, and multifamily property owners to finance energy-efficiency and renewable energy improvements. The repayment of qualified energy improvements takes place through a voluntary property tax assessment, allowing local governments to finance the up-front costs of the improvements. Responsibility for repayment transfers to the next owner if the property is sold. Although many states have passed legislation enabling C-PACE, a lack of model legislation has led to states setting their own loan terms, qualifying retrofits, and target markets (PACE, n.d.).

The lack of standardization has prevented C-PACE from scaling to its potential within the private lending community. States continue to modify their existing C-PACE statutes accordingly. In the 2014 state legislative session alone, for example, C-PACE statutory changes were made in Maryland (HB 202), New Hampshire (HB 532), and Oregon (HB 4041), all of which had preexisting C-PACE legislation that needed to be modified. The lack of expertise and mature financial standardization remains a barrier to taking C-PACE to scale. Detailed, independent analysis of existing C-PACE and multifamily PACE policies would help determine whether statutory changes would enable programs to reach greater scale and operate more efficiently.

The federal government could support states' efforts to overcome cost-of-capital barriers through green banks in a number of ways. DOE and the Treasury Department could undertake research to determine what role(s), if any, green banks could play. Depending on their findings, those departments could provide states with model legislation and regulations and key technical advice. Standardization of financial markets generally increases access and lowers costs. Thus, another possibility is to offer streamlined syndication assistance to help states design programs in ways that leverage private capital and low-cost financing mechanisms most effectively.

ADDRESSING BARRIERS THAT REMAIN AT THE DEPLOYMENT STAGE

As detailed in Chapter 2, the greatest historical barrier to the deployment of increasingly clean electric power technologies once they are technically ready[10] is that their price in the market has been higher, often significantly so, than that of conventional energy sources—a much more significant issue than the costs of capital associated with these technologies. As discussed in earlier chapters, a chief reason for this price differential is that the prices of conventional technologies do not reflect their full costs, particularly the "hidden" costs of pollution (externalities) (see also NRC, 2010b). As long as the first prices to purchasers for increasingly clean technologies remain high,

[10]That is, once the technologies have matured to a technology readiness level of 8 or 9.

investors will continue to seek other investment opportunities. This barrier, in other words, does not stem directly from financing or from financial markets but from market imperfections in the electric power sector.

A chief way to alleviate this barrier is to implement policies that price the pollution caused by various technologies so that market participants will receive appropriate signals regarding their value. In simpler terms, this approach would bring the prices of increasingly clean technologies closer to, and even below, those of incumbent technologies. Market demand would pull from there, and investors would find investing in the deployment of increasingly clean electric power technologies more attractive, accelerating their market deployment.

> **Finding 7-4:** Properly pricing pollution would cause market pull for increasingly clean energy technologies and attract more investors and investment capital to these technologies.

This finding aligns with and supports prior recommendations of the National Research Council (NRC, 2013c) and others that appropriate steps be taken to correct the market so it will give consumers appropriate price signals. It is important to bear in mind, though, that pollution prices would have their greatest impact on technologies that are technically developed enough to be ready for deployment. Pollution prices would have only small or modest effects on early-stage basic research and R&D.

One way to think about pollution pricing is as a very inexpensive insurance policy. Like any good insurance policy, it would diffuse risk to such a great number of people that costs borne by each person would be vanishingly small. The benefits, however, could be quite large. Focusing just on GHG emissions, the National Research Council committee that produced *America's Climate Choices* (NRC, 2010c, p. 5) recommended that U.S. policy makers "adopt a mechanism for setting an economy-wide carbon-pricing system" as an important element of a comprehensive national mitigation program. According to that report, "most economists and policy analysts have concluded…that putting a price on CO_2 emissions (that is, implementing a 'carbon price') that rises over time is the least costly path to significantly reduce emissions and the most efficient means to provide continuous incentives for innovation and for the long-term investments necessary to develop and deploy new low-carbon technologies and infrastructure" (NRC, 2011, p. 58). Further, "a carbon price designed to minimize costs could be imposed either as a comprehensive carbon tax with no loopholes or as a comprehensive cap-and-trade system that covers all major emissions sources" (p. 58).

This committee reconfirms those findings and the value of internalizing the cost of GHG emissions in the market price of fossil fuels. The committee takes no position on whether a carbon price would best be established as a tax or in a cap-and-trade regime; both approaches have advantages and disadvantages. A carbon tax is the most direct method for pricing carbon, but involves such issues as how to adjust the tax over time and the point in the supply chain where the tax should be collected. More important, it neither defines nor guarantees a specific level of GHG reductions (Marron and Toder, 2014). A cap-and-trade system sets a specific limit on emissions but can entail greater administrative complexity and cost. In addition, past cap-and-trade proposals in Congress, as well as the experience of the European Union's trading system, have demonstrated the tendency for lawmakers to include numerous concessions to stakeholders, which can reduce the effectiveness and transparency of the trading regime. An analysis of California's carbon market suggests that state-level trading schemes may have similar tendencies (Cullenward, 2014).

Both approaches lead to secondary policy issues, including how to estimate the full costs of GHG emissions to society and how to allocate the revenues from carbon pricing. Past proposals on revenue allocation have included deficit reduction, clean energy R&D, climate change mitigation and adaptation measures, and compensatory decreases in corporate or personal income taxes (or increases in public assistance to low-income families that do not pay taxes) to mitigate the impact of energy price increases. The great need for expanded technology options to address the climate and other problems due to pollution suggests that any future revenues generated by carbon pricing would be well invested in the research, development, and commercialization of increasingly clean energy resources and technologies, in measures to reduce any adverse and regressive impacts on energy prices, and in efforts to mitigate and adapt to the impacts of climate change. While recognizing the importance of deciding how to use any future revenues from pollution pricing, the committee notes that an analysis of revenue recycling is beyond the scope of this study. The committee notes further that future deliberations regarding a national carbon pricing policy would benefit greatly from assessing the experience of the two regional cap-and-trade systems that have emerged in the United States.[11]

In addressing the issue of carbon pricing, the committee was keenly aware of the political divide involved. Simply stated, pricing of carbon emissions on a national basis is unlikely to be quickly embraced or easily implemented. With that said, the committee believes it is necessary and appropriate to acknowledge what other committees of the National Academies have concluded as to the benefits of pricing carbon emissions, concurring that no other policy could be more important and no other may be more necessary to meet the daunting challenge facing the United States and the world.

---

[11]Nine northeastern states (Connecticut, Delaware, Maine, Maryland, Massachusetts, New Hampshire, New York, Rhode Island, and Vermont) participate in the Regional Greenhouse Gas Initiative, an emissions trading program. California launched its own emissions trading program in January 2013 and created a partnership with the Canadian Province of Quebec a year later, which is managed by the Western Climate Initiative. In 2013, the performance of both systems exceeded that of trading regimes in other countries.

Although Congress ultimately did not adopt cap-and-trade legislation in 2009, leading to a period in which global climate change ranked low on the national agenda, public discussion of market-based approaches appears to be intensifying as a result of the Environmental Protection Agency's proposed regulation of carbon pollution from power plants, increasingly definitive climate science, extreme weather events, and encouragement from several prominent experts from both political parties to establish a price on carbon. The committee finds that the analysis contained in *America's Climate Choices* (NRC, 2011) remains relevant and an important reference for a renewed national conversation about the most efficient ways to address climate change and to spur innovation in increasingly clean energy technologies.

References

ACEEE (American Council for an Energy-Efficient Economy). 2012. *The 2014 State Energy Efficiency Scorecard.* http://www.aceee.org/research-report/u1408.

Acemoglu, D., P. Aghion, L. Bursztyn, and D. Hemous. 2012. *The environment and directed technical change. American Economic Review* 102(1):131-166.

Acemoglu, D., U. Akcigit, D. Hanley, and W. Kerr. 2014. *Transition to clean technology.* Working paper 20743. Cambridge, MA: National Bureau of Economic Research.

Adams, S.P. 2006. Promotion, competition, captivity: The political economy of coal. *Journal of Policy History* 18(1):74-95.

Adams, S.P. 2010. Energy and politics. In *The Princeton encyclopedia of American political history,* edited by M. Kazin, R. Edwards, and A. Rothman. Princeton, NJ: Princeton University Press. Pp. 207-211.

Adler, J. 2011. Eyes on a climate prize: Rewarding energy innovation to achieve climate stabilization. *Harvard Environmental Law Review* 35(1):1-45.

AEE (Advanced Energy Economy). 2014. *Creating a 21st century electricity system for New York State: An energy industry working group position paper.* http://info.aee.net/21st-century-ny-energy-industry-wg-position-paper (accessed June 29, 2016).

AEIC (American Energy Innovation Council). 2010. *A business plan for America's energy future: Technical appendices.* Washington, DC: AEIC.

Alberini, A., and C. Towe. 2015. Information v. energy efficiency incentives: Evidence from residential electricity consumption in Maryland. *Energy Economics* 52:S30-S40.

Aldy, J.E. 2011. *A preliminary review of the American Recovery and Reinvestment Act's Clean Energy package.* Harvard Kennedy School Faculty Research Working Paper Series RWP11-048. Cambridge, MA: John F. Kennedy School of Government, Harvard University.

Allcott, H., and M. Greenstone. 2012. Is there an energy efficiency gap? *Journal of Economic Perspectives* 26(1):3-28.

Allcott, H., and S. Mullainathan. 2010. Behavior and energy policy. *Science* 327(5970):1204-1205.

Allcott, H., and T. Rogers. 2014. The short-run and long-run effects of behavioral interventions: Experimental evidence from energy conservation. *American Economic Review* 104(10):3003-3037.

Allcott, H., and D. Taubinsky. 2015. Evaluating behaviorally motivated policy: Experimental evidence from the lightbulb market. *American Economic Review* 105(8):2501-2538.

Anadon, L. D., M. Bunn, G. Chan, M. Chan, C. Jones, R. Kempener, A. Lee, N. Logar, and V. Narayanamurti. 2011. *Transforming U.S. energy innovation*. November. Cambridge, MA: Harvard Kennedy School. http://belfercenter.ksg.harvard.edu/files/uploads/energy-report-january-2012.pdf.

APPA (American Public Power Association). 2014. *2014-15 annual directory and statistical report.* Arlington, VA: APPA.

Apt, J., and P. Jaramillo. 2014. *Variable renewable energy and the electricity grid.* New York: RFF Press.

Argote, L., S.L. Beckman, and D. Epple. 1990. The persistence and transfer of learning in industrial settings. *Management Science* 36(2):140-154.

Arimura, T.H., S. Li, R.G. Newell, and K. Palmer. 2011. *Cost-effectiveness of electricity energy efficiency programs*. Working paper 17556. Cambridge, MA: National Bureau of Economic Research.

ARPA-E (Advanced Research Projects Agency-Energy). 2016. Department of Energy announces 14 new projects for window efficiency technologies. *ARPA-E Newsletter*, May 18. http://arpa-e.energy.gov/?q=news-item/department-energy-announces-14-new-projects-window-efficiency-technologies.

Arrow, K. 1962. The economic implications of learning by doing. *The Review of Economic Studies* 29(3):155-173.

ASCE (American Society of Civil Engineers). 2011. *Failure to act: The economic impact of current investment trends in electricity infrastructure.* Reston, VA: ASCE.

ASCE. 2013. *2013 report card for America's infrastructure.* http://www.infrastructurereportcard.org/a/documents/2013-Report-Card.pdf.

Auffhammer, M., C. Blumstein, and M. Fowlie. 2008. Demand-side management and energy efficiency revisited. *The Energy Journal* 29(3):91-104.

Austin, D. 2012. *Addressing market barriers to energy efficiency in buildings.* CBO working paper 2012-10. http://www.cbo.gov/sites/default/files/cbofiles/attachments/AddressingMarketBarriersToEnergyEfficiencyInBuildings_WorkingPaper_2012-10.pdf.

Ayre, J. 2014. Japanese manufacturer solar frontier mulling factory in New York. *Clean Technica*, September 30. http://cleantechnica.com/2014/09/30/japanese-manufacturer-solar-frontier-mulling-factory-new-york.

Baker, E., V. Bosetti, L.D. Anadon, M. Henrion, L.A. Reis. 2015. Future costs of key low-carbon energy technologies: Harmonization and aggregation of energy technology expert elicitation data. *Energy Policy* 80(2015):219-232.

Banerjee, A., and B.D. Solomon. 2003. Eco-labeling for energy efficiency and sustainability: A meta-evaluation of US programs. *Energy Policy* 31(2):109-123.

Barbose, G., C. Goldman, I. Hoffman, and M. Billingsley. 2013. *The future of utility customer-funded energy efficiency programs in the United States: Projected spending and savings to 2025.* LBNL-5803E. Berkeley, CA: Ernest Orlando Lawrence Berkeley National Laboratory. https://emp.lbl.gov/sites/all/files/lbnl-5803e.pdf.

Barbose, G., S. Weaver, and N. Darghouth. 2014. *Tracking the sun VII: An historical summary of the installed price of photovoltaics in the United States from 1998-2013.* http://emp.lbl.gov/publications/tracking-sun-vii-historical-summary-installed-price-photovoltaics-united-states-1998-20.

Battelle Memorial Institute. 2011. 2012 global R&D funding forecast. *R&D Magazine*, December. http://www.battelle.org/docs/default-document-library/2012_global_forecast.pdf.

Behr, P. 2011. Futuristic U.S. power reactor may be developed overseas. *The New York Times*, June 23. http://www.nytimes.com/cwire/2011/06/23/23climatewire-futuristic-us-power-reactor-may-be-developed-86684.html.

Bell, M.L., K. Ebisu, R.D. Peng, J. Walker, J.M. Samet, S.L. Zeger, and F. Dominici. 2008. Seasonal and regional short-term effects of fine particles on hospital admissions in 2020 U.S. counties, 1999-2005. *American Journal of Epidemiology* 168(11):1301-1310.

Benkard, C.L. 2000. Learning and forgetting: The dynamics of aircraft production. *American Economic Review* 90(4):1034-1054.

Best Energy Investment. n.d. *Renewable energy could be the largest source of electricity by 2050. Renewable energy market facts.* http://www.bester energycapital.com/enegyfact.php.

Biello, D. 2014. Can carbon capture technology be part of the climate solution. *Yale e360*, September 8. http://e360.yale.edu/feature/can_carbon_capture_technology_be_part_of_the_climate_solution/2800.

BLS (Bureau of Labor Statistics). 2005. *Consumer Price Index detailed reports. Annual Average Indexes 2005. Table 1A. Consumer Price Index for All Urban Consumers (CPI-U): U.S. city average, by expenditure category and commodity and service group*. Washington, DC: BLS.

BLS. 2007. *Consumer Price Index detailed reports. Annual Average Indexes 2007. Table 1A. Consumer Price Index for All Urban Consumers (CPI-U):*

U.S. city average, by expenditure category and commodity and service group. Washington, DC: BLS.

BNEF (Bloomberg New Energy Finance). 2014a. *Global trends in renewable energy investment 2014*. http://fs-unep-centre.org/system/files/globaltrends report2014.pdf.

BNEF. 2014b. Sustainable energy in America 2014 factbook. *Bloomberg New Energy Finance*, February 5. http://about.bnef.com/white-papers/sustainable-energy-in-america-2014-factbook.

BNEF. 2015. *Wind and solar boost cost-competitiveness versus fossil fuels*. Press release, October 5. http://about.bnef.com/press-releases/wind-solar-boost-cost-competitiveness-versus-fossil-fuels.

BNEF. 2016. 2016 sustainable energy in America factbook. *Bloomberg New Energy Finance*, February 4. http://about.bnef.com/white-papers/sustainable-energy-in-america-2016-factbook (accessed July 29, 2016).

Board of Governors of the Federal Reserve System. 2012. *Industrial production and capacity utilization*. Release date August 15. http://www.federal reserve.gov/releases/g17/20120815.

Böhringer, C., A. Cuntz, D. Harhoff, and E.A. Otoo. 2014. *The impacts of feed-in tariffs on innovation: Empirical evidence from Germany*. CESifo working paper no. 4680. http://www.cesifo-group.de/portal/page/portal/DocBase_Content/WP/WP-CESifo_Working_Papers/wp-cesifo-2014/wp-cesifo-2014-03/cesifo1_wp4680.pdf.

Bollinger, M., and J. Seel. 2015. *Utility-Scale Solar 2014: An empirical analysis of project cost, performance, and pricing trends in the United States*. LBNL-1000917. Berkeley, CA: Lawrence Berkeley National Laboratory.

Bonbright, J., A. Danielsen, and D. Kamerschen. 1961. *Principles of public utility rates* (1st ed.). Arlington, VA: Public Utilities Reports.

Bonneville Power Administration. n.d. *History*. https://www.bpa.gov/news/AboutUs/History/Pages/default.aspx.

Boomhower, J., and L.W. Davis. 2014. A credible approach for measuring inframarginal participation in energy efficiency programs. *Journal of Public Economics* 113:67-79.

Borenstein, S. 2012. The private and public economics of renewable electricity generation. *Journal of Economic Perspectives* 26(1):67-92.

Borenstein, S. 2014. Money for nothing? *Energy Institute at Haas*, May 12. https://energyathaas.wordpress.com/2014/05/12/money-for-nothing (accessed June 29, 2016).

Bosseboeuf, D., and C. Richard. 1997. The need to link energy efficiency indicators to related policies. *Energy Policy* 25(7-9):813-823.

BP. 2014. *BP Energy Outlook 2035*. http://www.bp.com/content/dam/bp/pdf/energy-economics/energy-outlook-2016/bp-energy-outlook-2014.pdf.

Breakthrough Energy Coalition. n.d. *Introducing the breakthrough energy coalition*. http://www.breakthroughenergycoalition.com/en/index.html.

Bresler, S., P. Centolella, S. Covino, and P. Sotkiewicz. 2013. Smarter demand response in RTO markets: The evolution toward price responsive demand in PJM. In *Energy efficiency: Towards the end of demand growth*, edited by F.P. Sioshansi. Waltham, MA: Academic Press. Pp. 419-442.

Bresnahan, T., and M. Trajtenberg. 1992. *General purpose technologies: "Engines of growth?"* Working paper 4148. Cambridge, MA: National Bureau of Economic Research.

Broderick, J. 2007. CFLs in America: Lessons learned on the way to market. *LD+A Magazine*, July. http://apps1.eere.energy.gov/buildings/publications/pdfs/ssl/cfls_july_lessons.pdf.

Brounen, D., and N. Kok. 2011. On the economics of energy labels in the housing market. *Journal of Environmental Economics and Management* 62:166-179.

Brunt, L., J. Lerner, and T. Nicholas. 2011. *Inducement prizes and innovation.* Discussion paper. Bergen, Norway: Norwegian School of Economics.

Buongiorno, J., J. Jurewicz, M. Golay, and N. Todreas. 2015. *Light-water reactors on offshore floating platforms: Scalable and economic nuclear energy*. Paper 15366. Proceedings of ICAPP '15, Nice, France, May 3-6.

C2ES (Center for Climate and Energy Solutions). 2013. *Leveraging natural gas to reduce greenhouse gas emissions.* http://www.c2es.org/publications/leveraging-natural-gas-reduce-greenhouse-gas-emissions (accessed December 18, 2014).

California Energy Commission. 2012. *California energy demand 2012-2022 final forecast* (Vol. 1). http://www.energy.ca.gov/2012publications/CEC-200-2012-001/CEC-200-2012-001-SF-V1.pdf.

California Energy Commission. 2016. *Research & development: The science of innovation.* http://www.energy.ca.gov/research (accessed June 29, 2016).

California ISO (Independent System Operator). 2013. *What the duck curve tells us about managing a green grid.* Folsom, CA: California ISO.

Campbell, R. 2012. *Weather-related power outages and electric system resiliency.* Washington, DC: Congressional Research Service.

Cappers, P., C.A. Spurlock, A. Todd, P. Baylis, M. Fowlie, and C. Wolfram. 2016. *Time-of-use as a default rate for residential customers: Issues and insights (smart grid investment grant-consumer behavior study analysis).* Berkeley, CA: Lawrence Berkeley National Laboratory.

CBO (Congressional Budget Office). 2009. *Potential impacts of climate change in the United States*. Washington, DC: CBO.

CBO. 2015. *Using ESPCs to finance federal investment in energy-efficient equipment.* https://www.cbo.gov/publication/49869.

CCSP (Climate Change Science Program). 2007. *Scenarios of greenhouse gas emissions and atmospheric concentrations.* Synthesis and assessment product 2.1a. Washington, DC: CCSP. http://digitalcommons.unl.edu/cgi/viewcontent.cgi?article=1011&context=usdoepub.

CEER (Council of European Energy Regulators). 2012. *5th CEER benchmarking report on the quality of electricity supply 2011.* https://www.energy-community.org/pls/portal/docs/1522177.PDF (accessed June 29, 2016).

Census Bureau. 2014. *2014 American Community Survey.* Washington, DC: Census Bureau.

Centolella, P. 2012. A pricing strategy for a lean and agile electric power industry. *Electricity Policy,* September. http://www.electricitypolicy.com/images/pdf/Centolella%20-9-26-12-final.pdf.

Centolella, P., and M. McGranahan. 2013. *Understanding the value of uninterrupted service.* Proceedings of the CIGRE 2013 Grid of the Future Symposium, Boston, Massachusetts, October 21.

Centolella, P., and A. Ott. 2009. *The integration of price responsive demand into PJM wholesale power markets and system operations.* https://www.hks.harvard.edu/hepg/Papers/2009/Centolella%20%20Ott%20PJM%20PRD%2003092009.pdf.

Chen, K. 2012. *Overview of Chinese Academy of Sciences thorium molten salt reactor licensing strategy.* Berkeley, CA: University of California Berkeley.

Chitnis, M., and S. Sorrell. 2015. Living up to expectations: Estimating direct and indirect rebound effects for UK households. *Energy Economics* 52(Suppl. 1):100-116.

Chittum, A., and T. Sullivan. 2012. *Coal retirements and the CHP investment opportunity.* Report no. IE123. Washington, DC: American Council for an Energy-Efficient Economy (ACEEE). http://aceee.org/research-report/ie123.

CIER (Center for Integrative Environmental Research). 2007. *The US economic impacts of climate change and the costs of inaction: A review and assessment by the Center for Integrative Environmental Research at the University of Maryland.* College Park, MD: University of Maryland.

City Energy Project. 2014. *The CEP theory of change & the data revolution.* https://ifmt.app.box.com/s/grhiejemeij2ptq7j54yvijf7dx2rmmz.

Clarke, L., J. Weyant, and A. Birky. 2006. On the sources of technological change: Assessing the evidence. *Energy Economics* 28(5-6):579-595.

Clarke, L., J. Weyant, and J. Edmonds. 2008. On the sources of technological change: What do the models assume? *Energy Economics* 30(2):409-424.

Clarke, L., J. Edmonds, V. Krey, R. Richels, S. Rose, and M. Tavoni. 2009. International climate policy architectures: Overview of the EMF 22 international scenarios. *Energy Economics* 31:S64-S81.

Clarke, L., A. Fawcett, J. McFarland, J. Weyant, Y. Zhou, and V. Chaturvedi. 2014. Technology and U.S. emissions reductions goals: Results of the EMF 24 modeling exercise. *The Energy Journal* 35, Special Issue.

CLASP (Collaborative Labeling and Appliance Standards Program). 2005. *A multi-country comparative evaluation of labelling research.* http://clasponline.org/en/Resources/Resources/PublicationLibrary/2005/Multi-country-comparative-eval-of-labeling-research.aspx.

Clean Edge. 2014. *Clean energy trends 2014.* http://cleanedge.com/reports/Clean-Energy-Trends-2014.

CNA Military Advisory Board. 2007. *National security and the threat of climate change.* Alexandria, VA: CNA Corporation. https://www.cna.org/cna_files/pdf/National%20Security%20and%20the%20Threat%20of%20Climate%20Change.pdf.

CNA Military Advisory Board. 2014. *National security and the accelerating risks of climate change.* Alexandria, VA: CNA Corporation. https://www.cna.org/cna_files/pdf/MAB_5-8-14.pdf.

Compete Coalition. 2014. *Annual baseline assessment of choice in Canada and the United States.* http://www.competecoalition.com/files/ABACCUS-2014-Executive-Summary-vf.pdf (accessed June 29, 2016).

Conchado, A., and Linares, P. 2012. The economic impact of demand-response programs on power systems: A survey of the state of the art. In *Handbook of networks in power systems*, edited by A. Sorokin, S. Rebennack, P.M. Pardalos, N.A. Iliadis, and M.V.F. Pereira. Berlin, Heidelberg: Springer-Verlag.

Constable, G., and B. Somerville. 2003. *A century of innovation: Twenty engineering achievements that transformed our lives.* Washington, DC: Joseph Henry Press.

Cossent, R., and T. Gómez. 2013. Implementing incentive compatible menu of contracts to regulate electricity distribution investments. *Utilities Policy* 27:28-38.

Cossent, R., L. Olmos, T. Gómez, C. Mateo, and P. Frías. 2011. Distribution network costs under different penetration levels of distributed generation. *European Transactions on Electrical Power* 21(6):1869-1888.

Costello, K. 2009. How should regulators view cost trackers. *The Electricity Journal* 22(10):20-33.

Cullenward, D. 2014. *Leakage in California's carbon market: Preliminary trading is consistent with expected impacts of regulatory changes.* http://papers.ssrn.com/sol3/papers.cfm?abstract_id=2458773.

Curtright, A., and J. Apt. 2008. The character of power output from utility-scale photovoltaic systems. *Progress in Photovoltaics* 16(3):241-247.

Dalberg Global Development Advisors. 2013. *The advance market commitment for pneumococcal vaccines.* http://www.gavi.org/results/evaluations/pneumococcal-amc-process---design-evaluation.

Dale, L.L., C. Antinori, M.A. McNeil, J.E. McMahon, and S.K. Fujita. 2009. Retrospective evaluation of appliance price trends. *Energy Policy* 37(2):597-605.

Darr, E.D., L. Argote, and D. Epple. 1995. The acquisition, transfer, and depreciation of knowledge in service organizations: Productivity in franchises. *Management Science* 41(11):1750-1762.

Davis, L.W. 2008. Durable goods and residential demand for energy and water: Evidence from a field trial. *RAND Journal of Economics* 39(2):530-546.

Davis, L.W. 2010. Evaluating the slow adoption of energy efficient investments: Are renters less likely to have energy efficient appliances? In *The design and implementation of U.S. climate policy*. Cambridge, MA: National Bureau of Economic Research.

Davis, L.W., and J. Davis. 2004. *How effective are prizes as incentives to innovation? Evidence from three 20th century contests*. Presented at DRUID Summer Conference 2004, Elsinore, Denmark, June 14-16.

Davis, L.W., and G.E. Metcalf. 2014. *Does better information lead to better choices? Evidence from energy-efficiency labels*. http://works.bepress.com /gilbert_metcalf/111.

Davis, L.W., A. Fuchs, and P.J. Gertler. 2014. Cash for coolers: Evaluating a large-scale appliance replacement program in Mexico. *American Economic Journal: Economic Policy* 6(4):207-238.

de Figueiredi, M.A., D.M. Reiner, and H.J. Herzog. 2005. Framing the long-term in-situability issue for geologic carbon storage in the United States. *Mitigation and Adaptation Strategies for Global Change* 10:547-657.

Defense Science Board Task Force. 2011. *Trends and implications of climate change for national and international security*. Washington, DC: Defense Science Board. http://www.acq.osd.mil/dsb/reports/ADA552760.pdf.

Dell, M., B.F. Jones, and B.A. Olken. 2012. Temperature shocks and economic growth: Evidence from the last half century. *American Economic Journal: Macroeconomics* 4(3):66-95.

Dell, M., B.F. Jones, and B.A. Olken. 2014. What do we learn from the weather? The new climate-economy literature. *Journal of Economic Literature* 52(3):740-798.

Deutch, J.M. 2011. *An energy technology corporation will improve the federal government's efforts to accelerate energy innovation*. Discussion paper 2011-05. http://www.hamiltonproject.org/assets/legacy/files/downloads_and_links/05_energy_corporation_deutch_paper_1.pdf.

DHS (Department of Homeland Security). 2013. Incident response activity. *ICS-CERT Monitor*, April-June. https://ics-cert.us-cert.gov/sites/default/files/Monitors/ICS-CERT_Monitor_Apr-Jun2013.pdf.

Dlouhy, J.A. 2013. Utility executives: Major cyberattack on power grid is inevitable. *Fuel Fix*, August 6. http://fuelfix.com/blog/2013/08/06/utility-executives-major-cyberattack-on-power-grid-is-inevitable.

DoD (Department of Defense). 2014. *Quadrennial defense review 2014*. Washington, DC: DoD. http://archive.defense.gov/pubs/2014_Quadrennial_Defense_Review.pdf.

DOE (Department of Energy). 2010. *Public benefit funds: Increasing renewable energy and industrial energy efficiency opportunities.* http://www1.eere.energy.gov/manufacturing/states/pdfs/publicbenefitfunds.pdf.

DOE. 2011a. *Buildings energy data book.* Washington, DC: DOE.

DOE. 2011b. *Innovation ecosystem development initiative.* Washington, DC: DOE.

DOE. 2012a. *Application of automated controls for voltage and reactive power management—initial results.* https://www.smartgrid.gov/files/VVO_Report_-_Final.pdf.

DOE. 2012b. *Demand reductions from the application of advanced metering infrastructure, pricing programs, and customer-based systems—initial results.* https://www.smartgrid.gov/files/peak_demand_report_final_12-13-2012.pdf.

DOE. 2012c. *Reliability improvements from the application of distribution automation technologies—initial results.* https://www.smartgrid.gov/files/Distribution_Reliability_Report_-_Final.pdf.

DOE. 2013. *U.S. energy sector vulnerabilities to climate change and extreme weather.* Washington, DC: DOE.

DOE. 2014a. *AEO2014 early release overview.* http://www.fossil.energy.gov/programs/gasregulation/authorizations/2013_applications/sc_exhibts_13_116_118/Ex._109_-_AEO_2014_Early_Release_Overvie.pdf.

DOE. 2014b. *Energy conservation standards activities report to Congress.* http://energy.gov/sites/prod/files/2014/02/f7/2014_feb_report_to_congress.pdf.

DOE. 2014c. *Energy Department invests $3.2 million to support clean energy small businesses and entrepreneurs.* http://energy.gov/articles/energy-department-invests-32-million-support-clean-energy-small-businesses-and (accessed June 20, 2014).

DOE. 2014d. *2014 SunShot Initiative portfolio book: Tackling challenges in solar energy.* http://energy.gov/sites/prod/files/2014/08/f18/2014_SunShot_Initiative_Portfolio8.13.14.pdf.

DOE. 2015a. *Net metering policies as of March 2015. DSIRE (Database of State Incentives for Renewables & Efficiency.* http://www.dsireusa.org/resources/detailed-summary-maps/net-metering-policies-2 (accessed July 26, 2016).

DOE. 2015b. *Quadrennial technology review 2015.* Washington, DC: DOE.

DOE. 2015c. *Quadrennial technology review: An assessment of energy technologies and research opportunities.* http://energy.gov/sites/prod/files/2015/09/f26/Quadrennial-Technology-Review-2015_0.pdf.

DOE. 2016a. *On the path to SunShot: Executive Summary.* DOE/EE 1412. http://energy.gov/sites/prod/files/2016/05/f31/OTPSS%20-%20Executive%20Summary-508.pdf (accessed July 23, 2016).

DOE. 2016b. *Energy Department announces up to $70 million for new REMADE in America Institute.* http://energy.gov/eere/articles/energy-department-announces-70-million-new-remade-america-institute.

DOE. n.d.-a. *Buildings energy data book.* http://buildingsdatabook.eren.doe.gov/TableView.aspx?table=1.4.1.

DOE. n.d.-b. *History and impacts.* http://energy.gov/eere/buildings/history-and-impacts.

DOE. n.d.-c. *Renewable electricity production tax credit (PTC).* http://energy.gov/savings/renewable-electricity-production-tax-credit-ptc (accessed August 9, 2016).

DOE and EPA (Environmental Protection Agency). 2012. *Combined heat and power: A clean energy solution.* http://www1.eere.energy.gov/manufacturing/distributedenergy/pdfs/chp_clean_energy_solution.pdf.

DOE EAC (Electricity Advisory Committee). 2012. *Recommendations on electricity workforce.* Memorandum. http://energy.gov/sites/prod/files/EAC%20Paper%20-%20Recommendations%20on%20Electricity%20Workforce%20-%20Final%20-%208%20Nov%202012.pdf.

DOE EAC. 2014a. *EAC recommendations regarding emerging and alternative regulatory models and modeling tools to assist in analysis (September 2014).* http://energy.gov/sites/prod/files/2014/10/f18/Recs-EmergAltRegModels-Tools.pdf.

DOE EAC. 2014b. *Status and recommendations for electricity delivery workforce.* Memorandum. http://energy.gov/sites/prod/files/2014/10/f18/Status-Recs-ElectricityDeliveryWorkforce-Sep2014.pdf.

Domingo, C., T. Gomez San Roman, A. Sanchez-Miralles, J. Pascual Peco Gonzalez, and A. Candela. 2011. A reference network model for large-scale distribution planning with automatic street map generation. *IEEE Transactions on Power Systems* 26(1).

DSIRE (Database of State Incentives for Renewables & Efficiency). 2016. *Renewable portfolio standard policies.* http://www.dsireusa.org/resources/detailed-summary-maps (accessed August 3, 2016).

Dubin, J., A. Miedema, and R. Chandran. 1986. Price effects of energy-efficient technologies: A study of residential demand for heating and cooling. *RAND Journal of Economics* 17(3):310-325.

Dumagan, J.C., and T.D. Mount. 1993. Welfare effects of improving end-use efficiency: Theory and application to residential electricity demand. *Resource and Energy Economics* 15(2):175-201.

EDA (Economic Development Administration). 2011. U.S. Commerce Department announces launch of i6 green challenge to promote clean energy innovation and economic growth. Press release. *Newsroom*, March 10. https://www.eda.gov/news/press-releases/2011/03/10/303.htm.

EEI (Edison Electric Institute). 2014. *2013 financial review: Annual report of the U.S. investor-owned electric utility industry.* http://www.eei.org/

resourcesandmedia/industrydataanalysis/industryfinancialanalysis/finreview/Documents/FinancialReview_2013.pdf.

EEI. 2016. *2016 Q1 credit ratings*. http://www.eei.org/Search/Pages/Results.aspx?k=credit%20ratings&r=write%3E%3D%221%2F25%2F2016%22 (accessed July 26, 2016).

EIA (Energy Information Administration). 2007. *State electricity profiles 2005*. Washington, DC: EIA. http://www.eia.gov/electricity/state/archive/062905.pdf.

EIA. 2009a. *Energy market and economic impacts of HR 2454, the American Clean Energy and Security Act of 2009*. SR-OIAF/2009-05. http://www.eia.gov/analysis/requests/2009/hr2454/pdf/sroiaf(2009)05.pdf.

EIA. 2009b. *2009 Residential Energy Consumption Survey*. http://www.eia.gov/consumption/residential/data/2009/#sf?src=‹ Consumption Residential Energy Consumption Survey (RECS)-b1.

EIA. 2010. *Manufacturing Energy Consumption Survey, 2010*. Washington, DC: EIA.

EIA. 2011. *Direct federal financial interventions and subsidies in energy in fiscal year 2010*. Washington, DC: EIA. https://www.eia.gov/analysis/requests/subsidy/archive/2010/pdf/subsidy.pdf.

EIA. 2012. *Annual Energy Outlook 2012*. http://www.eia.gov/forecasts/aeo/pdf/0383(2012).pdf.

EIA. 2013a. *Electric Power Annual 2012*. http://www.eia.gov/electricity/annual/archive/03482012.pdf.

EIA. 2013b. *Heating and cooling no longer majority of U.S. home energy use*. http://www.eia.gov/todayinenergy/detail.cfm?id=10271.

EIA. 2013c. *International Energy Outlook 2013*. http://www.eia.gov/forecasts/archive/ieo13/pdf/0484(2013).pdf.

EIA. 2013d. *Updated capital cost estimates for utility scale electricity generating plants*. Washington, DC: EIA. http://www.eia.gov/forecasts/capitalcost/pdf/updated_capcost.pdf.

EIA. 2014a. *Annual Energy Outlook 2014*. Washington, DC: EIA. http://www.eia.gov/forecasts/aeo/pdf/0383(2014).pdf.

EIA. 2014b. *Assumptions to the Annual Energy Outlook 2014*. http://www.eia.gov/forecasts/aeo/assumptions/pdf/0554(2014).pdf.

EIA. 2014c. California continues to set daily records for utility scale solar energy. *Today in Energy*, June 25. http://www.eia.gov/todayinenergy/detail.cfm?id=16851 (accessed July 25, 2016).

EIA. 2014d. *Carbon dioxide emission coefficients by fuel*. http://www.eia.gov/environment/emissions/co2_vol_mass.cfm (accessed December 19, 2014).

EIA. 2014e. *Electric Power Annual 2014*. http://www.eia.gov/electricity/annual.

EIA. 2014f. *Electric Power Monthly*. http://www.eia.gov/electricity/monthly/epm_table_grapher.cfm?t=epmt_5_6_a (accessed December 2014).

EIA. 2014g. *LED bulb efficiency expected to continue improving as cost decline*. http://www.eia.gov/todayinenergy/detail.cfm?id=15471.

EIA. 2014h. *Levelized cost and levelized avoided cost of new generation resources in the Annual Energy Outlook 2014*. http://www.eia.gov/forecasts/aeo/pdf/electricity_generation.pdf.

EIA. 2014i. Natural gas, solar, and wind lead power plant capacity additions in first-half 2014. *Today in Energy*, September 9. http://www.eia.gov/todayinenergy/detail.cfm?id=17891 (accessed June 28, 2016).

EIA. 2014j. *Short-term energy outlook*. Washington, DC: EIA.

EIA. 2015a. *Annual Energy Outlook 2015*. Washington, DC: EIA. http://www.eia.gov/forecasts/aeo/pdf/0383(2015).pdf.

EIA. 2015b. *Assumptions to the Annual Energy Outlook 2015*. http://www.eia.gov/forecasts/aeo/assumptions/pdf/0554(2015).pdf.

EIA. 2015c. *Direct federal financial interventions and subsidies in energy in fiscal year 2013*. https://www.eia.gov/analysis/requests/subsidy/pdf/subsidy.pdf.

EIA. 2015d. Hawaii and Vermont set high renewable portfolio standard targets. *Today in Energy*, June 29. http://www.eia.gov/todayinenergy/detail.cfm?id=21852 (accessed August 9, 2016).

EIA. 2015e. *January 2015: Monthly energy review*. https://www.eia.gov/totalenergy/data/monthly/archive/00351501.pdf (accessed July 22, 2016).

EIA. 2015f. *Levelized cost and levelized avoided cost of new generation resources in the Annual Energy Outlook 2015*. http://www.eia.gov/forecasts/aeo/pdf/electricity_generation.pdf.

EIA. 2015g. *Monthly Energy Review: 2015 data as of July 2016*. http://www.eia.gov/totalenergy/data/monthly.

EIA. 2016a. *Annual Energy Outlook 2016*. Washington, DC: EIA. https://www.eia.gov/forecasts/aeo/index.cfm.

EIA. 2016b. Chapter 7. Industrial sector energy consumption. In *International Energy Outlook 2016*. DOE/EIA-0484. Washington, DC: EIA. http://www.eia.gov/forecasts/ieo/industrial.cfm (accessed July 29, 2016).

EIA. 2016c. Demand trends, prices, and policies drive recent electric generation capacity additions. *Today in Energy*, March 18. http://www.eia.gov/todayinenergy/detail.cfm?id=25432 (accessed July 27, 2016).

EIA. 2016d. *Electric Power Annual 2014*. Washington, DC: EIA. http://wnew.www.eia.gov/electricity/annual/pdf/epa.pdf.

EIA. 2016e. *Electric Power Monthly with data for May 2016*. Washington, DC: EIA. https://www.eia.gov/electricity/monthly/pdf/epm.pdf.

EIA. 2016f. Extended policies case. In *Annual Energy Outlook 2016: Issues in Focus*, by M. Bowman, E. Boedecker, J. Maples, and K. Perl. Washington, DC: EIA. http://www.eia.gov/forecasts/aeo/policies.cfm.

EIA. 2016g. *Levelized cost and levelized avoided cost of new generation resources in the Annual Energy Outlook 2016*. http://www.eia.gov/forecasts/aeo/pdf/electricity_generation.pdf (accessed August 9, 2016).

EIA. 2016h. Total U.S. electricity sales projected to grow slowly as electricity intensity declines. *Today in Energy*, June 15. http://www.eia.gov/todayinenergy/detail.cfm?id=26672.

EIA. 2016i. Wind adds the most electric generation capacity in 2015, followed by natural gas and solar. *Today in Energy*, March 23. http://www.eia.gov/todayinenergy/detail.cfm?id=25492 (accessed July 25, 2016).

ELCON (Electric Consumers Resource Council). 2004. *The economic impacts of the August 2003 blackout.* http://www.elcon.org/Documents/Profiles%20and%20Publications/Economic%20Impacts%20of%20August%202003%20Blackout.pdf.

ELI (Environmental Law Institute). 2009. *Estimating U.S. government subisidies to energy sources: 2002-2008.* Washington, DC: ELI. https://www.eli.org/sites/default/files/eli-pubs/d19_07.pdf.

ELI. 2013. *Estimating U.S. government spending on coal: 2002-2010.* Washington, DC: ELI. https://www.eli.org/sites/default/files/eli-pubs/d23-10.pdf.

EPA (Environmental Protection Agency). 2015. *National emissions inventory air pollutant emissions trends data, 1970-2014 Average annual emissions, all criteria—February 6.* https://www.epa.gov/air-emissions-inventories/air-pollutant-emissions-trends-data (accessed July 26, 2016).

EPA. 2016. *Inventory of U.S. greenhouse gas emissions and sinks: 1990-2014.* Washington, DC: EPA.

EPRI (Electric Power Research Institute). 2011. *Estimating the costs and benefits of the Smart Grid: A preliminary estimate of the investment requirements and the resultant benefits of a fully functioning Smart Grid.* https://www.smartgrid.gov/files/Estimating_Costs_Benefits_Smart_Grid_Preliminary_Estimate_In_201103.pdf.

EPRI. 2012. *Volt-VAR control for electric distribution systems.* Briefing presented for the Public Utilities Commission of Ohio, January 18. Palo Alto, CA: EPRI.

EPRI. 2013. *Enhancing distribution resiliency: Opportunities for applying innovative technologies.* Palo Alto, CA: EPRI.

EPRI. 2014. *The integrated grid: Realizing the full value of central and distributed energy resources.* http://www.epri.com/Press-Releases/Pages/EPRI-Unveils-Study-on-the-Integrated-Electric-Grid-.aspx (accessed June 29, 2016).

Eto, J., and K. LaCommare. 2008. *Tracking the reliability of the U.S. electric power system: An assessment of publicly available information reported to state public utility commissions.* Berkeley, CA: Ernest Orlando Lawrence Berkeley National Laboratory. https://emp.lbl.gov/sites/all/files/lbnl-1092e.pdf.

Executive Office of the President. 2013. *Economic benefits of increasing electric grid resilience to weather outages.* http://energy.gov/sites/prod/files/2013/08/f2/Grid%20Resiliency%20Report_FINAL.pdf.

Faruqui, A. 2013. Surviving sub-one-percent growth. *Electricity Policy*, June. http://www.electricitypolicy.com/images/2013/06/Faruqui/Faruqui-1-percent-6-11-13%202.pdf.

Faruqui, A., and J. Palmer. 2012. *The discovery of price responsiveness—a survey of experiments involving dynamic pricing of electricity.* https://www.hks.harvard.edu/hepg/Papers/2012/The%20Arc%20of%20Price%20Responsiveness%20(03-18-12).pdf.

Faruqui, A., S. Sergici, and A. Sharif. 2010. The impact of informational feedback on energy consumption. A survey of the experimental evidence. *Energy* 35(4):1598-1608.

Fawcett, A.A., K.V. Calvin, F.C. de la Chesnaye, J.M. Reilly, and J.P. Weyant. 2009. Overview of EMF 22 U.S. transition scenarios. *Energy Economics* 31(Suppl. 2):S198-S211.

Fawcett, A.A., L. Clarke, S. Rausch, and J. Weyant. 2013. Overview of EMF 24 policy scenarios. *The Energy Journal* 35.

Fehrenbacher, K. 2014. SolarCity breaks ground on huge solar factory in New York, strikes deal with state. *GIGAOM*, September 24. https://gigaom.com/2014/09/24/solarcity-breaks-ground-on-huge-solar-factory-in-new-york-strikes-deal-with-state.

FEMP (Federal Energy Management Program). n.d. *Energy-efficient product procurement.* http://energy.gov/eere/femp/energy-efficient-product-procurement.

FERC (Federal Energy Regulatory Commission). 2013. *Assessment of demand response and advanced metering: Staff report.* http://www.ferc.gov/legal/staff-reports/2013/oct-demand-response.pdf.

FERC (Federal Energy Regulatory Commission). n.d. *Monthly report of cost and quality of fuels for electric plants (Form 423).* Washington, DC: FERC.

Fischer, C. 2010. Renewable portfolio standards: When do they lower energy prices? *The Energy Journal* 31(1):101-119.

Fischer, C., and R. Newell. 2008. Environmental and technology policies for climate mitigation. *Journal of Environmental Economics and Management* 55:142-162.

Fischer, C., and L. Preonas. 2010. Combining policies for renewable energy: Is the whole less than the sum of its parts? *International Review of Environmental and Resource Economics* 4:51-92.

Fischer, C., R. Newell, and L. Preonas. 2013. *Environmental and technology policy options in the electricity sector: Interactions and outcomes.* Washington, DC: Resources for the Future. http://www.rff.org/Publications/Pages/PublicationDetails.aspx?PublicationID=22362.

Forsberg, C., D. Curtis, J. Stempien, R. MacDonald, and P. Peterson. 2014. *Fluoride-salt-cooled High-temperature Reactor (FHR) commercial basis and commercialization strategy: A Fluoride-salt-cooled High-temperature Reactor (FHR) with a Nuclear Air-Brayton Combine Cycle (NACC) and firebrick resistance-heated energy storage.* MIT-ANP-TR-153. Cambridge, MA: MIT Center for Advanced Nuclear Energy Systems.

Fowlie, M., M. Greenstone, and C. Wolfram. 2015. *Do energy efficiency investments deliver? Evidence from the Weatherization Assistance Program.* https://nature.berkeley.edu/~fowlie/WAP.pdf.

Fox-Penner, P. 2010. *Smart power: Climate change, the smart grid, and the future of electric utilities* (1st ed.). Washington, DC: Island Press.

Fox-Penner, P. 2014. *Smart power aniversary edition: Climate change, the smart grid, and the future of electric utilities* (2nd ed.). Washington, DC: Island Press.

Frondel, M., and C. Vance. 2012. Heterogeneity in the effect of home energy audits. Theory and evidence. *Environmental and Resource Economics* 55(3):407-418.

FTC (Federal Trade Commission). n.d. *EnergyGuide labeling: FAQs for appliance manufacturers.* http://www.business.ftc.gov/documents/bus-82-energyguide-labels-faqs#products.

Gallagher, K.S., A. Grübler, L. Kuhl, G. Nemet, and C. Wilson. 2012. The energy technology innovation system. *Annual Review of Environment and Resources* 37(1):137-162.

Gans, W., A. Alberini, and A. Longo. 2013. Smart meter devices and the effect of feedback on residential electricity consumption: Evidence from a natural experiment in Northern Ireland. *Energy Economics* 36:729-743.

GAO (Government Accountability Office). 2010. *Energy Star program: Covert testing shows the Energy Star program certification process is vulnerable to fraud and abuse.* GAO-10-470. Washington, DC: GAO.

GAO. 2013. *Coal leasing- BLM could enhance appraisal process, more explicitly consider coal exports, and provide more public information.* Washington, DC: GAO. http://www.gao.gov/assets/660/659801.pdf.

GAO. 2014. *Advanced reactor research: DOE supports multiple technologies, but actions needed to ensure a prototype is built.* GAO-14-545. Washington, DC: GAO.

Garnett, A., C. Greig, and M. Oettinger. 2014. *Zerogen IGCC with CCS: A case history.* Brisbane, Australia: The University of Queensland. https://energy.uq.edu.au/files/1084/ZeroGen.pdf.

Garvey, T. 2009. *The Yucca Mountain litigation: Breach of contract under the Nuclear Waste Policy Act of 1982.* Washington, DC: Congressional Research Service.

Gayer, T., and W. K. Viscusi. 2013. Overriding consumer preferences with energy. *Journal of Regulatory Economics* 43:248-264.

Geller, H., and S. Attali. 2005. *The experience with energy efficiency policies and programmes in IEA countries: Learning from the critics*. Paris, France: IEA. https://www.iea.org/publications/freepublications/publication/IEAEnergyPolicies_Learning_from_critics.pdf.

Generac. 2014. *2013 annual report*. Waukesha, WI: Generac.

Gillingham, K., and K. Palmer. 2014. Bridging the energy efficiency gap: Policy insights from economic theory and empirical evidence. *Review of Environmental Economics and Policy* 8(1):18-38.

Gillingham, K., and J. Sweeney. 2010. Market failure and the structure of externalities. In *Harnessing renewable energy in electric power systems: Theory, practice, policy*, edited by B. Moselle, J. Padilla, and R. Schmalensee. Washington, DC: RFF Press. Pp. 69-91.

Gillingham, K., R. Newell, and K. Palmer. 2004. *Retrospective examination of demand-side energy efficiency policies*. Discussion paper 04-19. Washington, DC: Resources for the Future.

Gillingham, K., R. Newell, and K. Palmer. 2006. Energy efficiency policies: A retrospective examination. *Annual Review of Environment and Resources* 31:161-192.

Gillingham, K., R. Newell, and K. Palmer. 2009. Energy efficiency economics and policy. *Annual Review of Resource Economics* 1(1):597-619.

Gillingham, K., M. Harding, and D. Rapson. 2012. Split incentives in residential energy consumption. *The Energy Journal* 33(2):37-62.

Gillingham, K., D. Rapson, and G. Wagner. 2015. The rebound effect and energy efficiency policy. *Review of Environmental Economics and Policy* 10(1):68-88.

Gilmore, E., and L. Lave. 2007. *Increasing backup generation capacity and system reliability by selling electricity during periods of peak demand*. Presented at 26th USAEE/IAEE North American Conference, September 16-19. http://www.usaee.org/usaee2007/submissions/Presentations/Elisabeth%20Gilmore.pdf.

Global CCS Institute. 2016. *Large-scale CCS projects database*. http://www.globalccsinstitute.com/projects/large-scale-ccs-projects (accessed August 5, 2016).

Goellner, J., M. Prica, J. Miller, S. Pullins, J. Westerman, J. Harmon, T. Grabowski, H. Weller, B. Renz, S. Knudsen, and D. Cohen. 2011. *Demand dispatch: Intelligent demand for a more efficient grid*. Washington, DC: DOE. https://www.netl.doe.gov/File%20Library/Research/Energy%20Efficiency/smart%20grid/DemandDispatch_08112011.pdf.

Gosnell, G.K., J.A. List, and R. Metcalfe. 2016. *A new approach to an age-old problem: Solving externalities by incenting workers directly*. Working paper no. 22316. Cambridge, MA: National Bureau of Economic Research.

Goulder, L., and S. Schneider. 1999. Induced technological change and the attractiveness of CO_2 emissions abatement policies. *Resource and Energy Economics* 21:211-253.

Greenstone, M., and A. Looney. 2012. Paying too much for energy? The true costs of our energy choices. *Dædalus* 141(2):10-30.

Grubb, M., N. Haj-Hasan, and D. Newbery. 2008. Accelerating innovation and strategic deployment in UK electricity: Applications to renewable energy. In *Delivering a low-carbon electricity system: Technologies, economics and policy*, edited by M. Grubb, T. Jamasb, and M. Pollitt. Cambridge: Cambridge University Press.

Hardcastle, A. 2008. *Workforce challenges of electric sector employers in Washington and Oregon.* Olympia, WA: Washington State University Extension Energy Program. http://www.energy.wsu.edu/Documents/WSU_Workforce_Challenges_Final_Report_090311.pdf.

Harrington W., and R.D. Morgenstern. 2004. Economic incentives versus command and control- what's the best approach for solving environmental problems? *Resources* (Fall/Winter):13-17. http://www.rff.org/files/sharepoint/WorkImages/Download/RFF_Resources_152_ecoincentives.pdf.

Heal, G., and H. Kunreuther. 2010. Environment and energy: Catastrophic liabilities from nuclear power plants. In *Measuring and managing federal financial risk*, edited by D. Lucas. Chicago, IL: University of Chicago Press. Pp. 235-260.

Heeter, J., and L. Bird. 2012. *Including alternative resources in state renewable portfolio standards: Current design and implementation experience.* http://www.nrel.gov/docs/fy13osti/55979.pdf.

Heeter, J., G. Barbose, L. Bird, S. Weaver, F. Flores-Espino, K. Kuskova-Burns, and R. Wiser. 2014. *A survey of state-level cost and benefit estimates of renewable portfolio standards.* http://www.nrel.gov/docs/fy14osti/61042.pdf.

Hejzlar, P. 2013. Terrapower traveling wave reactor development program overview. *Nuclear Engineering and Technology* 45(6):731-744.

Hines, P., J. Apt, and S. Talukdar. 2009. Large blackouts in North America: Historical trends and policy implications. *Energy Policy* 37(12):5249-5259.

Horowitz, M.J., and H. Haeri. 1990. Economic efficiency vs. energy efficiency. *Energy Economics* 122-131.

Houde, S. 2014. *How consumers respond to environmental certification and the value of energy information.* Working paper 20019. Cambridge, MA: National Bureau of Economic Research.

Houde, S., and J.E. Aldy. 2014. *Belt and suspenders and more: The incremental impact of energy efficiency subsidies in the presence of existing policy instruments.* Working paper 20541. Cambridge, MA: National Bureau of Economic Research.

Houde, S., and A. Spurlock. 2015. *Do energy efficiency standards improve quality? Evidence from a revealed preference approach*. Working paper. College Park, MD: University of Maryland.

Houde, S., A. Todd, A. Sudarshan, J. Flora, and K. Armel. 2013. Real-time feedback and electricity consumption: A field experiment assessing the potential for savings and persistence. *Energy Journal* 34(1):87-102.

Howell, S. 2015. *Financing constraints as barriers to innovation: Evidence for R&D grants to energy startups*. http://economics.yale.edu/sites/default/files/howell_innovation_finance_jmp_jan7.pdf.

Hudson, F. 2012. *Sustainability on a smarter planet*. Presented at TEDx, April. http://tedxtalks.ted.com/video/TEDxNJIT-Florence-Hudson-Sustai.

Huntington, H.G. 2009. *Structural change and U.S. energy use: Recent patterns*. Stanford, CA: Energy Modeling Forum. https://web.stanford.edu/group/emf-research/docs/occasional_papers/op65.pdf.

IAEA (International Atomic Energy Agency). 2016. *Nuclear power reactors in the world.* Vienna, Austria: IAEA. http://www-pub.iaea.org/MTCD/Publications/PDF/RDS_2-36_web.pdf (accessed July 27, 2016).

IAEA. n.d. *Small and medium sized reactors development, assessment, and deployment.* http://www.iaea.org/NuclearPower/SMR.

IEA (International Energy Agency). 2012. *World Energy Outlook 2012*. Paris, France: IEA. http://www.worldenergyoutlook.org/weo2012.

IEA. 2013. *Technology roadmap for carbon capture and storage.* Paris, France: IEA.

IEA. 2014. *World Energy Outlook 2014*. Paris, France: IEA. http://www.worldenergyoutlook.org/weo2014.

IEE (Innovation Electricity Efficency). 2013. *Utility-scale smart meter deployments: A foundation for expanded grid benefits.* http://www.edisonfoundation.net/iee/Documents/IEE_SmartMeterUpdate_0813.pdf (accessed June 29, 2016).

IEEE (Institute of Electrical and Electronics Engineers). 2011. *IEEE benchmarking 2011 results.* Distribution Reliability Working Group Meeting, San Diego, CA, July 24. http://grouper.ieee.org/groups/td/dist/sd/doc/2012-07-01-Benchmarking-Results-2011.pdf.

Interagency Working Group on Social Cost of Carbon. 2013. *Technical update of the social cost of carbon for regulatory impact analysis—under Executive Order 12866*. http://www.whitehouse.gov/sites/default/files/omb/inforeg/social_cost_of_carbon_for_ria_2013_update.pdf.

IPCC (Intergovernmental Panel on Climate Change). 2007. *Climate change 2007.* Fourth Assessment Report (AR4). Geneva, Switzerland: IPCC.

IPCC. 2013. *Climate change 2013: The physical sciences basis*. New York: Cambridge University Press. http://www.climatechange2013.org.

IPCC. 2014a. *Climate change 2014: Impacts, adaptation, and vulnerability*. New York: Cambridge University Press. http://ipcc-wg2.gov/AR5/report.

IPCC. 2014b. *Climate change 2014: Mitigation of climate change*. New York: Cambridge University Press. http://www.ipcc.ch/report/ar5/wg3 (accessed December 16, 2015).

IRS (Internal Revenue Service). 2014. *Internal Revenue Bulletin 2014-23*. https://www.irs.gov/pub/irs-irbs/irb14-23.pdf.

IRS. n.d. *Q&A on tax credits for Sections 25C and 25D, Notice 2013-70*. http://www.irs.gov/pub/irs-drop/n-13-70.pdf (accessed August 5, 2016).

Jackson, R., O.C. Onar, H. Kirkham, E. Fisher, K. Burkes, M. Starke, O. Mohammed, and G. Weeks. 2015. *Opportunities for energy efficiency improvements in the U.S. electricity transmission and distribution system*. ORNL/TM-2015/5. Oak Ridge, TN: Oak Ridge National Laboratory.

Jacobsen, G.D., and M.J. Kotchen. 2013. Are building codes effective at saving energy? Evidence from residential billing data in Florida. *The Review of Economics and Statistics* 95(1):34-49.

Jaffe, A., and K. Palmer. 1997. Environmental regulation and innovation: A panel data study. *The Review of Economics and Statistics* 79(4):610-619.

Jaffe, A., and R. Stavins. 1994. The energy-efficiency gap. What does it mean? *Energy Policy* 22(10):804-810.

Jaffe, A., R.G. Newell, and R.N. Stavins. 2004. Economics of energy efficiency. *Encyclopedia of Energy* 2:79-90.

Jaffe, A., R. Newell, R. Stavins. 2005. A tale of two market failures: Technology and environmental policy. *Ecological Economics* 54(2-3):164-174.

Jamasb, T. 2007. Technical change theory and learning curves: Patterns of progress in electricity generation technologies. *The Energy Journal* 28(3):51-72.

Jamasb, T., and J. Kohler. 2008. Learning curves for energy technology: A critical assessment. In *Delivering a low carbon electricity system*, edited by M. Grubb, T. Jamasb, and M. Pollitt. Cambridge, UK: Cambridge University Press. http://www.dspace.cam.ac.uk/handle/1810/194736.

Jamasb, T., and M. Pollitt. 2008. Reference models and incentive regulation of electricity distribution networks: An evaluation of Sweden's Network Performance Assessment Model (NPAM). *Energy Policy* 36(5):1788-1801.

Janeway, W. 2012. *Doing capitalism in the innovation economy*. New York: Cambridge University Press.

Jenkins, C. 2011. *RIIO Economics: Examining the economics underlying Ofgem's new regulatory framework*. Working paper. http://www.rpieurope.org/publications/2011/Jenkins_RIIO%20Economics_FSR%20working%20paper_130611.pdf.

Joint Committee on Taxation. 2014. *Estimates of federal tax expenditures for fiscal years 2014-2018*. Washington, DC: The Joint Committee on Taxation.

Jones, J., and M. Leff. (2014, April 28). Implications of accelerated power plant retirements. In *Annual Energy Outlook 2014 with Projections to 2040.* Washington, DC: EIA. http://www.eia.gov/forecasts/aeo/index.cfm (accessed January 20, 2015).

Joskow, P. 2011. Comparing the costs of intermittent and dispatchable electricity generating technologies. *The American Economic Review* 101(3):238-241. http://www.jstor.org/stable/29783746.

Junginger, M., P. Lako, S. Lensink, W. van Sark, and M. Weiss. 2008. *Technological learning in the energy sector.* Report 500102017, NWS-E-2008-14, ECN-E--08-034. PBL Netherlands Environmental Assessment Agency. http://www.ecn.nl/docs/library/report/2008/e08034.pdf.

Junginger, M., W. van Sark, and A. Faaij (Eds.). 2010. *Technological learning in the energy sector: Lessons for policy, industry and science.* Northampton, MA: Edward Elgar Publishing, Inc.

Kahn, A.E. 1970. *The economics of regulation* (Vol. 1). New York: John Wiley & Sons.

Kahn, A.E. 1971. *The economics of regulation* (Vol. 2). New York: John Wiley & Sons.

Kahn, M.E., N. Kok, and J.M. Quigley. 2013. *Commercial building electricity consumption dynamics: The role of structure quality, human capital, and contract incentives.* Working paper 18781. Cambridge, MA: National Bureau of Economic Research.

Kahouli-Brahmi, S. 2008. Technological learning in energy-environment economy modeling: A survey. *Energy Policy* 36(1):138-162.

Kay, L. 2011. *Managing innovation prizes in government.* Washington, DC: IBM Center for the Business of Government.

Kay, L. 2013. The effect of inducement prizes on innovation: Evidence from the Ansari XPrize and the Northrop Grumman Lunar Lander Challenge. *R&D Management* 41(4):360-377.

Kerber, T. n.d. *Residential energy savings through data analytics.* http://www.ecofactor.com/wp-content/uploads/2013/10/Residential-Energy-Savings-through-Data-Analytics.pdf.zip (accessed June 29, 2016).

Ketels, C.H.M., and O. Memedovic. 2008. From clusters to cluster-based economic development. *International Journal of Technological Learning, Innovation and Development* 1(3):375-392.

Kim, K.K., W. Lee, S. Choi, H.R. Kim, and J. Ha. 2014. SMART: The first licensed advanced integral reactor. *Journal of Energy and Power Engineering* 8(2014):94-102.

Kind, P. 2013. *Disruptive challenges: Financial implications and strategic responses to a changing retail electric business.* http://www.eei.org/ourissues/finance/documents/disruptivechallenges.pdf.

Kobos, P.H., J.D. Erickson, and T.E. Drennen. 2006. Technological learning and renewable energy costs: Implications for U.S. renewable energy policy. *Energy Policy* 34(13):1645-1658.

Kotchen, M.J. 2016. *Longer-run evidence on whether building energy codes reduce residential energy consumption.* http://environment.yale.edu/kotchen/pubs/codesLR.pdf.

Kremer, M. 1998. Patent buyouts: A mechanism for encouraging innovation. *Quarterly Journal of Economics* 113(4):1137-1167.

Krey, V., G. Luderer, L. Clarke, and L. Kriegler. 2014. Getting from here to there—energy technology transformation pathways in the EMF27 scenarios. *Climatic Change* 123(3):369-382.

Kriegler, E., J.P. Weyant, G.J. Blanford, V. Krey, L. Clarke, J. Edmonds, A. Fawcett, G. Luderer, K. Riahi, R. Richels, S.K. Rose, M. Tavoni, and D.P. van Vuuren. 2014. The role of technology for achieving climate policy objectives: Overview of the EMF 27 study on global technology and climate policy strategies. *Climatic Change* 123(3-4):353–367.

Laden, F., J. Schwartz, F.E. Speizer, and D.W. Dockery. 2006. Reduction in fine particulate air pollution and mortality. *American Journal of Respiratory and Critical Care Medicine* 173(6):667-672.

Lantz, E., R. Wiser, and M. Hand. 2012. *IEA wind task 26: The past and future cost of wind energy.* Golden, CO: National Renewable Energy Laboratory.

Larsson, M.B.-O. 2005. *The Network Performance Assessment Model: A new framework for regulating the electricity network companies.* Stockholm, Sweden: Royal Institute of Technology of Stockholm. http://www.diva-portal.org/smash/get/diva2:8800/FULLTEXT01.pdf.

Lave, L., M. Ashworth, and C. Gellings. 2007. The aging workforce: Electricity industry challenges and solutions. *The Electricity Journal* 20(2):71-80.

Lazard. 2015. Levelized cost of energy analysis 9.0. *Lazard*, November 17. https://www.lazard.com/perspective/levelized-cost-of-energy-analysis-90 (accessed July 29, 2016).

Lerner, J. 2009. *Boulevard of broken dreams: Why public efforts to boost entrepreneurship and venture capital have failed—and what to do about it.* Princeton, NJ: Princeton University Press.

Lerner. J. 2010. The future of public efforts to boost entrepreneurship and venture capital. *Small Business Economics* 35(3):255-264.

Lester, R., and D. Hart. 2012. *Unlocking energy innovation: How America can build a low-cost low-carbon energy system.* Cambridge, MA: MIT Press.

Lester, R.K., and D.M. Hart. 2015. Closing the energy demonstration gap. *Issues in Science and Technology* xxxi(2):48-54.

Levinson, A. 2014a. California energy efficiency: Lessons for the rest of the world, or not? *Journal of Economic Behavior & Organization* 107:269-289.

Levinson, A. 2014b. *How much energy do building energy codes really save? Evidence from California.* Working paper 20797. Cambridge, MA: National Bureau of Economic Research.

Levinson, A. 2015. A direct estimate of the technique effect: Changes in the pollution intensity of U.S. manufacturing 1990-2008. *Journal of the Association of Environmental and Resource Economists* 2(1):43-56.

Levitt, S.D., J.A. List, and C. Syverson. 2013. Toward an understanding of learning by doing: Evidence from an automobile assembly plant. *Journal of Political Economy* 121(4):643-681.

Lindman, Å., and P. Söderholm. 2012. Wind power learning rates: A conceptual review and meta-analysis. *Energy Economics* 34(3):754-761.

Lockwood, M. 2013. *Governance, innovation and the transition to a sustainable energy system: Perspectives from economic theory.* http://projects.exeter.ac.uk/igov/wp-content/uploads/2013/07/WP5-Governance-Innovation-and-the-Transition-to-a-Sustainable-Energy-System-Perspecitves-from-Economic-Theory.pdf.

Lueken, R., and J. Apt. 2014. The effects of bulk electricity storage on the PJM market. *Energy Systems* 5(4):677-704.

Lyon, T., and Yin, H. 2009. Why do states adopt renewable portfolio standards?: An empirical investigation. *The Energy Journal* 31(3):133-157.

Madison Park Group. n.d. *Guide to venture capital.* http://upload.madisonparkgrp.com/news/gtvc.pdf.

Malkin, D., and P. Centolella. 2014. Results-based regulation: A more dynamic approach to grid modernization. *Fortnightly Magazine,* March.

Malm, E. 1996. An actions-based estimate of the free rider fraction in electric utility DSM programs. *The Energy Journal* 17(3):41-48.

Mankins, J. 1995. *Technology readiness levels—a white paper*. Washington, DC: National Aeronautics and Space Administration.

Manne, A., and R. Richels. 2004. The impact of learning-by-doing on the timing and costs of CO_2 abatement. *Energy Economics* 26(4):603-619.

Marron, D., and E. Toder. 2014. Tax policy issues in designing a carbon tax. *American Economic Review* 104(5):563-568.

Martin, L., and J. Jones. 2016. Issues in focus: Effects of the clean power plan. In *Annual Energy Outlook 2016*. Washington, DC: EIA. http://www.eia.gov/forecasts/aeo/section_issues.cfm (accessed June 20, 2016).

Massachusetts Department of Public Utilities. 2013. *Massachusetts Electric Grid Modernization Stakeholder Working Group process: Report to the Department of Public Utilities from the Steering Committee*. DPU 12-76. http://www.mass.gov/eea/docs/dpu/electric/grid-mod/ma-grid-mod-working-group-report-07-02-2013.pdf.

Massachusetts Department of Public Utilities. 2014. *Investigation by the Department of Public Utilities on its own Motion into Modernization of the Electric Grid.* DPU 12-76-B. http://www.mass.gov/eea/docs/dpu/orders/dpu-12-76-b-order-6-12-2014.pdf.

Mathieu, J.L. 2012. *Modeling, analysis, and control of demand response resources.* LBNL-5544E. Berkeley, CA: Ernest Orlando Lawrence

Berkeley National Laboratory. https://eaei.lbl.gov/sites/all/files/lbnl-5544e_0.pdf.

Mathieu, J.L., M. Dyson, and D.S. Callaway. 2012. *Using residential electric loads for fast demand response: The potential resource and revenues, the costs, and policy recommendations.* Proceedings of the 2012 ACEEE Summer Study on Energy Efficiency in Buildings. http://aceee.org/files/proceedings/2012/data/papers/0193-000009.pdf.

Mauch, B., J. Apt, P.M. Carvallo, and M. Small. 2013. An effective method for modeling wind power forecast uncertainty. *Energy Systems* 4(4):393-417.

Mazzucato, M. 2011. The entrepreneurial state. *RENEWAL: A Journal of Social Democracy* 19(3/4).

McDermott, K. 2012. *Cost of service regulation in the investor owned electric utility industry: A history of adaptation.* http://www.eei.org/issues andpolicy/stateregulation/documents/cosr_history_final.pdf.

McDonald, A., and L. Schrattenholzer. 2000. Learning rates for energy technologies. *Energy Policy* 29(4):255-261.

McKinsey & Company. 2009. *"And the winner is..." Capturing the promise of philanthropic prizes.* http://mckinseyonsociety.com/downloads/reports/Social-Innovation/And_the_winner_is.pdf.

Metcalf, G.E. 2008. An empirical analysis of energy intensity and its determinants at the state level. *The Energy Journal* 29(3):1-26. http://globalchange.mit.edu/files/document/MITJPSPGC_Reprint08-8.pdf.

Metcalf, G.E., and K.A. Hassett. 1999. Measuring the energy savings from home improvement investments: Evidence from monthly billing data. *The Review of Economics and Statistics* 81(3):516-528.

METI (Ministry of Economy, Trade and Industry) Agency for Natural Resources and Energy. 2010. *Top Runner Program: Developing the world's best energy-efficient appliances.* http://www.enecho.meti.go.jp/category/saving_and_new/saving/enterprise/overview/pdf/toprunner2011.03en-1103.pdf.

Meyers, R.J., E.D. Williams, and H.S. Matthews. 2010. Scoping the potential of monitoring and control technologies to reduce energy use in homes. *Energy and Buildings* 42:563-569.

Midwest ISO. 2010. *Transmission Expansion Plan 2010.* https://www.misoenergy.org/_layouts/MISO/ECM/Redirect.aspx?ID=19942.

Mills, E. 2011. Building commissioning: A golden opportunity for reducing energy costs and greenhouse gas emissions in the United States. *Energy Efficiency* 4(2):145-173.

MISI (Management Information Services, Inc.). 2011. *60 years of energy incentives: Analysis of federal expenditures for energy development.* Washington, DC: MISI. http://www.nei.org/corporatesite/media/filefolder/60_Years_of_Energy_Incentives_-_Analysis_of_Federal_Expenditures_for_Energy_Development_-_1950-2010.pdf.

MIT (Massachusetts Institute of Technology). 2007. *The future of coal.* http://web.mit.edu/coal/The_Future_of_Coal.pdf (accessed August 1, 2016).

MIT. 2016. *Carbon capture and sequestration project database.* http://sequestration.mit.edu/tools/projects/index.html (accessed July 20, 2016).

Montgomery, W., and A. Smith. 2007. Price, quantity, and technology strategies for climate change policy. In *Human-induced climate change: An interdisciplinary assessment*, edited by M.E. Schlesinger, H.S. Kheshg, J. Smith, F.C. de la Chesnaye, J.M. Reilly, T. Wilson, and C. Kolstad. Cambridge, UK: Cambridge University Press. Pp. 328-342.

Morgan, M.G., and S.T. McCoy. 2012. *Carbon capture and sequestration: Removing the legal and regulatory barriers*. Washington, DC: RFF Press.

Muller, N.Z., R. Mendelsohn, and W. Nordhaus. 2011. Environmental accounting for pollution in the United States economy. *The American Economic Review* 101(5):1649-1675.

NACAP (North American Carbon Atlas Partnership). 2012. *The North American Carbon Storage Atlas.* https://www.netl.doe.gov/File%20Library/Research/Carbon-Storage/NACSA2012.pdf.

Nadiri, M.I. 1993. *Innovations and technological spillovers.* http://www.nber.org.ezp-prod1.hul.harvard.edu/papers/w4423.pdf?new_window=1.

NAREIT (National Association of Real Estate Investment Trusts). n.d.-a. *Understanding the basics of REITs.* https://www.reit.com/investing/reit-basics/faqs/basics-reits.

NAREIT. n.d.-b. *U.S. REIT industry equity market cap, historical REIT industry market capitalization: 1972-2014.* https://www.reit.com/data-research/data/us-reit-industry-equity-market-cap.

NAS (National Academy of Sciences), NAE (National Academy of Engineering), and NRC (National Research Council). 2010. *Real prospects for energy efficiency in the United States*. Washington, DC: The National Academies Press.

Nathan, S. 2013. PRISM project: A proposal for the UK's problem plutonium, *The Engineer*, May 13.

NEI (Nuclear Energy Institute). 2015. *Second license renewal roadmap.* http://www.nei.org/CorporateSite/media/filefolder/Federal-State-Local-Policy/Regulatory-Information/Second-License-Renewal-Roadmap-small.pdf?ext=.pdf.

NEI. 2016. *Nuclear units under construction worldwide.* http://www.nei.org/Knowledge-Center/Nuclear-Statistics/World-Statistics/Nuclear-Units-Under-Construction-Worldwide (accessed August 1, 2016).

Nemet, G.F. 2006. Beyond the learning curve: Factors influencing cost reductions in photovoltaics. *Energy Policy* 34(17):3218-3232.

Nemet, G.F. 2012. Subsidies for new technologies and knowledge spillovers from learning by doing. *Journal of Policy Analysis and Management* 31(3):601-622.

Nemet, G.F., and Baker, E. 2008. *Demand subsidies versus R&D: Comparing the uncertain impacts of policy on a pre-commercial low-carbon energy technology*. La Follette School of Public Affairs Working Paper No. 2008-006. https://www.lafollette.wisc.edu/images/publications/workingpapers/nemet2008-006.pdf.

Nest Labs, Inc. 2013. Our first Rush Hour Rewards results. *Inside Nest*, July 18. https://nest.com/blog/2013/07/18/our-first-rush-hour-rewards-results (accessed June 29, 2016).

Nest Labs, Inc. 2014. *Energy savings from Nest: The impact of Nest learning thermostat.* https://nest.com/downloads/press/documents/efficiency-simulation-white-paper.pdf.

Net Power. 2014. *Net Power announces launch of breakthrough clean power project.* http://www.netpower.com/news-posts/net-power-announces-launch-of-breakthrough-clean-power-project (accessed December 12, 2014).

Neuhoff, K. 2008. Learning by doing with constrained growth rates: An application to energy technology policy. *The Energy Journal* 165-183. http://www.eprg.group.cam.ac.uk/wp-content/uploads/2008/11/eprg0809.pdf.

Neuhoff, K., and P. Twomey. 2008. Will the market choose the right technologies. In *Delivering a low carbon electricity system: Technology, economics and policy*, edited by M. Grubb, T. Jamasb, and M. Pollitt. Cambridge, UK: Cambridge University Press. Pp. 259-277.

New York State Department of Public Service. 2014. *Reforming the energy vision: NYS Department of Public Service staff report and proposal.* Case no. 14-M-0101. http://www3.dps.ny.gov/W/PSCWeb.nsf/96f0fec0b45a3c6485257688006a701a/26be8a93967e604785257cc40066b91a/$FILE/ATTK0J3L.pdf/Reforming%20The%20Energy%20Vision%20(REV)%20REPORT%204.25.%2014.pdf.

Newell, R., and N. Wilson. 2005. *Technology prizes for climate change mitigation*. Discussion paper 05-33. Washington, DC: Resources for the Future.

Newell, R.G., A.B. Jaffe, and R.N. Stavins. 1999. The induced innovation hypothesis and energy-saving technological change. *The Quarterly Journal of Economics* 941-975.

Newell, R.G., and J. Siikamäki. 2013. *Nudging energy efficiency behavior: The role of information labels*. RFF DP 13-17. Washington, DC: Resources for the Future.

Newsham, G.R., S. Mancini, and B.J. Birt. 2009. Do LEED-certified buildings save energy? Yes, but… *Energy and Buildings* 41(8):897-905.

Nicholas, T. 2011. What drives innovation? *Antitrust Law Journal* 77(3):787-809.

Nicholas, T. 2013. Hybrid innovation in Meiji, Japan. *International Economic Review* 54(2):575-600.

NIST (National Institute of Standards and Technology). 2014. *NIST framework and roadmap for smart grid interoperability standards, release 3.0*. NIST special publication 1108r3. http://www.nist.gov/smartgrid/upload/NIST-SP-1108r3.pdf.

Nordhaus, W. 2011. The economics of tail events with an application to climate change. *Review of Environmental Economics and Policy* 5(2):240-257.

Nordhaus, W. 2013. *The climate casino: Risk, uncertainty, and economics for a warming world*. New Haven, CT: Yale University Press.

Nordhaus, W. 2014. The perils of the learning model for modeling endogenous technological change. *The Energy Journal 35*(1):1-13.

NRC (National Research Council). 1999. *Concerning federally sponsored inducement prizes in engineering and science*. Washington, DC: National Academy Press.

NRC. 2001. *Effectiveness and impact of Corporate Average Fuel Economy (CAFE) standards*. Washington, DC: National Academy Press.

NRC. 2007. *Innovation inducement prizes at the National Science Foundation*. Washington, DC: The National Academies Press.

NRC. 2010a. *Electricity from renewable resources: Status, prospects, and impediments*. Washington, DC: The National Academies Press.

NRC. 2010b. *Hidden costs of energy: Unpriced consequences of energy production and use*. Washington, DC: The National Academies Press.

NRC. 2010c. *Limiting the magnitude of future climate change*. Washington, DC: The National Academies Press.

NRC. 2010d. *Modeling the economics of greenhouse gas mitigation: Summary of a workshop*. Washington, DC: The National Academies Press.

NRC. 2011. *America's climate choices*. Washington, DC: The National Academies Press.

NRC. 2012. *Rising to the challenge: U.S. innovation policy for the global economy*. Washington, DC: The National Academies Press.

NRC. 2013a. *An assessment of the prospects for inertial fusion energy*. Washington, DC: The National Academies Press.

NRC. 2013b. *Assessment of advanced solid-state lighting*. Washington, DC: The National Academies Press.

NRC. 2013c. *Effects of U.S. tax policy on greenhouse gas emissions*. Washington, DC: The National Academies Press.

NRECA (National Rural Electric Cooperative Association). 2016. *Electric co-op fact sheet 2016*. http://www.nreca.coop/about-electric-cooperatives/co-op-facts-figures/electric-co-op-fact-sheet-2016-03 (accessed July 25, 2016).

NREL (National Renewable Energy Laboratory). 2012. *Renewable Electricity Futures Study*. http://www.nrel.gov/analysis/re_futures.

NSF (National Science Foundation). 2010. U.S. businesses report 2008 worldwide R&D expense of $330 billion: Findings from new NSF survey. NSF 10-322. *InfoBrief*, May.

Nuclear Engineering International. 2014. B&W scales back small reactor development. *NEI Magazine*, April 15. http://www. neimagazine. com/news/newsbw-scales-back-small-reactor-development-4214703.

NuScale Power. 2013. *NuScale Power modular and scalable reactor.* https://aris.iaea.org/sites/..%5CPDF%5CNuScale.pdf.

Oates, D.L., and P. Jaramillo. 2013. *Production cost and air emissions impact of coal cycling in power systems with large-scale wind penetration.* Bristol, UK: IOP Publishing Ltd.

Ofgem (Office of Gas and Electric Markets [United Kingdom]). 2010a. *Handbook for implementing the RIIO model.* https://www.ofgem.gov.uk/ofgem-publications/51871/riiohandbook.pdf.

Ofgem. 2010b. *RIIO: A new way to regulate energy networks: Final decision.* https://www.ofgem.gov.uk/ofgem-publications/51870/decision-doc.pdf.

Ofgem. 2013. *Strategy decision for the RIIO-ED1 electricity distribution price control: Overview.* https://www.ofgem.gov.uk/sites/default/files/docs/2013/03/riioed1decoverview.pdf.

Ofgem. 2014. *Ofgem announces £46m of funding for eight innovation projects to improve Britain's energy networks.* Press Release, November 24. https://www.ofgem.gov.uk/publications-and-updates/ofgem-announces-%C2%A346m-funding-eight-innovation-projects-improve-britain%E2%80%99s-energy-networks.

Ordowich, C., J. Chase, D. Steele, R. Malhotra, M. Harada, and K. Makino. 2012. Applying learning curves to modeling future coal and gas power generation technologies. *Energy & Fuels* 26(1):753-766.

PACE (Property Assessed Clean Energy). n.d. *PACE industry publications.* http://www.pacenation.us/publications.

Pacific Economic Group. 2013. *Alternative regulation for evolving utility challenges: An updated survey.* http://www.eei.org/issuesandpolicy/stateregulation/Documents/innovative_regulation_survey.pdf.

Palmer, K.L., and D. Burtraw. 2005. *The cost effectiveness of renewable electricity policies.* Washington, DC: Resources for the Future.

Palmer, K.L., and M.A. Walls. 2015. What homeowners say about home energy audits. *Resources* 188:34-39.

Palmer, K.L., R. Sweeney, and M. Allaire. 2010. *Modeling policies to promote renewable and low-carbon sources of electricity.* Washington, DC: Resources for the Future.

Palmer, K.L., M.A. Walls, H. Gordon, and T. Gerarden. 2013. Assessing the energy-efficiency information gap: Results from a survey of home energy auditors. *Energy Efficiency* 6(2):271-292.

Palmer, K.L., M.A. Walls, and L. O'Keefe. 2015. *Putting information into action: What explains follow-up on home energy audits?* RFF DP 15-34. Washington, DC: Resources for the Future.

Paltsev, S., J.M. Reilly, H.D. Jacoby, and J.F. Morris. 2009. The cost of climate policy in the United States. *Energy Economics* 31:S235-S243.

Parry, I., A. Morris, and R. Williams (Eds.). 2015. *Implementing a U.S. carbon tax: Challenges and debates*. New York: Routledge.

Passey, R., T. Spooner, I. Macgill, M. Watt, and K. Syngellakis. 2011. The potential impacts of grid-connected distributed generation and how to address them: A review of technical and non-technical factors. *Energy Policy* 39(10):6280-6290.

PCAST (President's Council of Advisors on Science and Technology). 2010. *Report to the President on accelerating the pace of change in energy technologies through an integrated federal energy policy*. http://www.whitehouse.gov/sites/default/files/microsites/ostp/pcast-energy-tech-report.pdf.

Peters, M., M. Schneider, T. Griesshaber, and V.H. Hoffmann. 2011. *The quest for adequate technology-push and demand-pull policies: Country-level spillovers and incentives for non-incremental innovation*. Zurich, Switzerland: Swiss Federal Institute of Technology Zurich.

Pfund, N., and B. Healey. 2011. *What would Jefferson do? The historical role of federal subsidies in shaping America's energy future*. San Francisco, CA: DBL Investors. http://www.dblpartners.vc/wp-content/uploads/2012/09/What-Would-Jefferson-Do-2.4.pdf?597435&48d1ff.

Phillips, C.F., Jr. 1988. *The regulation of public utilities*. Arlington, VA: Public Utilities Reports.

Piette, M., J. Granderson, M. Wetter, and S. Kiliccote. 2012. *Responsive and intelligent building information and control for low-energy and optimized grid integration*. LBNL-5662E. http://simulationresearch.lbl.gov/wetter/download/2012-acee-controls.pdf.

Pindyck, R. 2011. Fat tails, thin tails, and climate change policy. *Review of Environmental Economics and Policy* 5(2):258-274.

PJM. 2009. *Demand response in the PJM markets*. Statement of Terry Boston, President and CEO, on behalf of the PJM board of managers, June 26. http://www.pjm.com/~/media/committees-groups/committees/mic/20100722/20100722-item-02b-statement-on-demand-response-in-the-pjm-markets.ashx (accessed June 29, 2016).

Pope, C.A. III, M. Ezzati, and D.W. Dockery. 2009. Fine-particulate air pollution and life expectancy in the United States. *New England Journal of Medicine* 360(4):376-386.

Popp, D., R.G. Newell, and A.B. Jaffe. 2010. Energy, the environment, and technological change. *Handbook of the Economics of Innovation* 2(10):873-937.

Porter, M.E. 2001. *Clusters of innovation initiative: Regional foundations of U.S. competitiveness*. http://www.compete.org/storage/images/uploads/File/PDF%20Files/CoC_Reg_Found_national_cluster.pdf.

Pritchard, G., and J. Evans. 2009. Using AMI for outage notification at PECO. *Utility Automation & Engineering/T&D* 14(4).

PSERC (Power Systems Engineering Research Center). 2016. *Academic programs*. http://pserc.wisc.edu/education/academic.aspx (accessed June 29, 2016).

Qiu, Y., and L.D. Anadon. 2012. The price of wind power in China during its expansion: Technology adoption, learning-by-doing, economies of scale, and manufacturing localization. *Energy Economics* 34(3):772-785.

Ramos, A, A. Gago, X. Labandeira, and P. Linares. 2015. The role of information for energy efficiency in the residential sector. *Energy Economics* 52:S17-S29.

RAP (Regulatory Assistance Project). 2011. *Electricity regulation in the United States: A guide*. http://www.raponline.org/docs/RAP_Lazar_ElectricityRegulationInTheUS_Guide_2011_03.pdf (accessed June 29, 2016).

Rawls, J., Z. Johal, and J. Parmentola. 2014. Improving the economics and long-term sustainability of nuclear power. *Energy Development Frontier* 3(2):20-29.

Rocky Mountain Institute. 2013. *e-Lab: New business models for the distribution edge*. Boulder, CO: Rocky Mountain Institute.

Rothwell, G.S. 2001. *Does the US subsidize nuclear power insurance?* Stanford, CA: Stanford Institute for Economic Policy Research.

Rubin, E.S. 2005. *The government role in technology innovation: Lessons for the climate change policy agenda*. Proceedings of the 10th Biennial Conference on Transportation Energy and Environmental Policy: Toward a Policy Agenda for Climate Change, Pacific Grove, CA.

Rubin, E.S. 2014. Will cutting carbon kill coal? Only if the industry fails to adapt. *Pittsburgh Post-Gazette*, November 24, p. E1.

Rubin, E.S., C. Chen, and A.B. Rao. 2007a. Cost and performance of fossil fuel power plants with CO_2 capture and storage. *Energy Policy* 35:4444-4454.

Rubin, E.S., S. Yeh, M. Antes, M. Berkenpas, and J. Davison. 2007b. Use of experience curves to estimate the future cost of power plants with CO_2 capture. *International Journal of Greenhouse Gas Control* 1(2):188-197.

Russell, B. 2010. Educating the Workforce for the modern electric power system university-industry collaboration. *The Bridge on the Electricity Grid* 40(1):35-41.

Ryan, L., S. Moarif, E. Levina, and R. Baron. 2011. *Energy efficiency policy and carbon pricing*. Working paper. Washington, DC: International Energy Agency.

Santen, N.R., and L.D. Anadon. 2014. *Electricity technology investments under solar RD&D uncertainty: How interim learning and adaptation affect the optimal decision strategy*. Discussion paper 2014-10. Cambridge, MA:

Harvard Kennedy School, Belfer Center for Science and International Affairs. http://belfercenter.ksg.harvard.edu/publication/24874.

Santen, N.R., M.D., Webster, D. Popp, and I. Pérez-Arriaga. 2014. *Inter-temporal R&D and capital investment portfolios for the electricity industry's low carbon future*. Working paper no. 20783. Cambridge, MA: National Bureau of Economic Research. http://www.nber.org/papers/w20783.

SBA (Small Business Administration). 2016a. *About Startup America: StartUp America—Empowering America's entrepreneurs*. https://www.sba.gov/about-sba/sba-initiatives/startup-america/about-startup-america.

SBA. 2016b. *Eligible impact investments*. https://www.sba.gov/sbic/general-information/key-initiatives/impact-investment-fund/eligible-impact-investments.

SBA. 2016c. *Program overview*. https://www.sba.gov/sbic/general-information/program-overview.

Schmalensee, R. 2013. *The performance of U.S. wind and solar generating plants*. Working paper 19509. Cambridge, MA: National Bureau of Economic Research.

Schneider, K., J. Fuller, F. Tuffner, R. Singh, C. Bonebrake, N.P. Kumar, and B. Vyakaranam. 2012. *Evaluation of representative Smart Grid investment grant project technologies: Summary report*. PNNL-20892. Richland, WA: Pacific Northwest National Laboratory.

Schneider, S., and Goulder, L. 1997. Achieving carbon dioxide emissions reductions at low cost. *Nature* 389:13-14.

Schwienbacher, A. 2008. Innovation and venture capital exits. *The Economic Journal* 118(533):1888-1916.

Sekar, S., and B. Sohngen. 2014. *The effects of renewable portfolio standards on carbon intensity in the United States*. Washington, DC: Resources for the Future.

Singleton, G., H. Herzog, and S. Ansolabehere. 2009. Public risk perspectives on the geologic storage of carbon dioxide. *International Journal of Greenhouse Gas Control* 3(1):100-107.

Sissine, F. 2014. *Renewable energy R&D funding history*. Washington, DC: Congressional Research Service.

Sissine, F. 2015. *DOE's Office of Energy Efficiency and Renewable Energy: FY2016 appropriations*. Washington, DC: Congressional Research Service. https://www.fas.org/sgp/crs/misc/R44004.pdf.

SmartGrid Consumer Collaborative. 2013. *Smart Grid economic and environmental benefits: A review and synthesis of research on Smart Grid benefits and costs*. http://smartgridcc.org/wp-content/uploads/2013/10/SGCC-Econ-and-Environ-Benefits-Full-Report.pdf.

Söderholm, P., and Sundqvist, T. 2007. Empirical challenges in the use of learning curves for assessing the economic prospects of renewable energy technologies. *Renewable Energy* 32(15):2559-2578.

Spurlock, C.A. 2013. *Appliance efficiency standards and price discrimination.* Berkeley, CA: Ernest Orlando Lawrence Berkeley National Laboratory.

Sullivan, M., M. Mercurio, and J. Schellenberg. 2009. *Estimated value of service reliability for electric utility customers in the United States.* LBNL-2132E. Berkeley, CA: Ernest Orlando Lawrence Berkeley National Laboratory. https://emp.lbl.gov/sites/all/files/REPORT%20lbnl-2132e.pdf.

Sullivan, P., W. Cole, N. Blair, E. Lantz, V. Krishnan, T. Mai, D. Mulcahy, and G. Porro. 2015. *2015 standard scenarios annual report: U.S. electric sector scenario exploration.* NREL/TP-6A20-64072. Golden, CO: National Renewable Energy Laboratory. http://www.nrel.gov/docs/fy15osti/64072.pdf.

Sun, Y. 2013. *HTR development status in China.* Paper presented at IAEA TWG-GCR meeting, Vienna, Austria, March 5-7.

Taylor, M., E.S. Rubin, and G.F. Nemet. 2006. *The role of technological innovation in meeting California's greenhouse gas emission targets.* 1-2006. http://repository.cmu.edu/cgi/viewcontent.cgi?article=1081&context=epp.

Taylor, M., C.A. Spurlock, and H.C. Yang. 2015. *Confronting regulatory cost and quality expectations: An exploration of technical change in minimum efficiency performance standards.* Berkeley, CA: Ernest Orlando Lawrence Berkeley National Laboratory.

Tennessee Valley Authority. n.d. *Our history.* https://www.tva.com/About-TVA/Our-History.

Thompson, P. 2010. Learning by doing. *Handbook of the Economics of Innovation* 1(10):429-476.

Tierney, S., and T. Schatzki. 2008. *Competitive procurement of retail electricity supply: Recent trends in state policies and utility practices.* Boston, MA: Analysis Group. http://www.analysisgroup.com/uploadedfiles/content/insights/publishing/competitive_procurement.pdf.

Tomić, J., and W. Kempton. 2007. Using fleets of electric-drive vehicles for grid support. *Journal of Power Sources* 168(2):459-468.

Tuladhar, S.D., S. Mankowski, and P. Bernstein. 2014. Interaction effects of market-based and command-and-control policies. *The Energy Journal* 35.

U.S. NRC (United States Nuclear Regulatory Commission). 2007. *Feasibility study for a risk-informed and performance-based regulatory structure for future plant licensing.* NUREG-1860. Washington, DC: U.S. NRC.

U.S. NRC. 2012. *Report to Congress: Advanced reactor licensing.* http://pbadupws.nrc.gov/docs/ML1215/ML12153A014.pdf.

U.S. NRC. n.d. *Radioactive waste.* http://www.nrc.gov/waste.html.

U.S.-Canada Power System Outage Task Force. 2004. *Final report on the August 14, 2003 blackout in the United States and Canada: Causes and recommendations.* http://energy.gov/sites/prod/files/oeprod/DocumentsandMedia/BlackoutFinal-Web.pdf.

UN (United Nations). 2015. *Adoption of the Paris Agreement.* Paris, France: UN.

UNFCC (United Nations Framework Convention on Climate Change). 2015. *Adoption of the Paris Agreement. Proposal by the President.* FCCC/CP/2015/L.9/Rev.1. Geneva, Switzerland: UN. http://unfccc.int/documentation/documents/advanced_search/items/6911.php?priref=600008831.

United States. 2015. *United Nations Framework Convention on Climate Change, U.S. cover note, INDC, and accompanying information, March 31.* http://www4.unfccc.int/submissions/INDC/Published%20Documents/United%20States%20of%20America/1/U.S.%20Cover%20Note%20INDC%20and%20Accompanying%20Information.pdf?Mobile=1&Source=%2Fsubmissions%2FINDC%2F_layouts%2Fmobile%2Fdispform%2Easpx%3FList%3D4a67b500-675c-4a0e-be83-649364ee36bb%26View%3D428ea47e-35b4-4eb0-bd45-71f71031d074%26RootFolder%3D%252Fsubmissions%252FINDC%252FPublished%2520Documents%252FUnited%2520States%2520of%2520America%252F1%26ID%3D386%26CurrentPage%3D1 (accessed July 29, 2016).

Valentino, L., V. Valenzuela, A. Botterud, Z. Zhou, and G. Conzelman. 2012. System-wide emissions implications of increased wind penetration. *Environmental Science and Technology* 46:4200-4206.

van Benthem, A., K. Gillingham, and J. Sweeney. 2008. Learning-by-doing and the optimal solar policy in California. *The Energy Journal* 29(3):131-152.

Verdolini, E., and M. Galeotti. 2011. At home and abroad: An empirical analysis of innovation and diffusion in energy technologies. *Journal of Environmental Economics and Management* 61(2):119-134.

Wald, M., and J. Broder. 2011. Utility shelves ambitious plan to limit carbon. *The New York Times*, July 13, p. A1.

Walls, M., K. Palmer, T. Gerarden, and X. Bak. 2016. *Is energy efficiency capitalized into home prices? Evidence from three U.S. cities*. RFF DP 13-18 Rev. Washington, DC: Resources for the Future.

Walsh, J., D. Wuebbles, et al. 2014. *2014 national climate assessment.* Washington, DC: U.S. Global Change Research Program. http://nca2014.globalchange.gov/report (accessed August 4, 2016).

WBDG (Whole Building Design Guide). n.d. *Federal high performance and sustainable buildings*. http://www.wbdg.org/references/fhpsb.php.

Weisenthal, T., P. Dowling, J. Morbee, C. Thiel, B. Schade, P. Russ, S. Simoes, S. Peteves, K. Schoots, and M. Londo. 2012. *Technology learning curves for energy policy support.* https://setis.ec.europa.eu/sites/default/files/reports/Technology-Learning-Curves-Energy-Policy-Support.pdf.

Weitzman, M. 2009. On modeling and interpreting the economics of catastrophic climate change. *Review of Economics and Statistics* 91(1):1-19.

Weitzman, M. 2011. Fat-tailed uncertainty in the economics of catastrophic climate change. *Review of Environmental Economics and Policy* 5(2):275-292.

White House. 2012. *President Obama Signs Executive Order promoting industrial energy efficiency.* https://www.whitehouse.gov/the-press-office/2012/08/30/president-obama-signs-executive-order-promoting-industrial-energy-effici (accessed June 28, 2016).

White House. 2015. *Executive Order: Planning for federal sustainability in the next decade.* March 19. https://www.whitehouse.gov/the-press-office/2015/03/19/executive-order-planning-federal-sustainability-next-decade.

Wiesenthal, T., P. Dowling, J. Morbee, C. Thiel, B. Schade, P. Russ, S. Simoes, S. Peteves, K. Schoots, and M. Londo. 2012. *Technology learning curves for energy policy support.* https://setis.ec.europa.eu/sites/default/files/reports/Technology-Learning-Curves-Energy-Policy-Support.pdf.

Williams, H. 2012. Innovation inducement prizes: Connecting research to policy. *Journal of Policy Analysis and Management* 31(3):752-776.

Williams, J.H., B. Haley, F. Kahrl, J. Moore, A.D. Jones, M.S. Torn, and H. McJeon. 2014. *Pathways to deep decarbonization in the United States.* http://unsdsn.org/wp-content/uploads/2014/09/US-Deep-Decarbonization-Report.pdf.

Wilson, C., and H. Dowlatabadi. 2007. Models of decision making and residential energy use. *Annual Review of Environment and Resources* 32:169-203.

Wiser, R., and M. Bolinger. 2014. *2013 wind technologies market report.* Washington, DC: DOE. http://energy.gov/sites/prod/files/2014/08/f18/2013%20Wind%20Technologies%20Market%20Report_1.pdf.

WNA (World Nuclear Association). 2015. *Nuclear industry stands ready to help tackle climate change.* Press statement, December 9. http://world-nuclear.org/press/press-statements/nuclear-industry-stands-ready-to-help-tackle-clima.aspx (accessed August 1, 2016).

Wolf, J. 2012. Oncor knows outage detection. *T&D World Magazine*, November 1.

Xcel Energy. 2013. Xcel Energy proposes adding economic solar, wind to meet future customer energy demands. *Xcel Energy*, September 10. https://www.xcelenergy.com/company/media_room/news_releases/xcel_energy_proposes_adding_economic_solar,_wind_to_meet_future_customer_energy_demands.

Yeh, S., and E.S. Rubin. 2012. A review of uncertainties in technology experience curves. *Energy Economics* 34(3):762-771.

Zachmann, G., A. Serwaah, and M. Peruzzi. 2014. *When and how to support renewables? Letting the data speak.* http://ideas.repec.org/p/bre/wpaper/811.html.

Appendix A

Committee Biographies

Charles O. Holliday, Jr. (*Chair*) is chairman of Royal Dutch Shell, PLC. He also is a director on the boards of HCA Holdings, Inc., and CH2M. He served as a director on the board of Deere & Company, 2007-2015; was a member of the board of directors for Bank of America, 2009-2015, serving as chairman, 2010-2014; and was a member of the board of directors for E. I. du Pont de Nemours and Company (DuPont), 1997-2009, serving as its chairman, 1999-2009, and as DuPont's chief executive officer, 1998-2009. Mr. Holliday is chairman emeritus of Catalyst, a leading nonprofit organization dedicated to expanding opportunities for women and business, and chairman emeritus of the board of the U.S. Council on Competitiveness, a nonpartisan nongovernmental organization working to ensure U.S. prosperity. He is a founding member of the International Business Council, and previously served as chairman of the Business Roundtable's Task Force for Environment, Technology and Economy; the World Business Council for Sustainable Development; the Business Council; and the Society of Chemical Industry American Section. Mr. Holliday has served on and chaired several committees of the National Academies, including the Committee on Research Universities, the Committee on America's Climate Choices, the Committee on Prospering in the Global Economy of the 21st Century, and the Roundtable on Scientific Communication and National Security. He is former chair of the National Academy of Engineering Council. Mr. Holliday received a B.S. in industrial engineering from the University of Tennessee and honorary doctorates from Polytechnic University in Brooklyn, New York, and Washington College in Chestertown, Maryland. He is the author of *Walking the Talk*, a book that makes the business case for sustainable development and corporate responsibility.

Jerome Apt is a professor at Carnegie Mellon University's Tepper School of Business and in the university's Department of Engineering and Public Policy. He is co-director of the Carnegie Mellon Electricity Industry Center and director of the RenewElec (renewable electricity) project. He has authored more than 100 papers in peer-reviewed scientific journals, and has published op-ed pieces in *The Wall Street Journal*, *The New York Times*, and *The Washington Post*. His recent publications address the mathematical characteristics and economics of wind, solar, and hybrid solar-fossil fuel power generation. Prior to his work at Carnegie Mellon, he was a planetary astronomer at the Jet Propulsion Laboratory, a National Aeronautics and Space Administration (NASA) astronaut on four Space Shuttle missions, director of the Carnegie Museum of Natural History, and managing director and chief technology officer of iNetworks LLC Venture Capital. Dr. Apt received the Metcalf Lifetime Achievement Award for significant contributions to engineering in 2002 and NASA's Distinguished Service Medal in 1997. He has served on two committees of the National Academies: the Committee on the Rationale and Goals of the U.S. Civil Space Program and the Panel on Earth Science Applications and Societal Needs. Dr. Apt received an A.B. from Harvard College in 1971 and a Ph.D. in experimental atomic physics from the Massachusetts Institute of Technology in 1976.

Frances Beinecke is former president of the Natural Resources Defense Council (NRDC), one of the United States' most influential environmental advocacy organizations, which uses law and science to advance solutions to the nation's environmental challenges. Under her leadership, the organization focused on establishing a clean energy future that curbs climate change, revives the world's oceans, defends endangered wildlife and wild places, and ensures safe and sufficient water. President Obama appointed Ms. Beinecke to the National Commission on the BP *Deepwater Horizon* Oil Spill and Offshore Drilling. She has played a leadership role in many environmental organizations and currently serves on the boards of the World Resources Institute, Climate Central, the Meridian Institute, and the NRDC Action Fund. Ms. Beinecke received a bachelor's degree from Yale College and a master's degree from the Yale School of Forestry and Environmental Studies (FES). She is a member of the Leadership Council of the Yale School of Forestry and the Yale School of Management's Advisory Board and is a former member of the Yale Corporation. She served as a McCluskey fellow at Yale FES in 2015. Ms. Beinecke has received the Rachel Carson Award from the National Audubon Society, the Aldo Leopold Award from Yale FES, and honorary degrees from Vermont Law School and Lehman College. She served on the National Academies' Committee on Outer Continental Shelf (OCS) Safety Information from 1982 until 1984.

Nora Mead Brownell co-founded ESPY Energy Solutions, LLC, a women-owned business providing innovative and highly skilled consulting services. The company offers strategic planning, marketing, business, regulatory, and technical expertise to energy utilities, energy equipment manufacturing and supply companies, smart-grid manufacturers and service providers, and financial institutions evaluating investments in the energy sector. In 2001, President George W. Bush nominated and the U.S. Senate confirmed Ms. Brownell to be a commissioner to the Federal Energy Regulatory Commission (FERC), where she served until 2006, with a focus on fostering competitive markets to serve the public interest and policies that promote investment in national energy infrastructure development. Previously, she served as a member of the Pennsylvania Public Utility Commission (PUC), 1997 to 2001, taking an active role in the rollout of electric choice in Pennsylvania. She also has actively supported Pennsylvania's pursuit of competition in the local markets for telecommunications, deployment of advanced services, enhancement of services to rural areas, protection of consumers, and advancement of special services, helping to craft unique solutions to a number of these industry issues. In addition, Ms. Brownell is former president of the National Association of Regulatory Utility Commissioners. She currently serves on the boards of National Grid PLC, Tangent, and Spectra Energy Partners, having previously served on the boards of numerous for-profit and nonprofit organizations. At present, Ms. Brownell is serving on the advisory boards of Morgan Stanley Infrastructure, New World Capital, and TerViva. In addition, she has lectured at the Vermont Law School's Center for Energy and the Environment, the Michigan State University Institute of Public Utilities, the University of Idaho, the H. John Heinz III College School of Public Policy and Management at Carnegie Mellon University, and the Wharton Energy Club, among others.

Paul Centolella is president of Paul Centolella & Associates and a senior consultant with Tabors Caramanis Rudkevich. In these roles, he advises electric utility and technology companies on business strategy and regulatory issues and government on emerging electric industry business and regulatory models. Mr. Centolella served as a commissioner on the Public Utilities Commission of Ohio (PUCO), 2007-2012, overseeing a broad range of utility services and pursuing a regulatory strategy designed to take advantage of efficient power markets, advance innovation and grid modernization, improve utility asset utilization, enhance reliability, and provide customers with new tools for managing their energy needs. He has both public- and private-sector experience in regulation, economic and energy consulting, and public utility and environmental law, and also has background working with standards development and emerging technologies. During his 35-year career, Mr. Centolella has performed economic assessments of energy markets for power system operators and has analyzed policies related to energy pricing, investments, innovation, system reliability, and security. He has served on a range of energy-related working groups and task forces, and is a member of the Ohio, California, and Washington State Bar Associations; the American Economic Association; and the International Association for Energy Economics. Mr. Centolella holds a B.A. in economics from Oberlin College and a JD from the University of Michigan Law School.

David K. Garman, now retired, was most recently a principal and managing partner in the consulting firm Decker Garman Sullivan and Associates, LLC, a company with a client base that includes Fortune 500 companies, national laboratories, universities, think tanks, and "greentech" startups. Previously, he served as under secretary of the Department of Energy, 2001-2007, overseeing a wide spectrum of applied energy research, development, and demonstration projects ranging from new types of nuclear power plants to clean coal technologies, hydrogen and fuel cell energy technologies, superconductivity, advanced vehicles, thin-film solar photovoltaic technologies, and others. In 2001, President George W. Bush nominated him to serve as assistant secretary of Energy Efficiency and Renewable Energy, which has the Department of Energy's largest energy research, development, demonstration, and deployment portfolio. As assistant secretary, Mr. Garman was instrumental in the development of the FreedomCAR cooperative automotive research partnership and the President's Hydrogen Fuel Initiative. In recognition of his role, he was awarded the National Hydrogen Association's 2002 Meritorious Service Award and the Electric Drive Vehicle Association's 2003 "E-Visionary" Award. He also served as chairman of the FreedomCAR Executive Steering Committee and as chairman of the Steering Committee for the 15-nation International Partnership for a Hydrogen Economy. He was twice awarded the Department of Energy's highest award, the Secretary's Gold Medal. Prior to joining the Department of Energy, Mr. Garman served in a variety of positions on the staff of two U.S. senators and two Senate committees during a career spanning nearly 21 years. He represented the Senate leadership at virtually all of the major negotiations under the United Nations Framework Convention on Climate Change, 1995-2000. Mr. Garman holds a B.A. in public policy from Duke University and an M.S. in environmental sciences from The Johns Hopkins University.

Clark W. Gellings, an independent energy consultant, recently retired as a fellow at the Electric Power Research Institute (EPRI), where he was responsible for technology strategy in areas concerning energy efficiency, demand response, renewable energy sources, and other clean technologies. He was named EPRI fellow in 2009, in recognition of his 28+ years of technical innovation and leadership. Mr. Gellings has made significant contributions to the development of demand-side management (DSM) and smart-grid research, among other technical areas. He pioneered smart-grid research when EPRI established its IntelliGrid research program in 1999. He has also conducted research in energy utilization, electrotechnologies, power quality, electric transportation, thermal and electrical energy storage, and renewables. From 1982 to 2009, Mr. Gellings served in seven vice president positions at EPRI. He has received a number of distinguished awards from various organizations. A licensed professional engineer, he has served on the National Academies' Panel on Redesigning the Commercial Building and Residential Energy Consumption Surveys of the Energy Information Administration and Committee on Enhancing the Robustness and Resilience of Future Electrical Transmission and Distribution in the United States to Terrorist Attack. Mr. Gellings earned his

B.S. in electrical engineering from Newark College of Engineering, then earned an M.S. in mechanical engineering from New Jersey Institute of Technology and an M.S. in management science from the Stevens Institute of Technology.

Barton J. Gordon joined K&L Gates as partner in its Washington, DC, office after 26 years representing the state of Tennessee in the U.S. House of Representatives. During his congressional career, he was known as a bipartisan leader in innovation policy. As dean of Tennessee's congressional delegation, he represented the Sixth District, 1985-2011. From 2007 through 2011, he was chairman of the House Science and Technology Committee, authoring the landmark bipartisan America COMPETES Act. That law created the Advanced Research Projects Agency-Energy (ARPA-E) within the Department of Energy, which is tasked with leveraging talent in private industry, universities, and government laboratories to develop next-generation energy sources and technologies. Additionally, he led the effort to enact the Energy Independence and Security Act of 2007, which increased mileage standards, improved vehicle technology, promoted alternative energy research, and improved energy efficiency in a variety of ways. Mr. Gordon also served as a senior member of the House Energy and Commerce Committee and of three subcommittees: the Health Subcommittee; the Commerce, Trade, and Consumer Protection Subcommittee; and, the Communications, Technology, and the Internet Subcommittee. Prior to his public service, he was a lawyer in private practice. He earned his B.S. from Middle Tennessee State University and his JD from the University of Tennessee.

William W. Hogan is Raymond Plank professor of global energy policy at Harvard's Kennedy School of Government and research director of the Harvard Electricity Policy Group (HEPG), which is exploring issues involved in the transition to a more competitive electricity market. His current research focuses on major energy industry restructuring, network pricing and access issues, market design, and energy policy in nations worldwide. Dr. Hogan has been a member of the faculty of Stanford University, where he founded the Energy Modeling Forum, and he is a past president of the International Association for Energy Economics. He has been actively engaged in the design and improvement of competitive electricity markets in many regions of the United States, as well as around the world. His activities include designing the market structures and market rules by which regional transmission organizations, in various forms, coordinate bid-based markets for energy, ancillary services, and financial transmission rights. This research is also part of the larger activities on the future of energy and energy policy research at Harvard University through the Environment and Natural Resources Policy Program, the Environmental Economics Program, Harvard University Center for the Environment, and Mossavar-Rahmani Center for Business and Government. Dr. Hogan received his undergraduate degree from the U.S. Air Force Academy and his MBA and Ph.D. from the University of California, Los Angeles.

Richard K. Lester is Japan Steel Industry professor and associate provost for international activities at the Massachusetts Institute of Technology (MIT), where he oversees the Institute's international engagements. From 2009 to 2015, he served as head of MIT's Department of Nuclear Science and Engineering. Dr. Lester's research is concerned with innovation strategy and management, focusing most frequently on the energy and manufacturing sectors. He has led major studies of national and regional competitiveness and innovation performance commissioned by governments and industry groups around the world. He is the founding director and faculty chair of the MIT Industrial Performance Center. Dr. Lester is also widely known for his teaching and research on nuclear technology innovation, management, and control. He has been a long-time advocate of advanced nuclear reactor and fuel cycle technologies to improve the safety and economic performance of nuclear power, and his studies in the field of nuclear waste management helped provide the foundation for new institutional and technological strategies for dealing with this long-standing problem. Dr. Lester is the author or co-author of eight books, most recently *Unlocking Energy Innovation: How America Can Build a Low-Cost, Low-Carbon Energy System* (written with David Hart). He obtained his undergraduate degree in chemical engineering from Imperial College and his Ph.D. in nuclear engineering from MIT. He has been a member of the MIT faculty since 1979. He serves as an advisor to governments, corporations, foundations, and nonprofit groups, and is chair of the National Academies' Board on Science, Technology, and Economic Policy.

August W. Ritter was elected Colorado's 41st governor in 2006. During his 4-year term, he established Colorado as a national and international leader in clean energy by building a new energy economy. After leaving the Governor's Office, he founded the Center for the New Energy Economy at Colorado State University. The Center works with state and federal policy makers to create clean energy policy throughout the country. Governor Ritter authored the recently published *Powering Forward: What Everyone Should Know about America's Energy Revolution.*

James Rogers served most recently as chairman of the board for Duke Energy, having been elected in 2007. He also served as Duke Energy's president and CEO from 2006 until his retirement in 2013, following the company's merger with Cinergy, where he had served as chairman and CEO. Previously, he was chairman, president, and CEO of PSI Energy. Mr. Rogers is currently a director of Cigna Corp. and Applied Materials Inc. In 2010 and 2011, he was named by the National Association of Corporate Directors' *Directorship* magazine to its annual Directorship 100, recognizing the most influential people in corporate governance. He has advocated for investing in energy efficiency, modernizing the electric infrastructure, and pursuing advanced technologies to grow the economy and transition to a low-carbon future. Mr. Rogers serves as vice chairman of the World Business Council for Sustainable Development. He was chairman of the Edison Electric Institute when it changed its position to support federal climate change legislation in 2007. He was also founding chairman of the Institute for Electric Efficiency, a board member of the Alliance to Save Energy, and co-chair of the National Action Plan for Energy Efficiency. He serves on the boards of directors of the Institute of Nuclear Power Operations and the World Association of Nuclear Operators and on the board of Duke University's Nicholas Institute for Environmental Policy Solutions. He is a lifetime member of the Council on Foreign Relations, and in September 2011, United Nations Secretary-General Ban ki-Moon named him to a blue ribbon commission of business and nongovernmental organization leaders known as the High-Level Group on Sustainable Energy for All. Mr. Rogers attended Emory University and earned bachelor of business administration and JD degrees from the University of Kentucky.

Theodore Roosevelt, IV is a managing director in investment banking at Barclays, based in New York. Currently, he serves as chairman of the firm's Clean Tech Initiative and is a co-chair of Barclays Military Services Network. Mr. Roosevelt joined Barclays in 2008 when it acquired the North American assets of Lehman Brothers, for which he began working as a general banker in domestic corporate finance in 1972. By 1984, he had been named a managing director, and in 1991, he was asked to focus on the development of the firm's international business. He was elected chairman of the board of directors of Lehman Brothers Financial Products Inc. in 1994 and chairman of the board of directors of Lehman Brothers Derivative Products Inc. in 1998. In February 2007, he was appointed chairman of the firm's Council on Climate Change. Mr. Roosevelt received his A.B. from Harvard in 1965. After serving in the U.S. Navy, he joined the Department of State as a foreign service officer. In 1972, he received his MBA from Harvard Business School. Mr. Roosevelt is board chair of the Center for Climate and Energy Solutions, secretary of The Climate Reality Project, a member of the Governing Council of the Wilderness Society, and a trustee for the American Museum of Natural History. He was an advisory committee member on the MIT study *The Future of Natural Gas* and served on the advisory committee for the Council on Foreign Relations special report *The Future of US Special Operations*. He is a member of the Council on Foreign Relations and The Economic Club of New York and a governor of the Foreign Policy Association. He is chairman emeritus of the National League of Conservation Voters and served as trustee for Trout Unlimited and World Resource Institute.

Peter Rothstein is president of NECEC, which combines two sister nonprofit organizations—the Northeast Clean Energy Council, the lead voice for hundreds of clean energy companies across the Northeast, and NECEC Institute, a leader of programs in innovation and entrepreneurship and in industry research and development across the region. NECEC members and partners cut across dozens of cleantech sectors and stages, from start-ups to emerging companies and market segment leaders. NECEC is widely recognized for its regional innovation cluster initiatives, as well as stakeholder initiatives that bring together the region's cleantech entrepreneurs, companies, and supporters to advance the regional clean energy economy. Mr. Rothstein has 30 years of experience in cleantech venture and high-tech markets. Previously, he was part of the Flagship Ventures team, a leading seed and early-stage venture capital firm. He also founded Allegro Strategy, serving as a consultant, advisor, and executive with early-stage cleantech start-ups. Mr. Rothstein has served in early-stage deal or executive roles with a number of cleantech companies and has been involved in a range of leading cleantech and entrepreneurial organizations. Earlier, he was an entrepreneur and executive in the software industry, including as a Lotus/IBM vice president of strategy and leader of an internal Lotus incubator accelerating knowledge management ventures. He holds a master's degree from the MIT Sloan School of Management with a concentration in energy economics and a bachelor's degree in environmental design from Clark University.

Gary Roughead served as the U.S. Navy's 29th chief of naval operations after holding six operational commands. He is one of only two officers in the history of the Navy to have commanded both the U.S. Atlantic and Pacific Fleets. In retirement, Admiral Roughead is an Annenberg distinguished fellow at the Hoover Institution at Stanford University and serves on the boards of directors of the Northrop Grumman Corporation; Maersk Line, Limited; and the Center for a New American Security. He is a trustee of Dodge and Cox Funds and is on the board of managers of the Johns Hopkins University Applied Physics Laboratory. He advises companies in the national security and medical sectors.

Maxine L. Savitz is retired general manager for technology partnerships at Honeywell, Inc., (formerly AlliedSignal) and previously was general manager of AlliedSignal Ceramics Components. She was employed at the Department of Energy (DOE) and its predecessor agencies from 1974 to 1983 and served as deputy assistant secretary for conservation. Dr. Savitz served two terms (2006-2014) as vice president of the National Academy of Engineering, and has served on numerous committees of the National Academies. She serves on the board of the American Council for an Energy Efficient Economy and on advisory bodies for Pacific Northwest National Laboratory, Sandia National Laboratories, and Jet Propulsion Laboratory. She also serves on the MIT visiting committee for sponsored research activities. In 2009, she was appointed to the President's

Council of Advisors for Science and Technology. In 2013, Dr. Savitz was elected a fellow of the American Academy of Arts and Sciences, and she is also a fellow of the California Council on Science and Technology. Past board memberships include the National Science Board, Secretary of Energy Advisory Board, Defense Science Board, Electric Power Research Institute, Draper Laboratories, and Energy Foundation. Dr. Savitz's awards and honors include the Orton Memorial Lecturer Award (American Ceramic Society) (1998), the DOE Outstanding Service Medal (1981), the President's Meritorious Rank Award (1980), recognition by the Engineering News Record for Contribution to Construction Industry (1979 and 1975), and the MERDC Commander Award for Scientific Excellence (1967). She is the author of about 20 publications. Dr. Savitz earned a Ph.D. in organic chemistry from MIT in 1961.

Mark Williams (deceased) served as a member of the executive committee and downstream director of Royal Dutch Shell PLC from 2009 to 2012. He had previously held the positions of executive vice president, global businesses, and vice president of strategy, portfolio and environment for oil products. In 2004, he was appointed executive vice president of supply and distribution in Shell Downstream Inc., a position he held through 2008. He joined Shell in 1979 for Shell Oil Exploration & Production in the United States. He also served as engineering manager for Shell Offshore Inc.; operations manager for Shell Western EP Inc.; head of EP staff planning and head of downstream strategy for Shell Oil Co.; and vice president transportation for Equilon Enterprises LLC, the Shell and Texaco joint venture in the United States. He was a member of the board of visitors of the McDonald Observatory, University of Texas, and a trustee of Carleton College. He also chaired the Downstream Committee of the American Petroleum Institute. Dr. Williams had been chairman of the board and independent director of Hess Corporation since May 2013. He had also served as chairman of the executive committee of the Athabasca Oil Sands Project and chairman of the Downstream Committee of the American Petroleum Institute. Dr. Williams held a Ph.D. in physics from Stanford University.

Appendix B

Benchmark Levelized Cost of Electricity Estimates

Evaluation of policies for market adoption of clean energy for electricity generation begins with a view of the economic competitiveness of alternative technologies. Although the details will vary for myriad reasons for any particular installation, benchmark numbers are available to indicate the relative magnitude of delivered energy costs. The Energy Information Administration (EIA) has long experience and provides substantial supporting information regarding cost estimates used in the National Energy Modeling System (NEMS) and applied in the *Annual Energy Outlook* (AEO). This appendix summarizes the essential elements for consistent cost estimates as applied in AEO2016.

EQUIVALENT GENERATION COSTS

The NEMS includes a series of components that utilize information about the various contributions to the cost of electricity generation. The levelized cost of electricity (LCOE) is a summary benchmark statistic based on the elements in the NEMS:

> Levelized cost of electricity (LCOE) is often cited as a convenient summary measure of the overall competiveness of different generating technologies. It represents the per-kilowatt hour cost (in real dollars) of building and operating a generating plant over an assumed financial life and duty cycle. Key inputs to calculating LCOE include capital costs, fuel costs, fixed and variable operations and maintenance (O&M) costs, financing costs, and an assumed utilization rate for each plant type. The importance of the factors varies among the technologies. For technologies such as solar and wind generation that have no fuel costs and relatively small variable O&M costs, LCOE changes in rough proportion to the

> estimated capital cost of generation capacity. For technologies with significant fuel cost, both fuel cost and overnight cost estimates significantly affect LCOE. The availability of various incentives, including state or federal tax credits, can also impact the calculation of LCOE. As with any projection, there is uncertainty about all of these factors and their values can vary regionally and across time as technologies evolve and fuel prices change. (EIA, 2015f, p. 1)

The AEO supporting information identifies the methodology and assumptions that affect the reported estimates of LCOE for utility-scale generation technologies. The reported estimates are for the years 2022 and 2040. The focus here is on the 2022 estimates as the benchmark for the "current" costs. The assumptions include choices regarding the effects of learning, capital costs, transmission investment, operating characteristics, and externalities. These choices are both important and appropriate for the benchmark comparison (e.g., learning rates), are important and require some adjustment (e.g., capital costs), or are supplemental to the EIA assumptions (e.g., externality costs).

LEARNING RATES

The cost estimates include detailed construction of the components of overnight capital cost of investment (EIA, 2013d). An important feature of the capital cost estimates reflects the maturity of the technology. For mature technologies, such as conventional coal, there is a minimum exogenous annual cost reduction. For newer technologies, such as wind and solar, the technologies have different learning rates in the range of 1-20 percent (EIA, 2015b, p. 107). For the first four units of a technology, there is an assumed increase in the costs above the base cost estimate to capture the effect of first-of-a-kind units. The learning rates applied to the base cost estimates decline after the first three and subsequent five doublings of installed capacity. The cost of competing technologies increases, particularly for natural gas. The result is a different ranking of the technologies in 2040 than in 2022, with renewables being more competitive in 2040.

CAPITAL CHARGES

The NEMS model supports many purposes, from projection to analysis. For projection applications, a constant issue arises as to what policies to assume for the future. In general, the practice is to assume that existing law would determine future policies. Hence, if there is a production tax credit for wind that is scheduled to expire before 2022, it is not included in the 2022 estimates, even though the tax credit has been renewed many times before.

This is an appropriate compromise regarding policy assumptions for projection. However, in constructing a benchmark for comparing the underlying competitiveness of technologies, the starting point should be the cost estimates without included policies that selectively target technologies. Ideally, the levelized cost estimates would be computed in a way that eliminates the impact of policies that cause market distortions, such as those that preferentially subsidize one technology or a class of technologies over others. A second-best option would be to include the effects only of policies that are technology-neutral, such as the income tax.

The cost estimates in AEO2016 include two important assumptions reflecting selective policies that affect the capital cost estimates. First, the weighted average cost of capital (WACC) is 5.6 percent in real terms. But for new coal plants, there is an additional 3 percent added to proxy for anticipated carbon restriction policies (EIA, 2015f, p. 3). Second, certain technologies, particularly wind and solar, use accelerated tax depreciation that is not available to other technologies. This produces substantially lower fixed-charge rates for renewable capital costs.[1]

Without these selective polices, the fixed-charge rates would be closer and reflect only real differences in component costs and lead times for production that affect capital costs. The assumed lead times for the technologies appear in the AEO assumptions (EIA, 2015b, p. 105). The capital charge rates applied here are for the comparable fossil fuel technology that has the same lead time. Thus the capital costs reported differ from those in EIA (2015f) in removing the 3 percent premium for coal and the accelerated depreciation for selected technologies.

TRANSMISSION INVESTMENT

In general, the technologies differ in terms of the transmission investment required to serve the final load. Potential wind and solar capacity factors can differ substantially by location. Often a wind or solar resource is geographically distant from the demand load and necessitates a transmission investment, such as an upgrade to existing infrastructure or a new transmission line. In contrast, a natural gas or coal power plant can usually be sited as close to the demand load as possible, given other factors such as the cost of land and social factors. Even in the case of local versus distant renewables, there is a well-recognized trade-off between optimal performance of the renewables and the cost of transmission investment (Midwest ISO, 2010). AEO2015 incorporates estimates of the

[1]Personal communication, EIA, spreadsheet AEO2014_financial.xls.

required transmission investment, and these costs are included in the LCOE (EIA, 2015f, p. 6).

DISPATCH PROFILES

EIA separates electricity generation technologies into categories of dispatchable and nondispatchable (EIA, 2015f, p. 6). The former include conventional fossil fuel plants that have a fairly consistent available capacity and can follow dispatch instructions to increase or decrease production. The latter consist of intermittent plants such as wind and solar, which depend on the availability of the wind and sunlight and typically cannot follow dispatch instructions easily or at all. It is generally recognized that the different operating profiles create different values for the technologies (Borenstein, 2012; Joskow, 2011). Empirical estimates for existing technologies show that the value of wind, which blows more at night when prices are low, can be 12 percent below the unweighted average price of electricity; and the value of solar, with the sun tending to shine when prices are higher, can be 16 percent greater than the unweighted average (Schmalensee, 2013).

One procedure utilized for putting nondispatchable technologies on an equivalent basis is to pair them with appropriately scaled dispatchable peaking technologies to produce an output that is like that of a conventional fossil fuel plant (Greenstone and Looney, 2012). Another approach, used by Schmalensee (2013), is to calculate the value of nondispatchable technologies based on spot prices. EIA provides a similar estimate based on its projected simulations, which is known as the levelized avoided cost estimate (LACE):

> Conceptually, a better assessment of economic competitiveness can be gained through consideration of avoided cost, a measure of what it would cost the grid to generate the electricity that is otherwise displaced by a new generation project, as well as its levelized cost. Avoided cost, which provides a proxy measure for the annual economic value of a candidate project, may be summed over its financial life and converted to a stream of equal annual payments. The avoided cost is divided by average annual output of the project to develop the "levelized" avoided cost of electricity (LACE) for the project. (EIA, 2015f, p. 2)

For purposes of equivalent comparison of the LCOE, the approach here combines these adjustments to provide an estimate of the net difference between the LACEs for the technology and for a conventional combined-cycle natural gas plant. The net differences are added to (e.g., for wind) or subtracted from (e.g., for solar) the other components of the LCOE.

EXTERNALITIES

The LCOE excluding marginal externalities provides a benchmark for comparison. However, a central point of the consideration of clean energy technologies is the impact of externalities. Hence, in addition to the equivalent cost estimates absent policies for pricing externalities, the estimates here incorporate separate components for the major externalities.

The principal externalities for electricity generation include the criteria pollutants and carbon dioxide (CO_2). The supporting documents for AEO2016 provide emission rates and associated heat rates (EIA, 2013d, pp. 2-10). For the criteria pollutants—such as sulfur dioxide, nitrogen oxide, and small particulates—the externality values are highly dependent on location because of different exposure effects, and are difficult to quantify with a benchmark value. However, the basic story for noncarbon externalities is relatively simple. The impacts are an order of magnitude larger for coal than for natural gas. By one estimate, the noncarbon health impacts for coal are greater than the value added of the coal sector (Muller et al., 2011). For illustrative purposes, however, the estimates here include the noncarbon damages for coal and natural gas plants from Greenstone and Looney (2012), using the emission rates of the criteria pollutants.

For carbon, the locational differentials do not pose any difficulty because of the nature of the global effects of emissions anywhere. Here emission rates by technology type come from the heat rate and emission data in AEO2016 (EIA, 2013d).[2] The price of carbon is set at the round number of \$15/ton of CO_2, taken as representative of the 3 percent discount case of the Interagency Working Group on Social Cost of Carbon (2013). Note that this is a global externality cost estimate. Looking only at the damages for the United States would reduce the social cost by about an order of magnitude. The \$15 figure is selected as a round number to make it easy to convert the graphic for other arguable values of the appropriate prices of carbon.

LEVELIZED COST OF ELECTRICITY

With the above assumptions and adjustments to obtain an approximation of equivalent LCOE, the results appear in Figure B-1 and Table B-1.

It is clear from Figure B-1 that new natural gas plants are the dominant technology. And without accounting for the costs of externalities, new IGCC

---

[2]The EIA data show carbon emissions for biomass and geothermal. The NEMS model assumes zero carbon emissions, presuming that biomass fuel recycles the carbon. Geothermal is treated here as zero-carbon.

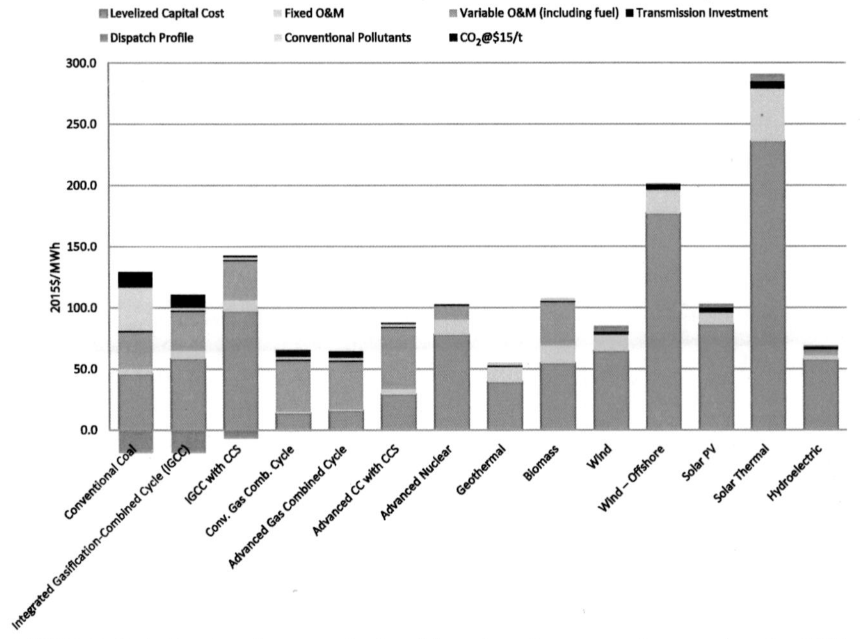

FIGURE B-1 Levelized cost of electricity for plants entering service in 2022 (2015 $/MWh).
SOURCE: EIA, 2015f, 2016g. Because *Annual Energy Outlook 2016* does not assess conventional coal and IGCC technologies, their values (in 2013 dollars) were sourced from *Annual Energy Outlook 2015* and then converted to 2015 dollars using the Bureau of Economic Analysis' gross domestic product (GDP) implicit price deflator.

coal plants are more competitive than even the best of the wind and solar. Onshore wind is the closest to being competitive. But the relative cost estimates shown here are similar to those in Greenstone and Looney (2012). The primary renewable technologies are not cost-competitive, and the differences are significant. This is for entry year 2022. Looking ahead to 2040, with some additional cost reductions for renewables and more substantial increased fuel costs for natural gas, the situation changes for wind but not for solar.

TECHNOLOGY ADOPTION

The EIA projections include many factors other than the implied LCOE. For example, although geothermal and conventional hydroelectric technologies are among the most cost-competitive, their penetration is limited because of constrained resources. The AEO2016 Reference Case shows the projected growth in electricity generation and the relative role of the various

TABLE B-1 Summary of Levelized Cost of Electricity (LCOE) for Year 2022 Entry (2015 $/MWh)

Plant Type	Levelized Capital Cost	Fixed Operations and Maintenance (O&M) Costs	Variable O&M Costs (including fuel)	Transmission Investment	Dispatch Profile	Criteria Pollutants	CO_2 @ $15/ton	Total System Average LCOE
Conventional Coal	45.9	4.3	30.2	1.2	-18.2	35.0	12.3	111.0
Integrated Gasification-Combined Cycle (IGCC)	58.4	7.1	31.5	1.2	-18.2	2.0	10.5	92.6
IGCC with Carbon Capture and Storage (CCS)	97.2	9.2	31.9	1.2	-6.5	2.0	1.2	136.2
Conventional Gas Combined Cycle	13.9	1.4	41.5	1.2	0.0	2.0	5.4	65.5
Advanced Gas Combined Cycle	15.8	1.3	38.9	1.2	0.0	2.0	5.1	64.3
Advanced Combined Cycle with CCS	29.2	4.3	50.1	1.2	0.0	2.0	0.6	87.5
Advanced Nuclear	78.0	12.4	11.3	1.1	-0.3	0.0	0.0	102.5
Geothermal	38.9	12.6	0.0	1.4	0.2	2.0	0.0	55.2
Biomass	54.7	14.9	35.0	1.2	-0.1	2.0	0.0	107.8
Wind	64.6	13.2	0.0	2.8	4.4	0.0	0.0	85.0
Wind—Offshore	177.0	19.3	0.0	4.8	0.2	0.0	0.0	201.3
Solar Photovoltaic (PV)	86.2	9.9	0.0	4.1	2.9	0.0	0.0	103.1
Solar Thermal	235.9	43.3	0.0	6.0	5.6	0.0	0.0	290.8
Hydroelectric	57.5	3.6	4.9	1.9	0.9	0.0	0.0	68.8

SOURCE: EIA, 2015f, 2016g. Because *Annual Energy Outlook 2016* does not assess conventional coal and IGCC technologies, their values (in 2013 dollars) were sourced from *Annual Energy Outlook 2015* and then converted to 2015 dollars using the Bureau of Economic Analysis' gross domestic product (GDP) implicit price deflator.

technology groupings (see Figure B-2). Importantly, the total growth of electricity is modest by historical standards, and this inhibits the introduction of new technology.

In part, the renewable growth is driven by implicit policies, such as the capital cost add-on for coal or explicit policies such as renewable portfolio standards. With the low cost for new facilities, natural gas shows the largest growth rate. Coal and nuclear remain essentially constant. There is modest growth in renewables.

The breakout of the Reference Case projections for renewables shows that hydroelectric stays about constant, but the other renewables grow at modest rates (see Figure B-3). The penetration of renewables is important but substantially below the levels that would be indicated by an aggressive program to address remaining health effects and the challenges of global warming.

CONCLUSION

Equivalent estimates of the LCOE are available from the supporting analyses of AEO2016. The data without the effect of selective policies indicate that existing technologies for clean energy are not competitive with new natural gas. And without accounting for the costs of externalities, the principal renewable technologies of wind and solar are not cost-competitive with new coal plants.

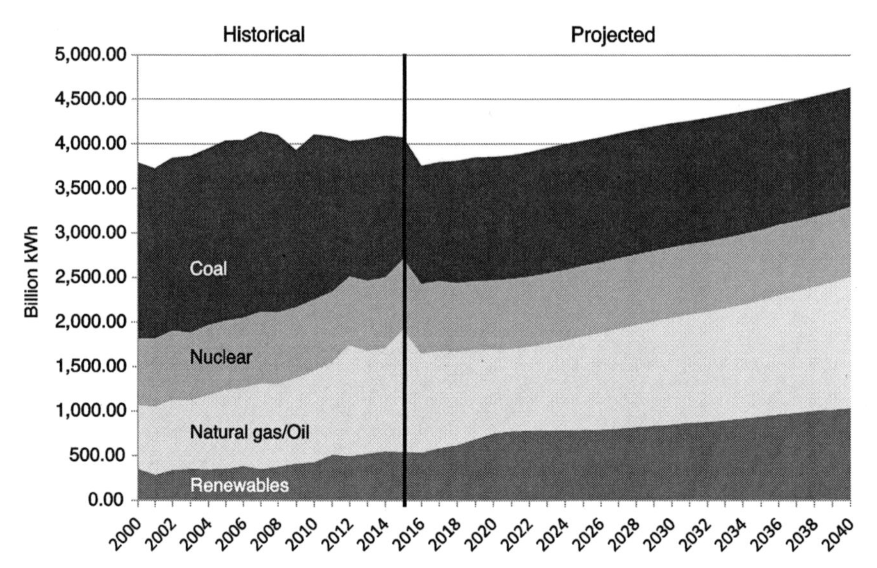

FIGURE B-2 Electric power generation by fuel (billions of kilowatt hours [kWh]) assuming No Clean Power Plan, 2000-2040.
SOURCE: EIA, 2016f, Figure IF3-6.

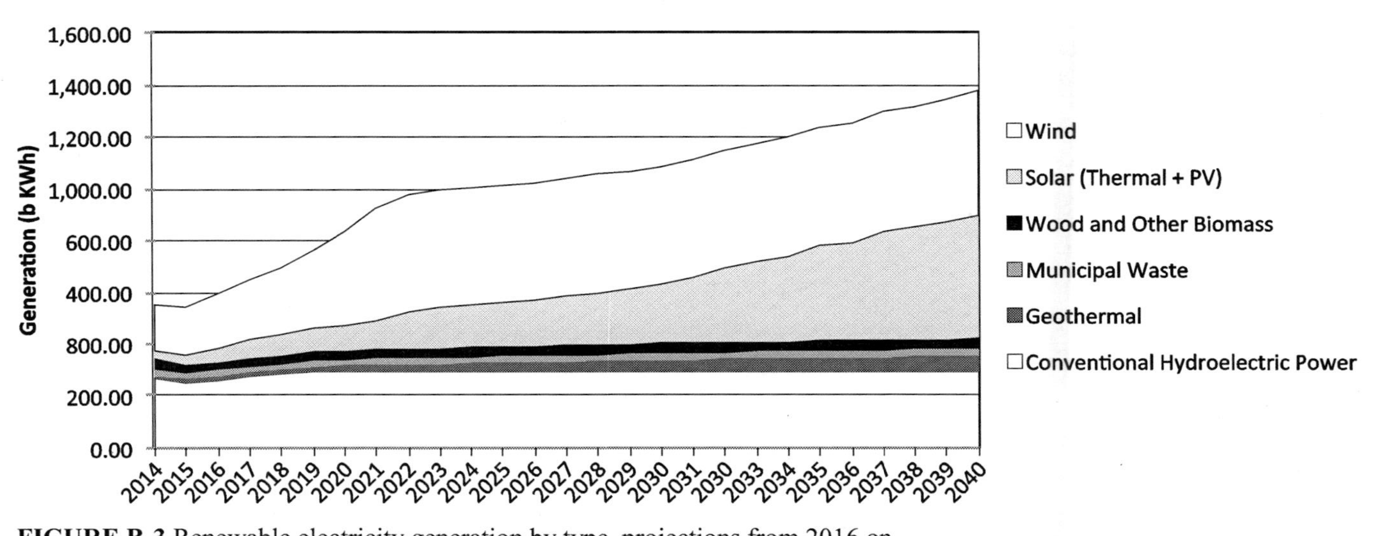

FIGURE B-3 Renewable electricity generation by type, projections from 2016 on.
SOURCE: Renewable Energy Generating Capacity and Generation, Reference Case Tables (EIA, 2016a, Table 16).

Appendix C

The Role of Research, Learning, and Technology Deployment in Clean Energy Innovation

"To avoid the most harmful effects of rising greenhouse gas concentrations while still meeting the growing demand for affordable and reliable energy services, nothing less than a fundamental transformation of current patterns of energy production, delivery, and use on a global scale will be required" (Lester and Hart, 2012, p. 2).

"The chances are slim that technology can do the job without government policies mainly because the required changes are so large....Achieving the emissions reductions contemplated by U.S. policy and consistent with the Copenhagen 2° C temperature target would require more rapid technological shifts than have been seen in almost any industry." (Nordhaus, 2013, p. 276)

"Accelerating innovation so as to decarbonize one of the least innovative sectors of the entire economy is a major policy challenge" (Grubb et al., 2008, p. 333).

SCALE OF THE CHALLENGE AND NEED FOR CORRESPONDINGLY SCALED INNOVATION EFFORT

The scale of the challenge to develop and deploy clean energy technology underscores the importance of investment in research and development (R&D) to discover and develop transformative innovations. Deployment of viable existing technologies is important, but will be far from enough to meet the global challenge, particularly in controlling the concentrations of greenhouse gases in the atmosphere.

A portfolio of public policies will be needed to support innovation at all stages of the process. An aspect of the necessary portfolio is the balance of effort among several types of initiatives to support innovation. These initiatives include everything from basic research on risky but potentially transformative technologies to support for deployment of clean technologies that may not yet be commercially viable. The purpose here is to address basic elements of the policy balance between R&D and large-scale uptake and deployment of more expensive technologies, and the interactions with pricing of carbon and other pollutant emissions.

MARKET FAILURES AND EXTERNALITIES

A primary argument for public policy in clean energy innovation is to overcome market failures and to internalize the costs of major externalities (Popp et al., 2010). There are many externalities in energy production and use. The most significant are the emissions of criteria pollutants, especially for small particulates, and the emissions of carbon dioxide and other greenhouse gases (Muller et al., 2011). The treatment of emission externalities interacts with the recognition of other market failures (NRC, 2010b). An example of an overview of market failures and the associated policy instruments appears in Table C-1 (Gillingham and Sweeney, 2010).

Although not exhaustive, this overview illustrates the interaction of policies and problems. On some dimensions, such as regulation or the treatment of market power, the policies may be considered relatively independently. On other dimensions, there is a strong interaction. For instance, the policy for internalizing the cost of carbon emissions has a major impact on the scope and design of policies for supporting technology innovation and deployment of clean energy technologies that address the problem of global warming.

A particular focus is on the benefits of early deployment. It is clear that investment in R&D, to include everything from discovering new technologies to deployment of first-of-a-kind plants, presents a requirement for public support to overcome the traditional market failures associated with spillover effects (Nadiri, 1993). What is not as clear, however, is how much public policy should go beyond this traditional realm of R&D to support large-scale deployment of clean energy technologies that would not be adopted absent public support. There is a widely accepted view that deployment produces the benefits of learning and subsequent cost reduction for new technologies, and this learning benefit can be difficult for investors to capture in the market (Popp et al., 2010, p. 895). In principle, the learning spillover is a type of externality that justifies a policy to support deployment. The issue is how much, and how the learning benefit interacts with other policies.

TABLE C-1 Sources of Market Failure and Some Illustrative Potential Policy Instruments

Market Failures \ Some Policy Instruments	Direct regulation	Direct government-sponsored R&D	Competions, such as X prize	R&D Tax Incentives	Excise taxes	Production Subsidies	Feed-in tariffs	Information programs	Product standards	Cap-and-trade	Marketable Market-Wide Standards	Transparency Rules	Macroeconomic Policy	Corporate Taxation Reform	Competition Policy/Laws	Restructured Regulation	Intellectual Property Law
Labor supply/demand imbalances						T							T				
Environmental externalities	P	P	P	P	P	P	P	P	P	P	P	P					
National security externalities	P	P	P	P	P	P	P	P	P	P	P	P					
Information market failures								P	P			P					
Regulatory failures																P	
Too High Discount Rates				P									P	P			
Imperfect foresight								P									
Economies of scale						T	T								P		
Market power									P						P	P	
R&D Spillovers		P	P	P											P		P
Learning-by-Doing						T	T										
Network Externalities						T									P		

NOTE: "P" indicates permanent change or instrument; "T" indicates transient instrument.
SOURCE: Gillingham and Sweeney, 2010, p. 81.

LEARNING AND INNOVATION

The extensive literature on learning incorporates a variety of pathways for innovation. These pathways touch the innovation stages in different ways and have different policy implications. Figure C-1 shows the stages of the innovation process and key obstacles to accelerating innovation at each stage that are important policy targets.

For example, at the earliest stages of fundamental research and proof of concept, there is a consensus that research has both very large spillover externalities and very large social returns (Popp et al., 2010, pp. 896-898). The policy implication calls for a variety of ways to provide public support for

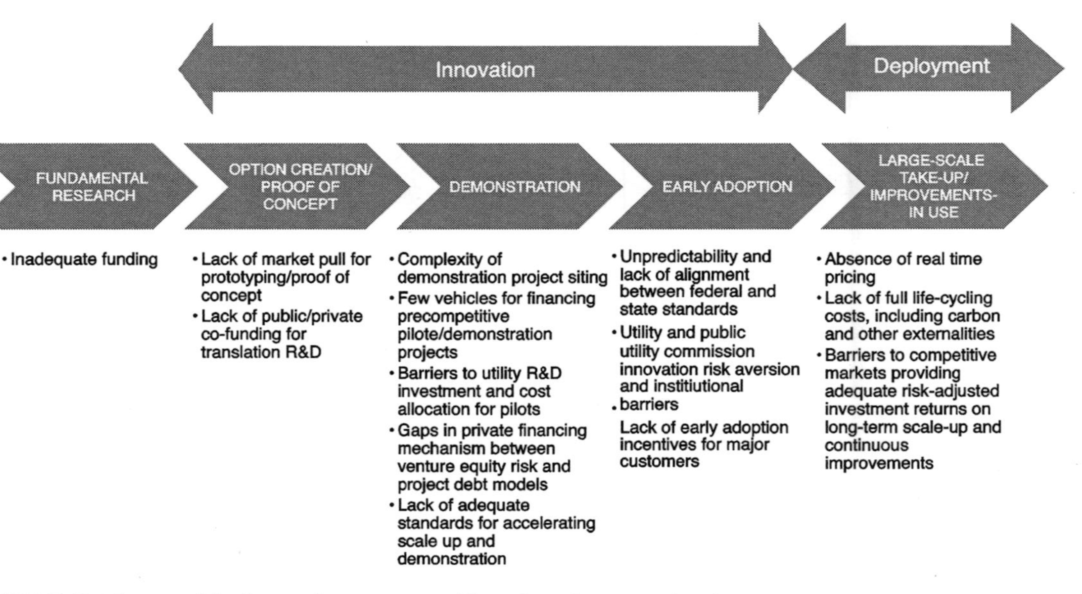

FIGURE C-1 Stages of the innovation process and key obstacles to acceleration.
SOURCE: Adapted from Lester and Hart, 2012, Figure 2.1, p.33.

fundamental research and proof of concept. In the energy context, government support of basic research and programs such as the Advanced Research Projects Agency-Energy (ARPA-E), pursuing risky innovations with potentially high payoffs and large spillovers, fits naturally at this early stage of R&D.

The more expensive "muddle in the middle" for demonstration and first-of-a-kind commercial projects presents a bigger challenge. The costs and risks of the individual efforts at this stage are much larger, and the inevitable failures can be more visible and controversial. Furthermore, the incentives for early adoption are in part connected to the incentives for large-scale take-up/improvements in use. Absent a sufficient ultimate market, even a market with very risky payoffs, the normal model for venture capital to take projects beyond proof of concept and demonstration would not apply. The venture capital firm is a for-profit activity, and without a high potential market, there is not enough profit to match the risk in innovation. Thus there is another area in which there may be high social returns where the spillovers and risks call for government support in the middle stages to include the first few plants of a new and unproven technology that have a low probability of success. For these stages before large-scale deployment, innovation is the product, and clean energy is the by-product.

A related but different question addresses the degree to which there should be public policy support, and the form of policy most likely to produce results, at the later stage of large-scale market penetration, where clean energy is the principal product, and innovation through cost reduction is a by-product. How large and of what types are the aggregate social returns for diffusion of innovative clean energy technologies? How much of this benefit can be captured by market participants, and how much spills over and leaves a market failure amenable to government policy intervention?

Abstracting from many of the details of the innovation process, the focus here is on models used to quantify the benefits of research and learning and the interaction with large-scale uptake of clean technology. The literature includes a variety of frameworks for addressing these questions. The terminology is not fully settled, but the basic ideas have been widely discussed (Junginger et al., 2010). For present purposes, the arguments for learning and the connection with large deployment make distinctions about different types of cost reductions or possible spillover effects:

- Learning by (re)searching (LBS): This is R&D broadly writ. The most important features include an intentional and costly effort to develop innovations. Typically, the R&D is risky, and there are large spillover effects. The goal is to develop the innovation until it is at or near the stage of large-scale deployment.
- Economies of scale: This represents the standard argument about decreasing unit production costs as a production plant reaches an efficient size. In the energy market, this is unlikely to be a market-

wide phenomenon—i.e., the case of natural monopoly—but may be relevant at the firm level (Gillingham and Sweeney, 2010).

- Learning by doing (LBD): Early production produces new information that reduces the cost of future production. Critical characteristics include that LBD is passive (Thompson, 2010) in the sense that it is treated as a free by-product of deployment rather than an explicit costly product of the deployment. There can be significant spillover effects.
- Learning by using (LBU): This is the demand-side counterpart of LBD, often not treated separately from LBD.
- Learning by waiting (LBW): The spillover effects from other industries, technologies, or countries are essentially exogenous—i.e., developed from outside—from the perspective of the firm in the present industry (Thompson, 2010). The resulting benefits from the innovative technology will appear over time and can be exploited by waiting.

This list is not exhaustive, but it covers enough of the aggregate pathways to characterize some of the key features of policy trade-offs. The focus is on the long-term character of the learning process. For example, there is no doubt that there can be a great deal of learning during the early shakeout period of any new production facility. There is evidence, for instance, of significant learning in the early stages of production of new cars, but the learning quickly reaches a sustainable cost level (Levitt et al., 2013; Thompson, 2010).

Similarly, economies of scale are not likely to produce the type of externality that justifies a policy intervention:

> The distinction between learning-by-doing and economies of scale may seem minor, but the implications for public policy are immense. If one firm can drive down its costs by producing at large scale in its factory or its installation operation, those benefits are highly appropriable by that large firm. Smaller firms are not likely to experience a cost decline because a competitor is enjoying economies of scale. Thus, significant economies of scale in any industry, short of creating a natural monopoly, are not generally seen as a basis for government intervention. (Borenstein, 2012, p. 83)

While an important benefit, the economies of scale apply only for very early deployment, and there may be little or no spillover effect that creates a market externality or a justification for public policy.

Hence, the relevant question is the degree of learning after the initial R&D stages, where the major benefits of initial shakeout of the first-of-a-kind plants have already been achieved, and production economies of scale are already being exploited. The immediate task is to integrate the roles of the other forms of learning and consider the implications for public policy.

LEARNING BY WAITING AND DOING

A key feature of LBD is the requirement of large-scale penetration of the market. LBS may occur with or without accompanying large-scale deployment. There may be important feedback effects between deployment and active research initiatives. For the present discussion, however, the focus is on the independent effects of LBD (and LBU). From this perspective, the impacts of LBS and LBW share the common feature that they do not depend on large-scale deployment of the innovative technology.

The importance of LBD has long been recognized (Arrow, 1962). "The empirical literature on firm progress curves is distressingly large…" (Thompson, 2010, p. 446). A standard model incorporates an experience curve that relates cumulative production to the cost of new production. As new production occurs, there is a knowledge by-product that reduces the cost of future capacity. The lower cost of the future production is the premium associated with LBD. If there is a great deal of learning, then the premium may justify early large-scale deployment.

A Learning Curve Model

In Nordhaus (2014), a representative version of an experience or learning curve model for a given technology, for a firm or an industry, illustrates the central components and key parameters. New production of clean energy in period t is Q_t. The associated cost of production is simplified to a constant marginal cost C_t. Cumulative production is Y_t, where

$$Y_t = \int_{v=-\infty}^{t} Q_v dv .$$

Nordhaus includes the effect of all learning that does not require deployment as represented in an exogenous trend where costs decrease at rate h. This exogenous trend provides benefits by waiting until costs come down before undertaking large-scale deployment.

The learning effect appears as the elasticity b that captures the percentage reduction in costs associated with an increase in experience, measured as cumulative production. Nordhaus effectively normalizes units so that $Y_0 = 1$; here it is convenient to retain Y_0 as a scaling factor. Hence, the stylized model of LBD represents the marginal cost of new production as

$$C_t = C_0 Y_0^b e^{-ht} Y_t^{-b}.$$

This is the experience or learning curve model, which has a long history. Without the effect of the exogenous rate h, this would be the original single-factor experience curve (Thompson, 2010). Including a possible nonzero exogenous rate makes this the simplest two-factor experience or learning model.

The model is linear in logarithms of cost, a time trend for exogenous change, and cumulative experience that drives LBD.

$$\ln C_t = \ln C_0 + b \ln Y_0 - ht - b \ln Y_t \ .$$

The description of learning can be summarized by the progress ratio (PR), which is the ratio of costs after a doubling of cumulative production.

$$PR = 2^{-b}.$$

The related learning rate (LR) is one minus the PR.

$$LR = 1 - PR = 1 - 2^{-b}.$$

Hence, with an LR of 20 percent, a doubling of cumulative production results in a 20 percent reduction in future production cost, or a PR of 80 percent. Table C-2 summarizes the connection between learning rates and the associated cumulative output elasticity b.

TABLE C-2 Learning Curve Rates

LR (%)	PR (%)	b
5	95	0.07
10	90	0.15
15	85	0.23
20	80	0.32
25	75	0.42
30	70	0.51
35	65	0.62
40	60	0.74
45	55	0.86
50	50	1.00

SOURCE: Nordhaus, 2014.

Single-factor models across a range of industries yield LRs dispersed around 20 percent (Thompson, 2010). However, the dispersion of LRs across industries is wide, and within industries and technologies, the LR can vary over time and be quite different for technologies of different types and stages of development. Early adoption may even show increasing costs for first-of-a-kind plants (Rubin et al., 2007b). Consistent with this literature, EIA assumes cost increases for the first-four-of-a-kind plants (EIA, 2014b). Figure C-2 provides the best graphic illustration of both phenomena from a study that examined LRs for a range of energy technologies, including the prominent sources of clean energy. LRs can appear to be negative at certain stages. And the relative cost of a prominent competing fossil fuel technology creates a receding target for clean energy technologies.

Nordhaus describes the problem of separating out the effect of accumulating experience from independent factors that change over time (Nordhaus, 2014). In an important and oft-cited paper (Nemet, 2006), Nemet provides a disaggregated study of the development of photovoltaic (PV) technology. Nemet separates PV costs into several components and addresses the evidence for cost reduction in each of the components: "The evidence presented here indicates that a much broader set of influences than experience alone contributed to the rapid cost reductions in the past" (Nemet, 2006, p. 3230). A similar result for Chinese experience in producing wind generators

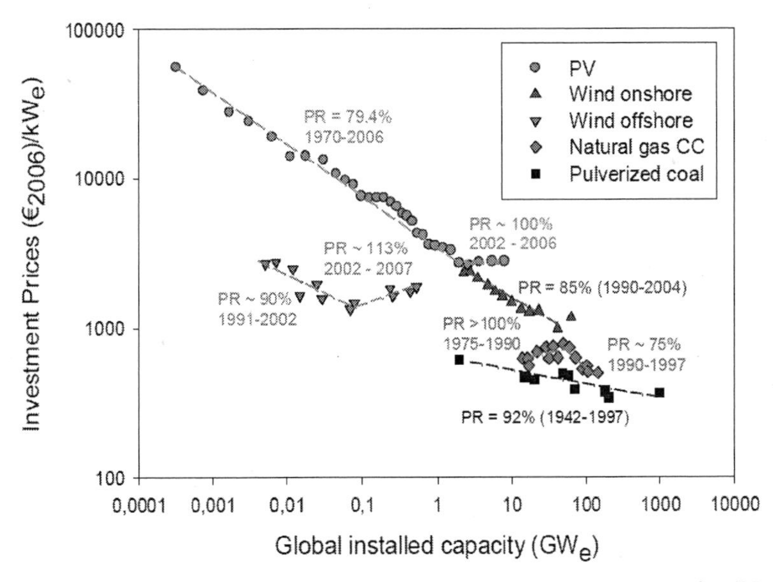

FIGURE C-2 Comparison of historical experience curves and progress ratios (PR = 1 - learning rate) of energy supply technologies.
SOURCE: Junginger et al., 2008, p. 10.

appears in Qiu and Anadon (2012), where there were large cost reductions over time, but the LBD rate estimate is only 4 percent.

As is usual with effects of compound growth, extrapolations of learning benefits are sensitive to small initial rate errors. "However, especially for long-term forecasts, small variations in PRs can lead to significantly deviating cost reductions in scenarios or completely different model outcomes in energy and climate models" (Junginger et al., 2008, p. 13; see also Wiesenthal et al., 2012). Nordhaus builds on related observations to make the point that small errors in the estimated LR can produce large biases in the forecast of future costs and the associated premium that might be the target of a deployment subsidy motivated by LBD. Furthermore, and importantly for present purposes, Nordhaus finds that reasonable LRs imply relatively small learning premiums.

The problem of empirical estimation complicates any application of the resulting model for the forecast of costs of deployment (Söderholm and Sundqvist, 2007). Furthermore, even if the estimation problems are set aside, translating LR estimates into policy prescriptions requires a better understanding of the impacts of the key assumptions. How large is the learning premium? And how does the learning premium interact with other policies?

Fixed Clean Deployment

Following Nordhaus (2014) in an application of simplicity to improve insight, a basic model provides a pedagogical tool. The purpose is to understand the primary elements of an estimate of any learning premium. The basic model employed by Nordhaus starts with an exogenous forecast of future clean technology production, Q_t. Implicitly, the analysis treats the cost of capacity increases as the present value of all future output of that capacity. Given this fixed trajectory of output, with discount rate (r), the present value of the future clean production is given by

$$V_0 = \int_{t=0}^{\infty} Q_t C_t e^{-rt} dt = \int_{t=0}^{\infty} Q_t \left[C_0 Y_0^b e^{-ht} Y_t^{-b} \right] e^{-rt} dt.$$

Hence, with an immediate production increment of θ,

$$V_0(\theta) = (Q_0 + \theta) C_o + \int_{t=0}^{\infty} Q_t \left[C_0 Y_0^b e^{-ht} (Y_t + \theta)^{-b} \right] e^{-rt} dt.$$

With exogenous Q_t, the resulting marginal cost of the incremental production is

$$V_0'(\theta) = C_0 - b\int_{t=0}^{\infty} Q_t C_t Y_t^{-1} e^{-rt}\, dt.$$

The resulting difference between the current marginal cost (C_0) and the total marginal cost is the learning premium:

$$\pi_I = b\int_{t=0}^{\infty} Q_t C_t Y_t^{-1} e^{-rt}\, dt$$

This result shows that the premium is declining in cumulative production and increasing in the volume of future production. Using a further simplification by assuming a constant growth in the rate of deployment, the model yields a convenient closed-form solution. See Nordhaus (2014) for details. With a 3 percent deployment growth rate and zero exogenous reduction in cost, the Nordhaus solution provides a learning premium summarized here as a percentage of the marginal cost of clean production, C_0. The Nordhaus learning premium as a function of the assumed discount rate and LR is shown in Table C-3.

Table C-3 illustrates that the LBD premium is sensitive to the LR and discount rate assumptions, and the premium is relatively low. Clean energy subsidies justified as necessary to jump start deployment and bring the cost down have often been much higher. Subsidies set high enough to make clean technologies competitive with dirty technologies that cost half as much or less would require a premium of more than 50 percent of the clean technology cost.

Variable Clean Deployment

The Nordhaus model provides an important set of insights, but raises immediate issues that appear to be relevant to an evaluation of the future role of LBD. First, the illustrative calculation assumes zero exogenous technological change or other improvements that are independent of cumulative production, which is not consistent with both conventional logic and available empirical

TABLE C-3 Learning by Doing (LBD) Premium

	Discount Rate (%)		
LR	**3%**	**5%**	**10%**
0	0	0	0
10	13	8	4
20	24	16	9
30	34	24	13

SOURCE: Nordhaus, 2014.

work. Second, the single-factor experience curve does not incorporate the lesson that different components of the technology may have different LRs. Third, the model does not connect to the other benefits of clean technology, most notably in reducing or eliminating the environmental externality of the competing dirty technologies. Fourth, the assumption of constant growth of future production is convenient analytically but not likely to be a good description of the rate of diffusion, which could have faster growth rates once the technology became commercial. Fifth, the difference between social cost and current marginal cost may in part be subject to capture by firms; only the spillover that cannot be captured is an externality that could be the target for a policy intervention. Each of these issues could have a material impact on the estimation of the LBD premium.

Addressing these issues allows for a minimal extension of the Nordhaus model without losing too much transparency. The premium calculations can be done with a nonzero assumption about the rate of exogenous technological change. This is straightforward but has an important interaction with the analysis of the rate of clean technology diffusion. In particular, with a nonzero rate of exogenous technological change (or R&D investments in LBS), the best policy might be to learn by waiting; that is, wait until the costs have declined enough to make the other premiums justify large-scale deployment (Montgomery and Smith, 2007; Santen et al., 2014).

The simple learning model implies that with enough cumulative production, the unit cost of new production will be driven to zero. Although going this far would be extreme, as will be seen, this is not an innocuous assumption. Clearly there is some positive lower bound on the cost. This could be included as a simplified version of a component model where the total cost is divided into two parts, one of which is amenable to technological change and learning, and the other of which is the long-run lower bound on the total cost, $C_{\min}$ (Manne and Richels, 2004; Neuhoff, 2008).

The treatment of the benefits of substitution for dirty technologies could be incorporated by assuming an exogenous growth rate for the total new technologies and allowing introduction of the clean technology when it is competitive, including accounting for the effect of the negative dirty technology externality and the positive clean technology learning externality. This could be seen as a cost-effectiveness analysis for the electricity sector, without accounting for any impacts of higher electricity prices and reduced electricity consumption.

Some of the benefits of learning can be captured by those deploying the technology. For example, the common argument about improved international competitiveness through early deployment is implicitly an argument that a large portion of the benefits is captured and does not spill over to others. The capture of learning benefits can be set as a parameter ρ. The component that is a spillover is $1-\rho$, following Fischer and Newell (2008).

With these extensions, the experience curve for clean technologies would modify the Nordhaus model to yield

$$C_t = \left(C_0 - C_{\min}\right) Y_0^b e^{-ht} Y_t^{-b} + C_{\min}.$$

The present value of future costs is

$$V_0 = \int_{t=0}^{\infty} Q_t C_t e^{-rt} dt = \int_{t=0}^{\infty} Q_t \left[\left(C_0 - C_{\min}\right) Y_0^b e^{-ht} Y_t^{-b} + C_{\min}\right] e^{-rt} dt.$$

Therefore, the marginal cost of an immediate increment of production becomes

$$V_0'(\theta) = C_0 - b \int_{t=0}^{\infty} Q_t \left(C_t - C_{\min}\right) Y_t^{-1} e^{-rt} dt.$$

Incorporating the uncaptured spillover effect, the corresponding LBD premium becomes

$$\pi_{II} = (1-\rho) b \int_{t=0}^{\infty} Q_t \left(C_t - C_{\min}\right) Y_t^{-1} e^{-rt} dt$$

The assumption of exogenous deployment of clean technology can be replaced with a dynamic optimization model. For instance, let the exogenous growth of total power demand be D_t. Although learning is not confined to new or clean technologies (Ordowich et al., 2012), for simplicity the dirty technology represented here has no learning and is described by an exogenous price P_t and an externality cost E_t. Again, for simplicity, assume that there is no emission externality cost for the clean technology. Ignoring uncertainty and all the other constraints in the market, the minimalist deterministic discrete dynamic optimization finds the deployment of clean and dirty technology to minimize the total social costs of a given growth path for new capacity production.

$$\begin{aligned} \underset{Q_t, Y_t \geq 0}{Min} \quad & \sum_{t=1}^{T} \frac{\left[C_t\left(Y_{t-1}\right) Q_t + \left(P_t + E_t\right)\left(D_t - Q_t\right)\right]}{(1+r)^t} \\ s.t. \quad & Y_t = Y_{t-1} + Q_t, \\ & Q_t \leq D_t. \end{aligned}$$

This model allows for learning by waiting, wherein clean technology production is deferred until the costs have declined enough or the dirty externality cost has increased enough to make the clean technology efficient. Given the resulting now-variable deployment of clean technology, the LBD premium π_{II} can be calculated. The externality cost of the dirty technology is a parameter that allows for sensitivity analysis.

The optimization model has a straightforward implementation. The advantage of a closed-form solution, as in the Nordhaus model, is lost. In exchange, at the cost of a modest computation, the model addresses the interactions of LBD, waiting, and the role of dirty technology externalities. Different sets of the key input parameters needed to benchmark the model can provide alternative sensitivity analysis of the learning premium and optimal waiting times before deployment.

Learning Premium

Application of the simple benchmark model produces estimates of the immediate size of the learning premium, the optimal waiting time before large-scale deployment of the clean technology, and the maximum size of the learning premium at the time of first deployment. Figure C-3 illustrates the results for a representative wind-type technology (where the initial cost is 75 percent higher than the price of the dirty technology, and the learning rate is 10 percent) and a representative solar PV-type technology (where the initial cost is 140 percent higher than the price of the dirty technology, and the learning rate is 20 percent). A "no fossil externality policy" case assumes the dirty externality cost is zero. A "fossil externality policy" case assumes an initial dirty externality cost of 20 percent of the price of the dirty technology that grows at the discount rate. The results appear in Figure C-3.

Without a fossil externality price, an assumed 5 percent rate of exogenous technological change means a very long time is required to make the clean technologies competitive. The wait is essentially 35 years for wind and 40 years for solar PV before costs come down enough and deployment begins. Hence, without material deployment, the initial LBD premium is essentially zero. The premium becomes larger in the distant future and reaches its maximum when deployment begins, but the maximum learning premium is still relatively small. The maximum LBD premium is 1.96 percent of the cost for wind and 4.22 percent for solar PV.

With a fossil externality, it is still optimal to wait. The immediate premium is larger and the optimal waiting period is shorter for the wind-type technology than for the PV-type technology. The waiting period reduces to 9 and 12 years for wind and solar PV, respectively. The maximum premium rises to 4.76 percent for wind and 11.01 percent for solar PV. But in all cases, the size of the learning premium is small compared with the benefits of internalizing the dirty technology externality.

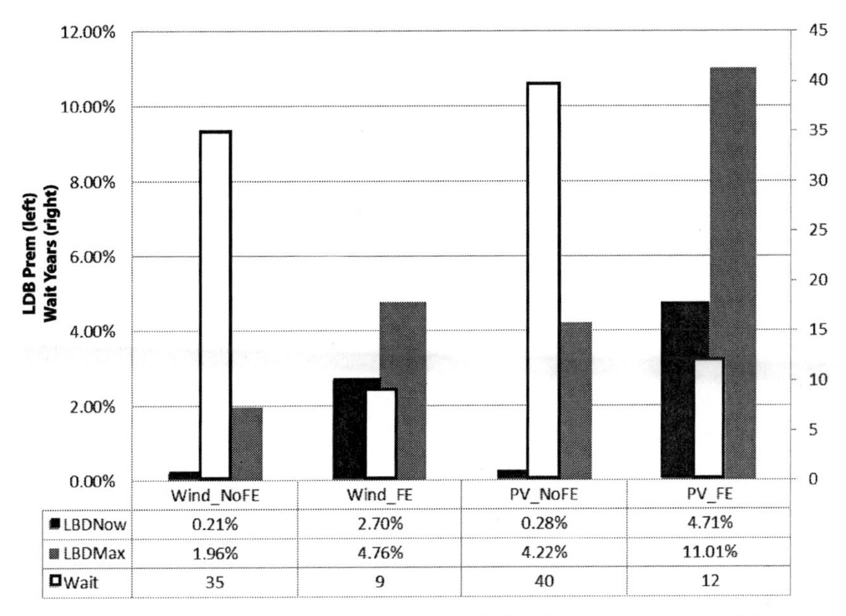

	Wind_NoFE	Wind_FE	PV_NoFE	PV_FE
LBDNow	0.21%	2.70%	0.28%	4.71%
LBDMax	1.96%	4.76%	4.22%	11.01%
Wait	35	9	40	12

FIGURE C-3 Learning by doing and by waiting, wind and solar photovoltaic (PV), with and without a fossil externality premium.

These illustrative calculations relate directly to an interpretation of the definition of the learning premium. Repeating the definition from above:

$$\pi_{II} = (1-\rho) b \int_{t=0}^{\infty} Q_t (C_t - C_{\min}) Y_t^{-1} e^{-rt} dt$$

The level of learning benefit captured by firms (ρ) has a proportional effect in reducing the learning premium. The premium is directly proportional to the learning elasticity (b), but a higher elasticity tends to reduce the cost disadvantage ($C_t - C_{\min}$) and make the premium correspondingly smaller. When costs are high and declining because of other factors, such as a trend, it pays to wait, which further discounts the immediate learning premium. Higher levels of deployment (Q_t) occur when costs are lower, which dampens the learning premium, and the higher deployment depresses the premium by the increase in cumulative output (Y_t). The large indirect effect of the dirty technology externality (E_t) increases the value of the clean technology and therefore increases deployment and makes future cost savings more valuable. With the exception of the important dirty externality impact, most of the elements of the learning premium interact in offsetting ways, and this helps explain why the value of the learning premium is relatively low.

Sensitivity Cases

Looking at the PV case with a carbon policy, eliminating the exogenous growth rate almost doubles the waiting period until the dirty technology externality costs have increased enough to justify deployment. But in this instance, the costs are so high that the learning premium increases by a factor of 1.5.

Setting the minimum cost bound C_{min} to zero reduces the PV deployment waiting period from 12 to 3 years and changes the deployment learning premium from 11 percent to 20 percent. Apparently, the assumption about the minimum cost level is important.

Surprisingly, changing the discount rate over the range 3 percent to 10 percent has only a modest effect on the size of the learning premium or the waiting time to deployment. The implementation here assumes the dirty technology externality grows at the rate of discount. The higher discount rate lessens the value of the future, but the costs in the future are correspondingly higher. The same offset occurs in the opposite direction with a lower discount rate.

The assumption of constant real dirty technology prices, albeit with increasing externalities, is an assumption distinctly influenced by the shale revolution. For example, in a preshale revolution model of LBD for prospective PV installations in California, van Bentham and colleagues (2008) assume high and increasing real natural gas and conventional electricity prices such that LBD dominates even the environmental benefits. The forecast implies PV prices below those of conventional dirty technology even without a carbon policy (van Benthem et al., 2008). With 3 percent growth of costs of dirty technologies due to fossil fuel shortages, the relative competitiveness of clean technologies would improve, the waiting period would fall, and the immediate learning premium for PV would nearly double. Essentially, the higher fossil fuel cost assumptions mean that in a few years, the PV technology would reach the holy grail of the Google mantra "RE <C"—clean renewable technology that costs less than carbon-based fuels.

POLICY MODELING FOR ENDOGENOUS TECHNOLOGICAL CHANGE

The importance of identifying the multiple factors driving innovation is clear from the empirical literature (Lindman and Söderholm, 2012). The single-factor experience models confound the effect of LBD with the separate impacts of LBS and exogenous change that support LBW.

> In relation to the invention-innovation-diffusion paradigm, single-factor learning curves amount to leaving out the effect

> of R&D on technical change as well as the main aspect of technology diffusion—i.e., the effect of cost reduction on higher technology adoption. The effect of cumulative capacity on the unit cost of technology as in single-factor learning by doing models is only a secondary effect of technology diffusion. Therefore, single-factor learning curves are not suitable for analysis of technologies that are in early stages of progress. A possible response to these shortcomings is to extend learning models to include R&D expenditures in addition to capital investment. (Jamasb and Kohler, 2008)

"Multi-factor models of this type offer improved explanations of the processes that contribute to cost reductions for the technology under study, and thus arguably provide more accurate assessment about the magnitude of investments or subsidies needed to bring down the cost of technology" (Yeh and Rubin, 2012, p. 768). EIA uses multifactor models with different LRs for different technology components and declining LRs with increased deployment (EIA, 2014b).

Distinct treatment of the factors is important because they each have different policy implications. Factors such as scale economies, when fully internalized by the private sector, may be important for the economy but do not require any public policy solution. A high level of exogenous technological change, whether from general technology trends or spillover effects from other industries, would imply waiting until costs decline before investing heavily in large-scale deployment. A high premium for LBS would imply a policy of investing in upstream research and small-scale deployment to niche markets to provide feedback and interaction with the directed research. A high LBD premium could imply policies to support large-scale take-up of otherwise expensive clean technologies. And, of course, the benefits of reduced dirty emissions would imply direct use of policies to create a market price for emissions.

In Fischer and Newell (2008), the analysis of technology change incorporates a learning premium within a two-period model that includes a representation for both intentional research (LBS) and learning through deployment of the clean technologies (LBD). The model includes multiple types of energy supply technologies and an aggregate model of electricity demand. There is an explicit analysis of subsidy policies for new technologies, learning capture, research subsidies, and caps on carbon emissions with associated prices. The model is simple enough to see and explain what is happening but includes more complicated trade-offs and benchmark calculations.

The Fischer and Newell (2008) technology learning model is a single-component model, i.e., $C_{min} = 0$, and treatment of future production and learning benefits includes output from both new and existing capacity, which should increase the implied learning premium. However, the main conclusions are consistent with the discussion above. "The underlying process of technological change, be it through learning by doing or R&D, turns out to be far less important than the incentives to use technology efficiently to reduce emissions. Nonetheless, the nature of technological change and the degree of knowledge spillovers do have discernable effects on the relative cost of alternative policies, which have differential effects on knowledge investment and how it occurs" (Fischer and Newell, 2008, p. 160).

An extension of this model to include energy efficiency appears in Fischer et al. (2013). For this expanded model, the authors conclude that "...it is telling that even with more refined representations of electricity generation options and market failures, emissions pricing still remains the single most cost-effective option for meeting emissions reduction goals. Technology policies are very poor substitutes, and when they overreach, they can be poor complements too" (p. 31).

The importance of active research (LBS), rather than passive learning (LBD), is intuitively plausible. Other things being equal, direct is better than indirect learning. Both can be valuable, and both types of investment face diminishing marginal returns. There is a consistent conclusion with two-factor models that represent LBS and LBD, and that "...examine the relative importance of R&D and capacity deployment for different technology categories. The results generally show higher learning by research than learning by doing rates. We do not find any technological development stage where learning by doing is the dominant driver of technical change" (Jamasb, 2007, p. 52). "An implication of devising policies based on overestimated learning by doing rates is that they can shift the scarce resources earmarked for innovation resources from more productive R&D activities to less productive and more costly capacity deployment policies" (Jamasb, 2007, p. 66). "The lesson from these observations is to be cautious in interpreting the policy conclusions of models that assume only a single source of technological progress or that neglect critical factors such as spillovers" (Clarke et al., 2006, p. 593).

The logical extension of these arguments points to multifactor models that include at least LBS, LBD, and exogenous technological change (Qiu and Anadon, 2012). The importance of controlling for the many factors is obvious for empirical work. However, the same conclusions apply to models of innovation used to prepare forecast simulations to estimate learning premiums and evaluate policies going forward (Clarke et al., 2008; DOE, 2014a).

A natural and relatively simple extension of current applications would be to modify the cost model to include a representation of a time trend, cumulative production, and knowledge (t, Y_t, K_t). This would yield a simulation model of the form

$$C_t = \left(C_0 - C_{\min}\right) Y_0^b K_0^k e^{-ht} Y_t^{-b} K_t^{-k} + C_{\min}.$$

The accumulation of knowledge could include a representation of the depreciation of knowledge found important in some studies (Nemet, 2012). Benchmarking the multiple components of the model would give attention to the differences between experience in and outside the market of interest (van Benthem et al., 2008). A more detailed simulation model would allow for more detailed representation of the technologies to consider portfolio effects (Neuhoff and Twomey, 2008) and constraints on rates of penetration (Manne and Richels, 2004; Neuhoff, 2008).

In forecasts over the long run, the two-period model of Fischer and colleagues (2013) provides insight. But a multiperiod model would be necessary to track vintages of technologies and the trade-offs of waiting for costs to decline or externality benefits to grow.

Over a long time horizon, and with opportunities to change policies given new information, the use of deterministic models cannot fully capture the dynamic paradigm of act-learn-act that incorporates uncertainty and hedging strategies. Although a full dynamic optimization model under uncertainty is challenging, both to implement and to interpret, the application of Santen and Anadon (2014) includes use of approximate dynamic programming techniques to investigate optimal innovation policies. The treatment of uncertainty affects the innovation strategy. "Results show that under a carbon constraint, the optimal investment strategy includes lower solar PV RD&D spending upfront but more RD&D spending later—and sometimes higher spending overall—when compared to a strategy under perfect foresight about RD&D outcomes, or based on single-shot decision-making under uncertainty without learning" (Santen and Anadon, 2014, p. v).

POLICY FOR INNOVATION

This overview of different types of learning supports a view that LBD is relevant, but the impact implies a relatively small premium that could be the focus of public policy. More important is LBS, through R&D, and waiting to get the costs for clean energy technology close to being commercially competitive. The horizon for deployment depends critically on the size of the fossil fuel externality cost. A policy needs to focus on the interactions of the market externalities and market failures. The learning premium is material, but cannot carry much of the burden of supporting large-scale deployment of expensive clean energy technologies.

Public policy needs to address the carbon price and the costs of other emissions, then invest in upstream R&D, the muddle in the middle, and initial limited deployment. The remaining LBD incentives for early large-scale uptake are likely to be small. This view of the proper balance of public policy is at odds with current conditions. There is too little attention to fundamental innovation and too much emphasis on deploying clean energy technologies that cannot meet the challenge (Zachmann et al., 2014). "The fundamental problem for cost-effective GHG [greenhouse gas] controls keeps coming back to the externality associated with R&D, and the unique degree to which solving the GHG externality depends on effective technological change. Until the R&D externality is resolved, there is almost no case to be made for starting to control the environmental externality" (Montgomery and Smith, 2007, p. 339).

As found by many others, the implications for the form of public support lean toward LBS and away from reliance on LBD. Furthermore, the basic model discussed above employs a simplifying assumption that a subsidy for clean technology leads to substitution for the dirty technology. In reality, this is not true when there are many different technologies. For example, the effect of the U.S. production tax credit for wind has been to substantially impair the economics of nuclear power rather than to solely substitute for fossil fuel production. Compared with the direct effect of a carbon price, indirect deployment subsidies for clean energy technology tend to be ineffective in addition to being unsupported by the small LBD premium:

> Subsidies for green power (or mandated utility offer prices for power generated in this way, known as "feed-in tariffs") have been portrayed as nearly equivalent to pricing externalities, but more politically acceptable. This approach, however, is very problematic for three closely related reasons.
>
> First, subsidizing green power for reducing pollution (relative to some counterfactual) is not equivalent to taxing "brown" power to reflect the marginal social damage. If end-use electricity demand were completely inelastic and green and brown power were each completely homogeneous, they would have the same effect; the only effect of the subsidy would be to shift the production share towards green and away from brown power. But the underlying market failure is the underpricing of brown power, not the overpricing of green power, so subsidizing green power from government revenues artificially depresses the price of power and discourages efficient energy consumption. As a result, government subsidies of green power lead to overconsumption of electricity and disincentives for energy efficiency. In addition, for any given level of reduction, it will be achieved more efficiently by equalizing the marginal price of the pollutant across sectors as well as within sectors. This is not achievable through ad hoc subsidies to activities that displace certain sources of emissions....

> Second, subsidizing green power generally fails to recognize the heterogeneity within the green power sector and among the brown power sources that are being displaced. Solar power that reduces coal-fired generation lowers greenhouse gas emissions by about twice as much on average as if it reduces natural-gas-fired generation. Assuming that the marginal generation displaced is equal to the average generation mix in the system can be a poor approximation....The problem arises because subsidizing green power is an indirect approach to the pollution problem, and the relationship between green power and emissions avoided is not uniform. It would not arise with a direct tax (or pricing through tradable permits) on pollution (Borenstein, 2012, pp. 79-80).

Nordhaus (2013, p. 266) makes a similar point:

> Subsidies pose a more general problem in this context. They attempt to discourage carbon-intensive activities by making other activities more attractive. One difficulty with subsidies is identifying the eligible low-carbon activities. Why subsidize hybrid cars (which we do) and not biking (which we do not)? Is the answer to subsidize all low-carbon activities? Of course, that is impossible because there are just too many low-carbon activities, and it would prove astronomically expensive. Another problem is that subsidies are so uneven in their impact. A recent study by the National Academy of Sciences looked at the impact of several subsidies on GHG emissions. It found a vast difference in their effectiveness in terms of CO_2 removed per dollar of subsidy. None of the subsidies were efficient; some were horribly inefficient; and others such as the ethanol subsidy were perverse and actually increased GHG emissions. The net effect of all the subsidies taken together was effectively zero! So in the end, it is much more effective to penalize carbon emissions than to subsidize everything else.

The optimal policy portfolio includes many instruments. It is important to note, however, that not all policies are equal:

> Some clear principles emerge. We find that when the ultimate goal is to reduce emissions, policies that create incentives for fossil-fueled generators to reduce emissions intensity, and for consumers to conserve energy, perform better than those that rely on incentives for renewable energy producers alone. Overall, we find that the nature of knowledge accumulation is

> far less important than the nature of the policy incentives....For the type of moderate emissions targets we explore, a renewable energy R&D subsidy turns out to be a particularly inefficient means of emissions reduction, since it postpones the vast majority of the effort to displace fossil-fueled generation until after costs are brought down....This requires very large R&D investments and forgoing near-term cost-effective abatement opportunities. While climate change is a long-term problem, the results for mid-term strategies emphasize the important role for policies that encourage abatement across all available forms and timeframes, as well as the limitations of narrowly targeted policies—particularly those focused solely on R&D.
>
> Nonetheless, given the presence of more than one market failure—an emissions externality and knowledge spillovers—no single policy can correct both simultaneously; each poses different trade-offs. The presence of knowledge spillovers means that separate policy instruments are necessary to optimally correct the climate externality and the externalities for both learning and R&D. In fact, we find that an optimal portfolio of policies can achieve emissions reductions at a significantly lower cost than any single policy, although the emissions reductions continue to be attributed primarily with the emissions price.
>
> Together, these results illuminate some of the arguments in Montgomery and Smith that R&D is the key for dealing with climate change and that an emissions price high enough to induce the needed innovation cannot be credibly implemented. We show that an emissions price alone, although the least costly of the single policy levers, is significantly more expensive alone than when used in combination with optimal knowledge subsidy policies. Although a high future emissions price may not be credible, with the combination policy the required emissions price is much more modest. However, if one believes that even a modest emissions price is not politically feasible, an R&D subsidy by itself is not the next best policy, and the costs of that political constraint are likely to be quite large and increasing with restrictions on the remaining policy options. It should be kept in mind, however, that we focus on reductions over the near-to-mid-term and incremental improvement of existing technology, rather than breakthrough technologies that might achieve deep reductions.

> It seems likely that R&D policies have greater salience in the latter context, although this lies beyond the scope of the current paper. (Fischer and Newell, 2008, pp. 143-144)

The optimal portfolio will address both the structure and the targets for policy support (Kriegler et al., 2014). A real carbon price on carbon emissions is better than searching for an equivalent subsidy for the right clean energy technology, particularly when the right technology may not yet be known. Estimating the appropriate price of carbon is a challenge, but it is a challenge that has already been undertaken by the U.S. government (Interagency Working Group on Social Cost of Carbon, 2013). And not all subsidies are the same. For example, an investment tax credit can affect the economics of wind without the perverse collateral effects of a production tax credit in lowering the perceived variable cost of wind from zero to minus the value of the credit. Given the low value of the LBD premium and the high value of reducing costs before large-scale deployment, a direct expansion of government support for upstream transformational R&D would be better than a broad subsidy for deploying existing clean energy technology.

The spillover effects of LBD can be invoked as a reason for public support for deploying expensive clean energy technologies. There are clear empirical difficulties in estimating the size of the appropriate LRs and implied premiums. The collective results support the view that the LBD premium is small. Much more important is the price on carbon and related fossil fuel externalities. Given the scale of the clean energy technology challenge and the state of current technology, greater emphasis is warranted on the earlier stages of the innovation process in the search for truly transformative technologies that would be cheap enough to be deployed with the market incentive of a price on carbon.

The implication for clean energy innovation policy is that the most important priorities are identifying and creating new options, demonstrating the efficacy of these options, and setting the stage for early adoption of those that are most promising. Although there are policies that would improve the conditions for eventual large-scale take-up and improvements in use, these policies are likely to be expensive and ineffective without a substantial investment in the earlier stages of the innovation process. The emphasis needs to be on developing technologies that can truly compete with incumbent energy sources. These technologies are not in hand today, and efforts to create these technologies for the future need to be expanded and accelerated. There is no guarantee of success, but the effort is worth a major investment with a clear view of the difficulties ahead. This challenge creates an opportunity and a need for governments at all levels, keeping an eye on the prize of expanding the innovation machine.

Appendix D

Technology Readiness

The technology readiness level (TRL) taxonomy is the most commonly utilized method for determining a given technology's readiness for ultimate application in electricity generation, energy storage, and power delivery, or utilization in power systems. The National Aeronautics and Space Administration (NASA) developed the TRL taxonomy as an aid to managing its space-related research and development. TRLs also are a convenient means of describing the stage of development of increasingly clean electric power technologies because they are intended to enable a consistent comparison of technological maturity across disparate technologies. However, the complexity of power systems makes the TRL assessment imperfect since components of a given system in development are usually at differing levels of technology readiness, meaning that some components are at high TRLs, while others are at low TRLs.

The committee assessed the technology readiness of a variety of increasingly clean electric power technologies; this appendix presents the results of that assessment. Table D-1 provides an approximate guide to how each TRL number corresponds to a specific stage of technological development. As the table indicates, TRLs encompass basic (blue-sky) research in new technologies and concepts (targeted identified goals, but not necessarily specific systems), focused technology development addressing specific technologies for one or more potential applications, technology development and demonstration for each application, system development, and commercialization.

To conduct this analysis, the committee had to reduce the extensive number of individual increasingly clean energy technologies to a manageable size. A review of currently available technologies prompted the committee to focus on a broad spectrum of technology options for achieving the transition to an increasingly clean electrical system while leaving the door open for potentially game-changing technical innovations. The committee used the

TABLE D-1 Technology Readiness Levels

Technology Readiness Level	Description
1	Exploratory research transitioning basic science into laboratory applications
2	Technology concepts and/or application formulated
3	Proof-of-concept validation
4	Subsystem or component validation in a laboratory environment to simulate service conditions
5	Early system validation demonstrated in laboratory or limited field application
6	Early field demonstration and system refinements completed
7	Complete system demonstration in an operational environment
8	Early commercial deployment (serial nos. 1, 2, etc.)
9	Wide-scale commercial deployment

SOURCE: Mankins, 1995.

following process to identify and categorize the technologies with the greatest potential:

- The committee created a master list of all technologies known to its members, including those referenced in the literature.
- The committee then reduced that list by selecting for technologies expected to have the greatest potential to reduce emissions of greenhouse gases (GHGs) and other pollutants and eliminating those with few technical or market prospects. These conclusions were based on an extensive literature review. The resulting reduced list reflects the committee's assignment of the highest priority to technologies that can both reduce energy consumption and accelerate the generation of power with no or low emissions. This reduced list, with detailed explanations, is included in this appendix and summarized in Table D-2.

As discussed in Chapter 2, the committee's review of available technologies indicated that there does not yet exist a suite of clean power technologies that can meet global demand at reasonable cost. Continued innovation, with particular attention to bridging the so-called "valleys of death" (see Chapter 3), is imperative. Therefore, policies need to address not only the deployment of clean energy technologies that are currently available but also the development of the technologies that are needed.

TABLE D-2 Promising Technologies for Increasingly Clean Electric Power

Technology Category	Technology Readiness Level[a]								
	1	2	3	4	5	6	7	8	9
Renewable Power Generation									
1: Electric energy storage			■	■	■	■	■	■	■
2: Hydro and marine hydrokinetic power[b]				■	■				■
3: Advanced solar photovoltaic power[c]					■	■			
4: Advanced concentrating solar power						■	■	■	■
5: Advanced solar thermal heating					■	■	■	■	
6: Advanced biomass power				■	■			■	■
7: Engineered/enhanced geothermal systems			■	■		■			
8: Advanced wind turbine technologies		■			■	■			
9: Advanced integration of distributed resources at high percent						■	■	■	■
Advanced Fossil Fuel Power Generation									
10: Carbon capture, transport, and storage							■	■	■
11: Advanced natural gas power and combined heat and power (CHP)[c]				■	■	■	■	■	■
12: Water and wastewater treatment				■	■				■
Nuclear Power Generation									
13: Advanced nuclear reactors	■	■	■	■	■	■	■	■	■
14: Small modular nuclear reactors	■	■	■	■	■	■	■	■	■
15: Long-term operation of existing nuclear plants	■	■	■	■	■	■	■	■	■

(Continued)

TABLE D-2 Continued

Technology Category	Technology Readiness Level[a]								
	1	2	3	4	5	6	7	8	9
Electricity Transmission and Distribution									
16: Advanced high-voltage direct current (HVDC) technologies							■		
17: Reducing electricity use in power systems									■
18: Smart-grid technologies (grid modernization)				■	■	■	■	■	■
19: Increased power flow in transmission systems			■	■	■	■	■	■	■
20: Advanced power electronics						■	■		
Energy Efficiency									
21: Efficient electrical technologies for buildings and industry	■	■	■	■	■	■	■	■	■

[a]Technology readiness levels are shown on a scale of 1 to 9, where 1 is the least ready. Most of the technology categories shown include technologies with varying readiness levels. A shaded box below a TRL number indicates there is at least one technology at that TRL.

[b]The committee identified barriers at lower TRLs for hydropower technologies but was unable to make specific level assignments.

[c]For concepts beyond three junctions.

Technology Category: 1. Electric Energy Storage

Description: Electric energy storage technologies for electric power applications with benefits for renewables integration; ancillary services; time arbitrage of on- and off-peak energy; and capital deferral at the grid connected, distribution, and customer levels are becoming better understood. Pumped hydro storage (generation from hydro sources is described under category 2) is the most prevalent storage technology at present, with 40 plants operating in the United States and capacity totaling more than 22 gigawatts (GW). Compressed air energy storage (CAES) technologies store ambient air at pressure in underground caverns. CAES produces electricity by releasing the air through a turbine-driven generator. Adiabatic CAES can achieve higher efficiency by recovering the heat of compression. Battery technologies vary tremendously in their underlying design and performance characteristics but hold great promise to allow for increased penetration of variable and distributed power resources. They also are used to provide other services including peak shaving, ramping, spinning reserve, and backup for specific uses such as data centers.

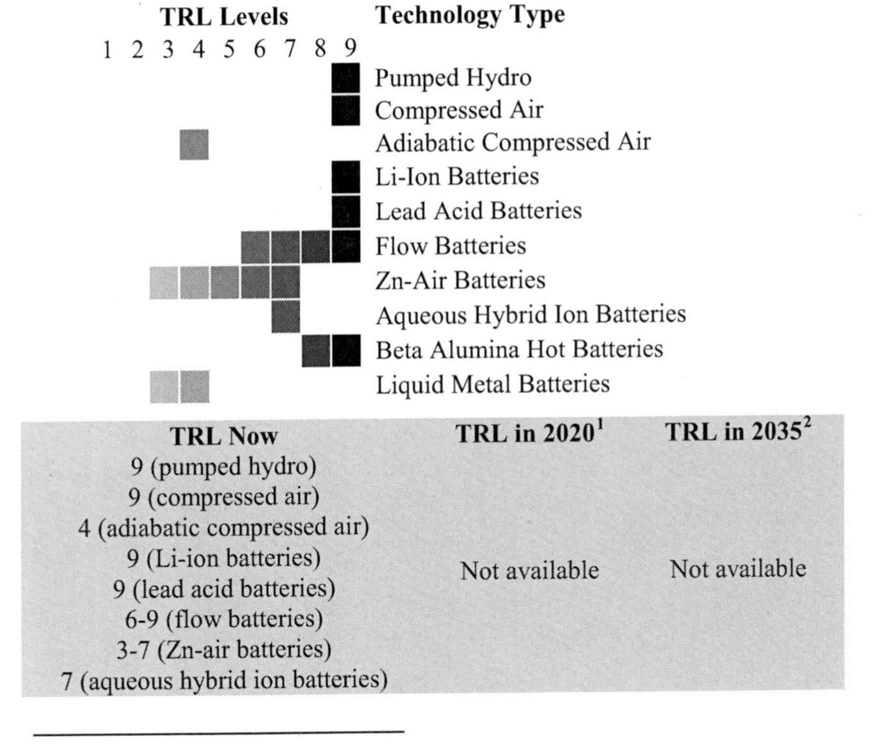

TRL Now	TRL in 2020[1]	TRL in 2035[2]
9 (pumped hydro) 9 (compressed air) 4 (adiabatic compressed air) 9 (Li-ion batteries) 9 (lead acid batteries) 6-9 (flow batteries) 3-7 (Zn-air batteries) 7 (aqueous hybrid ion batteries)	Not available	Not available

[1]The committee did not assess the 2020 TRLs for electric energy storage.
[2]The committee did not assess the 2035 TRLs for electric energy storage.

3-4 (liquid metal batteries)

Technology Barriers: There is a need for cycle-testing protocols for grid-scale storage. Once such protocols are developed, there will be a need to test single-cell and multi-cell systems under real-world conditions. Key needs are to reduce response times to demand and increase the total storage capability in order to make stored electricity dispatchable. Decreasing internal losses and improving calendar life are also important.

Commercialization Barriers: Electricity storage is not a mainstream technology considered in planning, building, and operating electric power infrastructure. Several regulatory, policy, financial, and awareness issues will have to be addressed before it can be accepted and exploited as part of the electricity supply chain. The most effective technology for large-scale electric energy storage at this time continues to be pumped hydro. Although that technology is relatively mature, the availability of new sites is extremely limited.

Technology Category: 2. Hydro and Marine Hydrokinetic (MHK) Power

Description: Large conventional hydro generation (greater than 30 megawatts [MW]) had an installed capacity in the United States of approximately 79,000 MW as of 2014, with the technical potential to double large (as well as small) hydro capacity. However, this expansion will likely not be realized by 2035 because of regulatory and financing constraints. MHK power technologies are still in various stages of development. Technologies to utilize ocean currents are in the proof-of-concept and laboratory demonstration phases. Wave, tidal, and ocean thermal technologies have components that have gone as far as open-water operation, although none have undergone array testing, and many wave and tidal technologies are still in the demonstration phases.

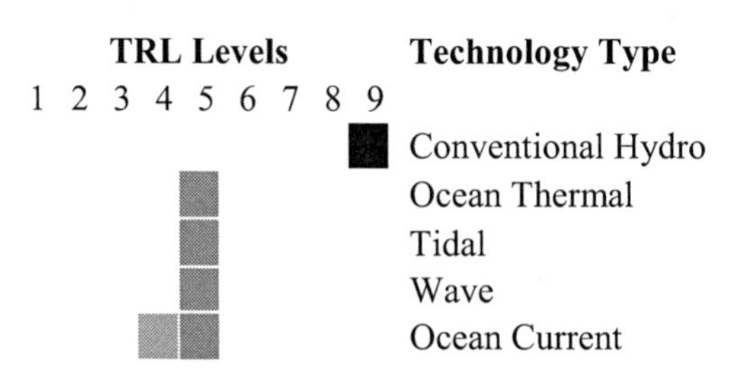

TRL Now	TRL in 2020[3]	TRL in 2035[4]
9 (conventional hydro; however, barriers identified at lower	9 (conventional hydro; however, barriers identified at lower	9 (conventional hydro; however, barriers identified at lower

[3]The committee did not assess the 2020 TRLs for MHK generation technologies.
[4]The committee did not assess the 2035 TRLs for MHK generation technologies.

TRLs)	TRLs)	TRLs)
5 (ocean thermal) 5 (tidal) 5 (wave) 4-5 (ocean current)	Not available for MHK technologies	Not available for MHK technologies

Technology Barriers: Improved operational performance of turbine runner and major components; improved flow measurement and control to reduce turbulence and increase energy conversion; fish passage/protection and environmental management; enhanced dam safety; development of room temperature superconductors (RTSs), precisely above 0° C. Some MHK technologies still require significant technology development. None have yet undergone array testing. Critical barriers include developing advanced controls and power take-off technologies, and optimizing device structures to improve energy capture, decrease mass, and improve system reliability.
Commercialization Barriers: Long-term financing for capital projects; long timeline for licensing and relicensing projects; low natural gas prices for competing generation; financial markets for hydro to benefit from providing ancillary services support.

Technology Category: 3. Advanced Solar Photovoltaic Power

Description: Triple-junction photovoltaic (PV) devices exist and have achieved efficiencies of ~43 percent under concentration with very advanced fabrication technology (the highest efficiencies were obtained with structures based on stacks of epitaxial III-V compounds), but represent the first actual devices to demonstrate very high efficiency potential.

TRL Levels									Technology Type
1	2	3	4	5	6	7	8	9	
				■	■				Advanced Solar PV

TRL Now	TRL in 2020	TRL in 2035
5-6	7-8	9

Technology Barriers: Multijunction cells (e.g., the incremental gain from adding another cell to a stack of N junctions is theoretically proportional to 1/N(squared); therefore, after including the per-junction losses (electrical and optical) in a practical device, the expected net gain from adding cells is close to zero after the 4th).
Commercialization Barriers: Needed cost reductions associated with each cell junction addition; utility-scale solar power generation assets depreciated as 5-year property.

Technology Category: 4. Advanced Concentrating Solar Power

Description: Concentrating solar power (CSP) encompasses a variety of configurations, including parabolic troughs, heliostats, and linear Fresnel reflector systems that range in size from a few kilowatts to 50 MW or more. The maturity of the technologies also varies. Approximately 4.8 GW of parabolic trough capacity is installed worldwide. Heliostats account for about 560 MW of installed capacity globally, while there is less than 50 MW of linear Fresnel systems installed.

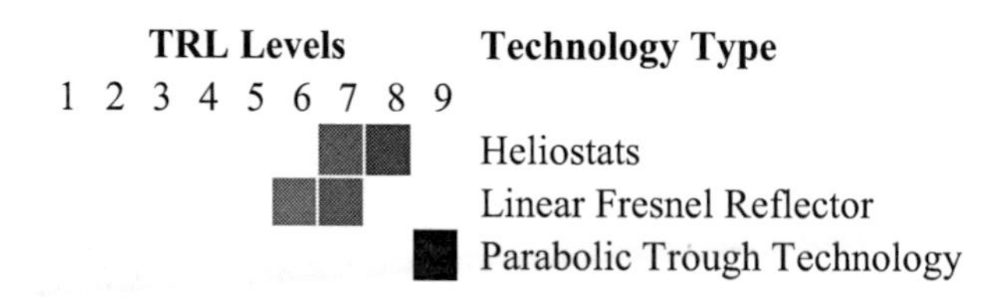

TRL Now	TRL in 2020	TRL in 2035
7-8 (heliostats) 6-7 (linear Fresnel reflector) 9 (parabolic trough)	9 (heliostats) 7-8 (linear Fresnel reflector) 9 (parabolic trough)	9 (heliostats) 7-8 (linear Fresnel reflector) 9 (parabolic trough)

Technology Barriers: Cost-effective thermal energy storage; low-cost solar fields; high-temperature receivers; advanced power block technologies; high-temperature heat transfer fluids with low melting points.

Commercialization Barriers: Financing; land use; siting issues in environmentally sensitive areas; transmission; regulatory framework; manufacturing; utility-scale solar power generation assets currently depreciated as 5-year property; supply chain.

Technology Category: 5. Advanced Solar Thermal Heating

Description: Solar thermal heating is used primarily for producing residential hot water, for space heating, and for heating pools. Use in North America has declined because of the high cost compared with other technologies, although use for pool heating remains quite cost-competitive. Solar thermal technologies continue to do well outside of North America where conventional fuels such as natural gas cost more. China remains the largest solar thermal market, with a preponderance of low-cost thermosiphon-type systems. Solar thermal for process applications is another area of great potential.

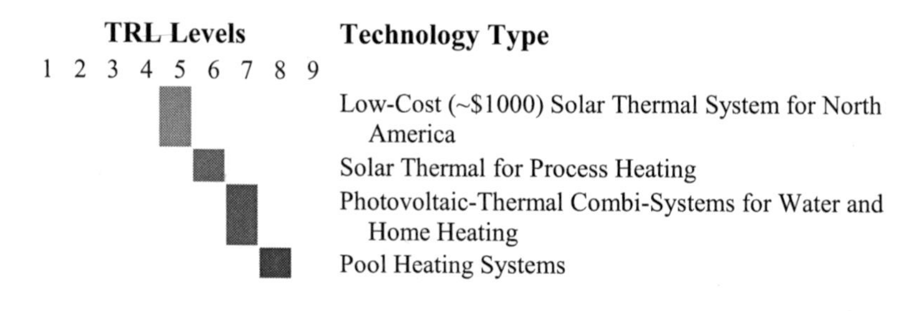

TRL Now	TRL in 2020	TRL in 2035
5 (low-cost solar thermal system for North America) 6 (solar thermal for process heating) 7 (PV-solar thermal combi-systems for water and home heating) 8 (pool heating systems)	6 (low-cost solar thermal system for North America) 7 (solar thermal for process heating) 7 (PV-solar thermal combi-systems for water and home heating) 8 (pool heating systems)	8 (low-cost solar thermal for North America) 8 (solar thermal for process heating) 8 (PV-solar thermal combi-systems for water and home heating) 9 (pool heating systems)

Technology Barriers: Low-cost systems with plug-and-play installation for residential and commercial use; large field integration for industrial applications; measurement of solar thermal output.

Commercialization Barriers: Incomplete value chain; lack of knowledge among building owners and/or operators; insufficient incentives to adopt new technology; split incentives between building owners and operators.

Technology Category: 6. Advanced Biomass Power

Description: Biomass power production, frequently referred to as biopower, refers to power generation from biomass sources such as grasses, straws, forest products, and energy crops. Pretreatment processes such as leaching and torrefaction help eliminate deleterious components from biomass and increase the energy density of biomass, making it more suitable as a fuel whether direct- or co-fired. Wood is the most common biopower fuel, generating more than 42 gigawatt hours (GWh) of electricity in 2015 (nearly twice the electricity produced by utility-scale solar PV).

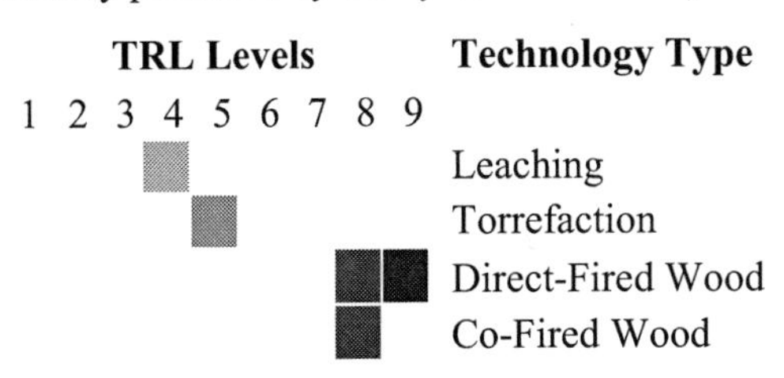

TRL Now	TRL in 2020	TRL in 2035
4 (leaching) 5 (torrefaction) 8-9 (direct-fired wood) 8 (co-fired wood)	7 (leaching) 7 (torrefaction) 9 (direct-fired wood) 9 (co-fired wood)	8 (for availability of commercial integrated leaching + torrefaction plants) 9 (direct-fired

		wood) 9 (co-fired wood)

Technology Barriers: Leaching/torrefaction plant demonstration projects; pilot burning tests using leached plus torrefied biomass in existing boilers; power production (e.g., distributed bipower systems, high-efficiency conversion technologies); feedstock development (e.g., efficient forest-thinning techniques, higher-yield crops/trees, improved biomass upgrading technology).

Commercialization Barriers: Cost of leached + torrefied biomass 3 times higher than that of coal on a per million British thermal units (MMBtu) basis; lack of leaching + torrefaction demonstration plants large enough to support pilot burning tests; high cost of delivered feedstock.

Technology Category: 7. Engineered/Enhanced Geothermal Systems

Description: Margin stimulation is being examined for the purpose of converting dry in-field wells that were originally deemed failures, while a hot dry rock method is also being considered to access existing subsurface heat in a wide geographic area by using water or supercritical carbon dioxide (CO_2).

TRL Levels **Technology Type**

1 2 3 4 5 6 7 8 9

Hot Dry Rock

Margin Stimulation

TRL Now	TRL in 2020	TRL in 2035
3-4 (hot dry rock) 6 (margin stimulation)	5-7 (hot dry rock) 8-9 (margin stimulation)	7-8 (hot dry rock) 9 (margin stimulation)

Technology Barriers: Cost-effective deep drilling technologies; high-temperature subsurface drilling instrumentation.

Commercialization Barriers: Cost and risk (e.g., cost for deep well completion can be tens of millions of dollars per well); ability to stimulate sufficiently large reservoir per well drilled; ability to create reservoir as designed and manage reservoir growth during operation; utility-scale geothermal power generation assets currently depreciated as 5-year property.

Technology Category: 8. Advanced Wind Turbine Technologies

Description: New wind generator technologies include advanced direct-drive permanent magnet generators (ADDPMGs), high-temperature superconducting generators (HTSCGs), and room-temperature superconducting generators (RTSCGs). Their development could reduce the levelized cost of electricity, increase capacity factors, reduce generator weight, and support the search of the wind industry for larger-scale wind platforms (in the 10-15 MW range), especially for off-shore wind.

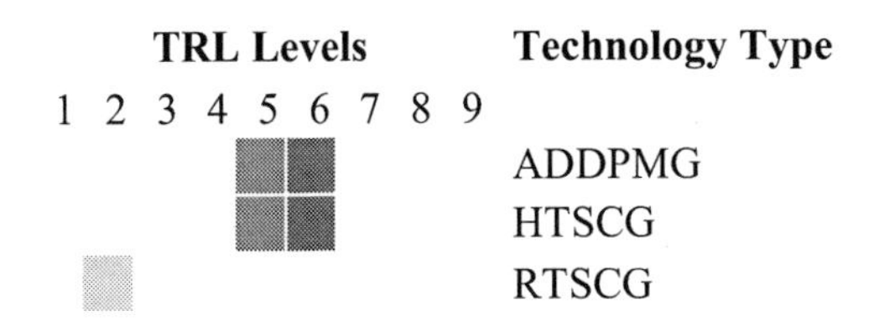

TRL Now	TRL in 2020	TRL in 2035
5-6 (ADDPMG)	7-8 (ADDPMG)	8 (ADDPMG)
5-6 (HTSCG)	7-8 (HTSCG)	8 (HTSCG)
2 (RTSCG)	5 (RTSCG)	7 (RTSCG)

Technology Barriers: ADDPMGs—structural robustness, long-term reliability, scale-up to 10 megawatts thermal (MWt); HTSCGs—industrialization of cryocoolers/low maintenance/higher efficiencies; RTSCGs—development of RTSs, precisely above 0° C.

Commercialization Barriers: Competitive capital cost and efficiency of commercial units; no evaluation of independent demonstration projects, including transportation, field assembly and operating performance; high perceived risk increase financial costs.

Technology Category: 9. Advanced Integration of Distributed Resources at Higher Rates

Description: Integrating a large amount of distributed resources into the power grid in an economical and sustainable way while ensuring system reliability will require new tools and methods. Understanding of the impacts on the rest of the power system, planning to ensure that the power system infrastructure can accommodate high penetrations of variable generation, and development of the operational tools needed to manage some of the unique aspects of wind and solar PV are needed. The scale of integration matters for both large-scale and distributed systems. Integrating distributed resources to supply more than 15 percent of the load will require smart inverters that enable distributed energy resources to provide voltage and frequency support and to communicate with energy management systems. It will also require distribution management systems and ubiquitous sensors so operators can reliably integrate distributed generation, storage, and end-use devices while also interconnecting those systems with transmission resources in real time.

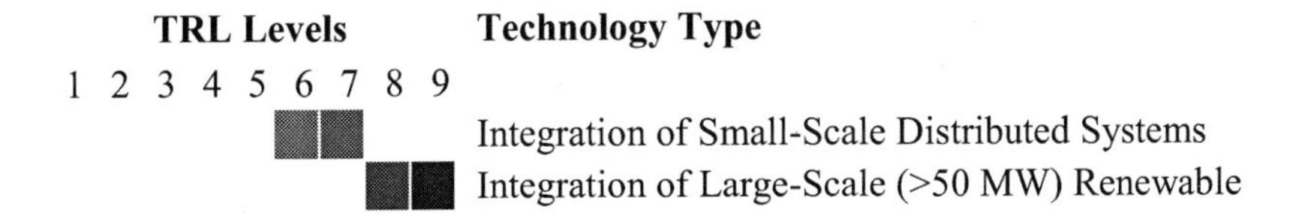

Systems

TRL Now	TRL in 2020	TRL in 2035
6-7 (integration of small-scale distributed systems) 8-9 (integration of large-scale renewable systems)	8-9 (integration of small-scale distributed systems) 8-9 (integration of large-scale renewable systems)	8-9 (integration of small-scale distributed systems) 9 (integration of large-scale renewable systems)

Technology Barriers: Variability and uncertainty of production, limiting the penetration in certain areas because of a lack of system flexibility. Inverter-based nature of the technology, limiting instantaneous penetration in the system (more than 50 percent not currently possible in a synchronous system). Distributed nature resulting in less visibility and control and potential reliability impacts.

Commercialization Barriers: Distributed PV may reduce utility revenue while still requiring significant transmission and distribution upgrades; lack of additional revenue streams for variable generation; large balancing costs imposed on variable generation.

Technology Category: 10. Carbon Capture, Transport, and Storage

Description: Although it is not cost-competitive at present, CO_2 capture and storage can work in fossil fuel power plants. There is also room for substantial improvement. Current technologies use three times the theoretical minimum energy to capture and compress CO_2, and efforts to prove and improve CO_2 capture and storage are in the early stages. Pipeline transportation of CO_2 in the United States is quite mature, with 50 individual pipelines spanning more than 4,500 miles.

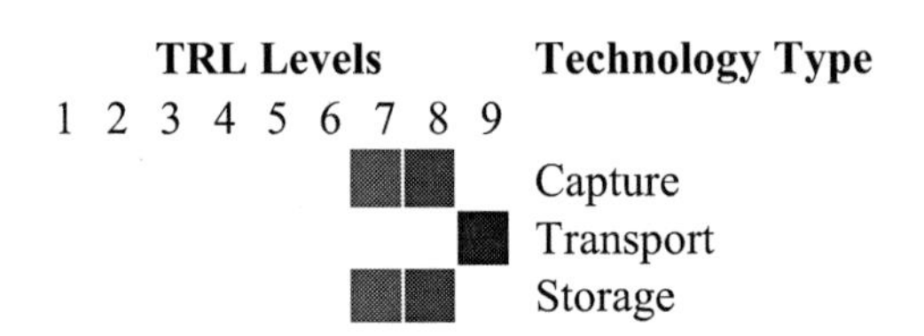

TRL Now	TRL in 2020	TRL in 2035
7-8 (capture) 9 (transport) 7-8 (storage)	8-9 (capture) 9 (transport) 8-9 (storage)	9 (capture) 9 (transport) 9 (storage)

Technology Barriers: Need to better understand long-term issues related to storage: impact on water tables, caprock, injection operations, liability (see Chapter 5 section on "Key Nonmarket Barriers").

Commercialization Barriers: Lack of regulatory and economic drivers (see

Chapter 5 section on "Key Nonmarket Barriers").

Technology Category: 11. Advanced Natural Gas Power and Combined Heat and Power (CHP)

Description: Advanced natural gas technologies, such as a new power generation concept based on the "Allam Cycle," could provide power at a thermal efficiency exceeding 50 percent. Heat rates for other advanced technologies are approaching 5,400 Btu/kilowatt hour (kWh). Combined heat and power (CHP), a technology of interest to large industrial organizations with a significant demand for thermal energy (steam or hot water), is another promising technology. Current installations are almost universally large and custom designed. Small-scale systems are not as well developed. Small-scale users would benefit from the development of modular "plug and play" thermal appliances.

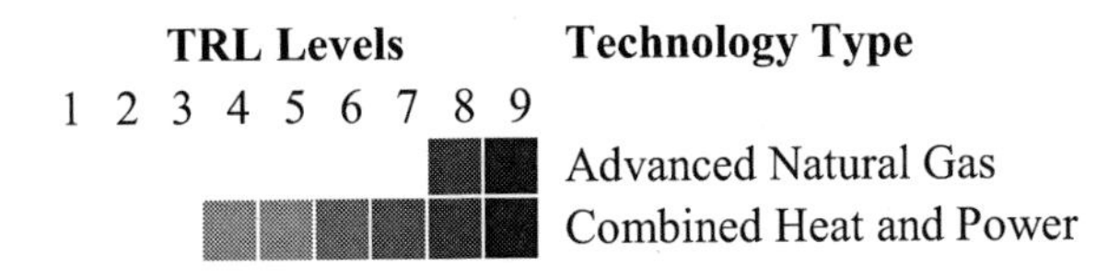

TRL Now	TRL in 2020	TRL in 2035
8-9 (advanced natural gas) 4-9 (CHP)	9 (advanced natural gas) 9 (CHP)	9 (advanced natural gas) 9 (CHP)

Technology Barriers: Need to prove the system (e.g., that it can operate as a whole while responding to typical demands of a natural gas-fired power plant); for micro-CHP, lack of availability of "plug and play" thermal appliances (most CHP installations need to be custom engineered).

Commercialization Barriers: In the long term, a continued preference for natural gas as a fuel source will pose a barrier to lowering GHG emissions.

Technology Category: 12. Water and Wastewater Treatment

Description: Water withdrawals and the treatment of wastewater are important limiting factors in the construction and operation of power plants. Conventional processes for generating desalinated (and deionized) water include reverse osmosis (RO), multistage flash distillation (MSF), and multiple effect desalination (MED). RO involves the use of membrane to generate pure water from salt water by applying a pressure higher than the osmotic pressure. MSF and MED involve thermal evaporation of water. In membrane distillation (MD), a heated aqueous solution passes through a hydrophobic membrane and is partially transformed to water vapor and collected as pure water. Electrodialysis (ED) transports salt ions through ion-exchange membranes under the influence of an applied electric potential difference, and forward osmosis (FO) is an osmotic process using a semipermeable membrane to effect separation of water from dissolved solutes under an osmotic pressure gradient.

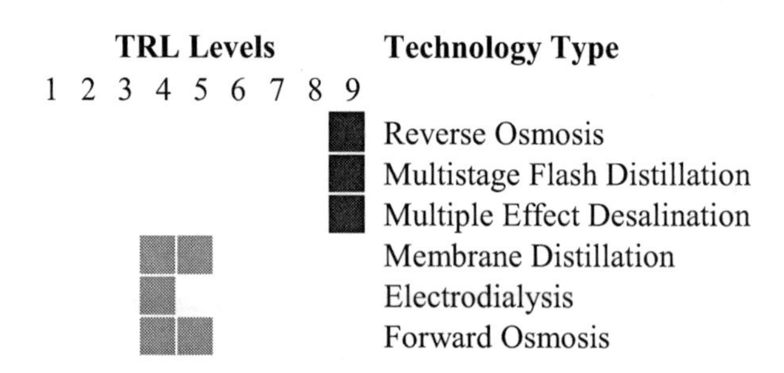

TRL Now	TRL in 2020	TRL in 2035
9 (reverse osmosis)	9 (reverse osmosis)	9 (reverse osmosis)
9 (multistage flash distillation)	9 (multistage flash distillation)	9 (multistage flash distillation)
9 (multiple effect desalination)	9 (multiple effect desalination)	9 (multiple effect desalination)
4-5 (membrane distillation)	7-9 (membrane distillation)	9 (membrane distillation)
4 (electrodialysis)	7 (electrodialysis)	9 (electrodialysis)
4-5 (forward osmosis)	7 (forward osmosis)	9 (forward osmosis)

Technology Barriers: RO—high power consumption and limitations of high salt concentration and fouling; MSF and MED—high investment, corrosion, energy cost; MD—availability of membrane with high flux; ED—energy-intensive, high treatment cost (which depends on salt concentration), competitive with RO in some cases (particularly for brackish water applications), fouling; FO—separation of draw solutes and high-flux membranes. The economics of FO are as yet unclear. Some studies argue that FO is economically/technologically less attractive than RO, while others argue the opposite. FO membranes are of insufficient permeability, and higher-permeability membranes are needed. Moreover, solute crossover limits use of FO for potable water production. In addition, draw solution requires regeneration, which adds to the overall cost. As in the case of RO, fouling and mineral scaling are of concern, and experience with FO systems in this regard is currently limited.

Commercialization Barriers: RO—development of membrane with less fouling; thermal distillation processes—predominantly large investment and energy costs; MD—need for significant improvements in pure water flux, probably by an order of magnitude, and need to prove cost and advantages through extensive field demonstration; ED—more complex to deploy for very large-scale systems, opportunities for integration of RO and ED for high-recovery water desalination applications; FO—must be proven superior to RO technology to gain a foothold in the commercial world, although niche applications are expected, especially where technical limitations prevent the use of RO (e.g., in treatment of high-salinity produced water).

Technology Category: 13. Advanced Nuclear Reactors

Description: Advanced reactor designs are intended to provide increased safety margins, reduce costs, and extend the length of useful life for nuclear power plants. A large number of systems are under development, including several that use gas, molten salts, or liquid metals for cooling instead of light water. There has been much development activity in the field of advanced nuclear power plant systems over the past 15 years, but a great deal of additional work will be needed for commercialization of these systems. Given the likely importance of very low-carbon or zero-carbon dispatchable power sources, the continued development of these systems is of high priority. The committee recognizes developments around the world that are under way that employ various technologies.

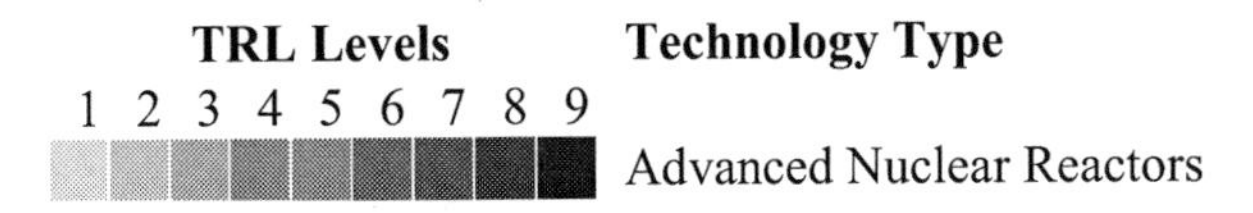

TRL Now	TRL in 2020	TRL in 2035
1-9	None assigned	None assigned

Technology Barriers: Need to develop materials capable of withstanding high neutron flux densities; no demand pull; spent fuel issue (see Chapter 5 section on "Nuclear Innovation Prospects and Obstacles").

Commercialization Barriers: Commercializing nuclear-related innovations is an expensive, lengthy, and risky process; need to develop regulations tailored to new technology systems (see Chapter 5 section on "Nuclear Innovation Prospects and Obstacles").

Technology Category: 14. Small Modular Nuclear Reactors

Description: Small modular reactors (SMRs) are smaller in size (300 MW or less) than current-generation baseload plants (typically 1,000 MW or larger). There are several systems under development across the world based on both light water and advanced designs. The committee recognizes developments around the world that are under way that employ various technologies.

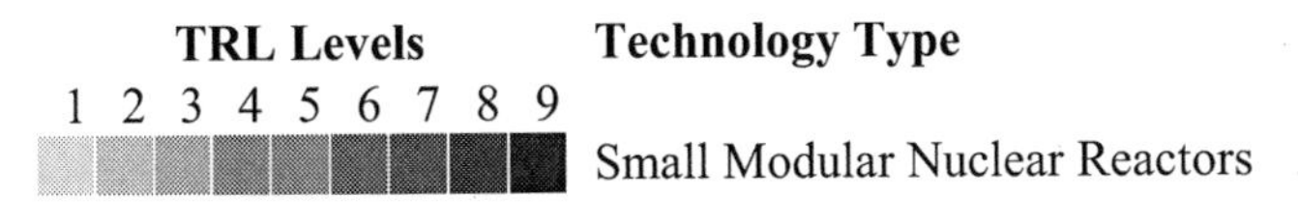

TRL Now	TRL in 2020	TRL in 2035

1-9	None assigned	None assigned

Technology Barriers: Designs need to be tested and proven (e.g., development of assessment methods for evaluating advanced SMR technologies and characteristics; development and testing of materials, fuels, and fabrication techniques; development of advanced instrumentation and controls and human-machine interfaces) (see Chapter 5 section on "Nuclear Innovation Prospects and Obstacles").

Commercialization Barriers: Cost and lack of experience; commercializing nuclear-related innovations is an expensive, lengthy, and risky process; need to develop regulations tailored to new technology systems (see Chapter 5 section on "Nuclear Innovation Prospects and Obstacles").

Technology Category: 15. Long-Term Operation of Existing Nuclear Power Plants

Description: Research is needed to address the technical bases for decisions regarding the continued high-performance operation of nuclear power plants. The research will need to address aging and life-cycle management, refurbishment and uprate decisions, and opportunities for modernization and performance improvement—especially needed to understand materials degradation and aging. Spent fuel continues to be a challenge, although local solutions—mainly onsite storage—are emerging in the absence of a single, central repository at the national level. The committee recognizes developments around the world that are under way that employ various technologies.

TRL Levels (1 2 3 4 5 6 7 8 9)	Technology Type
1–9	Long-Term Operation of Existing Nuclear Plants

TRL Now	TRL in 2020	TRL in 2035
1-9	None assigned	None assigned

Technology Barriers: Little research on degradation and aging of materials, including concrete; new technologies for online monitoring of critical equipment; new safety and risk analysis tools; integrated life-cycle management data, methods, and tools; enhanced nuclear fuel designs and analysis.

Commercialization Barriers: Development of repair and mitigation tools/technologies; development of accident-tolerant fuels and technologies; plant demonstrations to assess the new technologies; code and regulatory acceptance.

Technology Category: 16. Advanced High-Voltage Direct Current (HVDC) Technologies

Description: HVDC technology uses two types of converters: line commutated converters (LCCs) and voltage sourced converters (VSCs). LCCs use thyristors, can operate at ultra-high voltages of up 800 kilovolts (kV) to 1,000 kV, and can transmit power in the range of 6,000-8,000 MW. In use for the past 40 years, LCC is considered to be a relatively mature technology with high reliability and dependability. VSCs use integrated gate bipolar transistors (IGBTs) and can operate at voltage levels of 320 kV and transmit power levels of 1000-1200 MW. However, VSC ratings increase continuously over time, and the technology has strong potential to take a major share of new HVDC applications, especially DC grids and multiterminal DC systems.

TRL Levels 1 2 3 4 5 6 7 8 9	Technology Type
7	Advanced Line Commutated Converters
7	Advanced Voltage Sourced Converters

TRL Now	TRL in 2020	TRL in 2035
7 (advanced LCC)	9 (advanced LCC)	9 (advanced LCC)
7 (advanced VSC)	9 (advanced VSC)	9 (advanced VSC)

Technology Barriers: Increase operating voltages and levels of power transmission for VSCs.

Commercialization Barriers: No U.S. companies developing HVDC systems as power grid is almost entirely AC. Investments in R&D.

Technology Category: 17. Reducing Electricity Use in Power Systems (Production and Delivery)

Description: The electricity industry is the second largest electricity-consuming industry in the United States. The use of electrical energy in the production of electricity, as well as the uses or losses in power delivery (transmission and distribution), contribute to this total. There are opportunities to reduce electricity use in power production and delivery. These opportunities may include advances in control systems for auxiliary power devices and the use of adjustable-speed drive mechanisms (ASDs).

TRL Levels 1 2 3 4 5 6 7 8 9	Technology Type
9	Reducing Use in Power Systems

TRL Now	TRL in 2020	TRL in 2035
9	9	9

Technology Barriers: Power system designers seldom consider electrical losses in the design of power plants or transmission and distribution systems.

Commercialization Barriers: Retrofitting fossil or nuclear power plants requires regulatory approval (Environmental Protection Agency or U.S. Nuclear Regulatory Commission), often necessitating a complete review of the plant and resulting in many compliance requirements. State regulators are reluctant to consider distribution energy efficiency as part of energy-efficiency goals. Transmission operators pass-through losses, so have no incentive to reduce losses.

Technology Category: 18. Smart-Grid Technologies (Grid Modernization)

Description: Encompasses meters, appliances, power sources, phasor measurement units, power flow controls, and system automation. Smart-grid technologies permit systematic and reliable communication between suppliers and users, allowing for time-of-use pricing, peak load curtailment/leveling, smoother demand response, and greater penetration of variable and distributed generation sources.

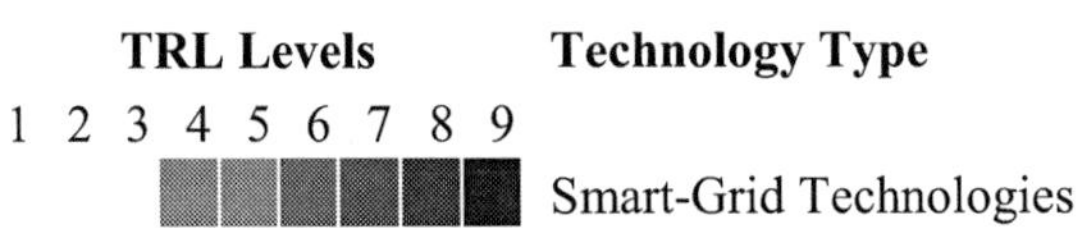

TRL Now	TRL in 2020	TRL in 2035
4-9	7-9	9

Technology Barriers: Mitigation of natural disaster impacts; cyber security of resources; accommodate and optimize the use of intelligent devices that reside at different points within the grid; integration of new and legacy technologies.

Commercialization Barriers: Savings may not be directly visible to consumers. Rate-based rate-of-return regulation may not allow for cost of infrastructure. Regulation may not allow for failed investment.

Technology Category: 19. Increased Power Flow in Transmission Systems

Description: Increasing power flow on existing and new transmission lines and corridors can facilitate greater use of renewable power generation options, enhance reliability, reduce control station power plant emissions, and reduce costs. Several technology options are commercially available, although some would benefit from additional advances. Others are in relatively early TRL stages and in need of continued development.

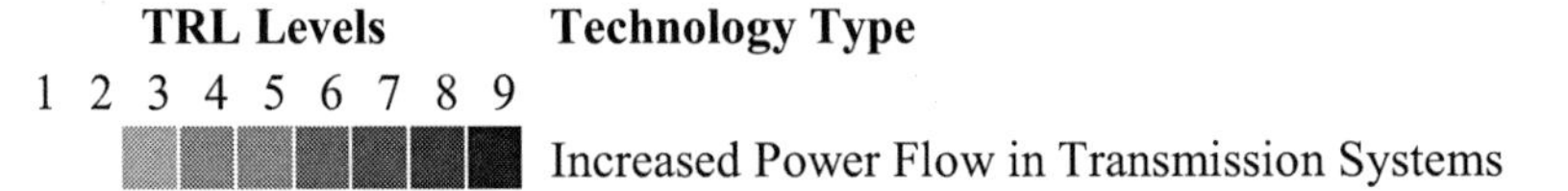

TRL Now	TRL in 2020	TRL in 2035
3-9	4-9	9

Technology Barriers: Standards need to be developed.

Commercialization Barriers: Savings may not be directly visible to consumers. Rate-based rate-of-return regulation may not allow for cost of infrastructure. Regulation may not allow failed investment.

Technology Category: 20. Advanced Power Electronics—Smart-Grid-Ready Inverters for Distributed Power Resources

Description: Increasing penetration of variable distributed energy resources (DER), especially solar PV systems on the distribution power grid, is creating grid integration challenges for utility engineers. Over voltage, reverse power flow, and excessive switching of capacitor banks and/or line tap changers often occur in circuits with higher penetration of variable generation sources such as solar PV. Some of these technical challenges can be resolved, or at least minimized, by employing the full potential of power electronics inside the inverters interfacing these sources with the electric grid. Inverters with grid supportive functionality, including reactive power support, low/high-voltage ride-through, watt-frequency, watt-voltage, and real power curtailment, can contribute to grid stability and hence help allow a higher adoption rate of variable DER technologies.

TRL Levels	Technology Type
1 2 3 4 5 6 7 8 9	
	Smart-Grid-Ready Inverters

TRL Now	TRL in 2020	TRL in 2035
Mostly 6 and 7 (early demonstration); in Europe, especially in Germany: 8 (early commercial deployment)	9	9

Technology Barriers: Need for common communication protocols.

Commercialization Barriers: Absence of widely accepted grid interconnection standards (e.g., Institute of Electrical and Electronics Engineers [IEEE] 1547) and testing standards (e.g., UL1741) to refer to regarding these smart-grid functionalities. There are also some open questions, such as the utility having access to customer-owned inverters, whether PV plant owners will be compensated for providing grid services, and whether a uniform grid code will be enforced.

Technology Category: 21. Efficient Electrical Technologies for Buildings and Industry

Description: Technologies are emerging that improve the efficiency of electricity use in buildings and industry, including heating, ventilation, and air conditioning (HVAC); lighting; water heating; plug loads, such as LED lighting; variable-speed HVAC systems; heat pumps; water heaters; smart thermostats; and even industrial processes. New efficient industrial technologies are emerging that can reduce electricity use. These include automation; controls; process heating; process cooling; motive power; compressed air; and other processes, such as 3-D printing, sensor networks, microwave processing, and the use of ultraviolet and other electromagnetic processing. Developments include enabling load devices to be demand-responsive. Still other technologies and processes are being electrified. If the electric power system evolves to significantly reduce GHG and other pollution emissions, then electrification of technologies and processes holds promise for reducing emissions. For example, electric vehicles that draw power from low- or no-emissions electricity sources should have no or significantly reduced emissions compared with internal combustion-powered vehicles. Technologies were identified across the entire range of TRLs.

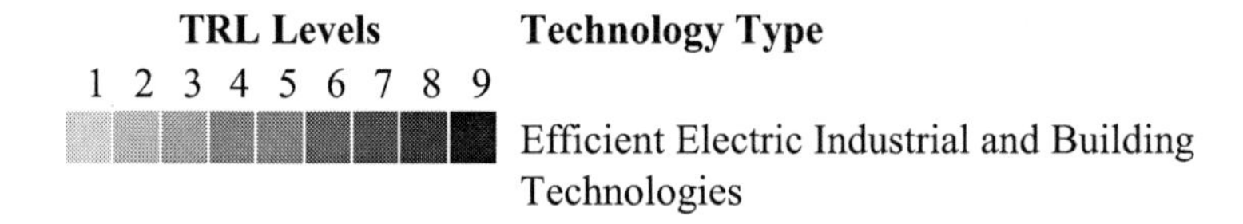

TRL Now	TRL in 2020	TRL in 2035
1-9	None assigned	None assigned

Technology Barriers: Because of the range of technologies, a full accounting of the technology barriers is difficult to summarize. Most technologies require additional development and refinement to improve their performance profiles. Further electrification would require distribution system upgrades (see categories 18 and 19 and Chapter 6).

Commercialization Barriers: Incomplete value chain; lack of knowledge or disconnected incentives between building owners and operators.

SOURCES

Category 1: Electric energy storage

1. Based in part on information from: Apt, J., and P. Jaramillo. 2014. *Variable renewable energy and the electricity grid.* New York: RFF Press.
2. EPRI (Electric Power Research Institute). 2014. *Bulk energy storage technologies: Performance potential, gird services and cost expectations.* EPRI report 3002003966. Palo Alto, CA: EPRI.
3. EPRI. 2012. *Coal technologies with CO_2 capture—status, risks, and markets 2012.* EPRI report 1023863. Palo Alto, CA: EPRI.
4. EPRI. 2008. *Operation experience, risk and market assessment of clean coal technologies.* EPRI report 1015679. Palo Alto, CA: EPRI.
5. DOE EAC (Department of Energy Electricity Advisory Committee). *2012 storage report: Progress and prospects. Recommendations for the U.S. Department of Energy.* Washington, DC: DOE EAC. http://energy.gov/sites/prod/files/EAC%20Paper%20-%202012%20Storage%20Report%20-%2015%20Nov%202012.pdf.

Category 2: Hydro and marine hydrokinetic (MHK) power

1. DOE. 2012. *An assessment of energy potential at non-powered dams in the United States.* Oakridge, TN: Oakridge National Laboratory. http://www1.eere.energy.gov/water/pdfs/npd_report.pdf.

Category 3: Advanced solar photovoltaic power

1. EPRI. 2012. *Engineering and economic evaluation of central-station solar photovoltaic power plants.* EPRI report 10025005. Palo Alto, CA: EPRI.

Category 4: Advanced concentrating solar power

1. EPRI. 2012. *Field assessment and optimization of the Enel Archimede Concentrating Solar Power Plant.* EPRI report 1026478. Palo Alto, CA: EPRI.

Category 5: Advanced solar thermal heating

1. Individual correspondence with the Tennessee Valley Authority.

Category 6: Advanced biomass power

1. EPRI. 2010. *Engineering and economic evaluation of biomass power plants.* EPRI report 1019762. Palo Alto, CA: EPRI.

Category 7: Engineered/Enhanced geothermal systems

1. EPRI. 2010. *Geothermal power: Issues, technologies, and opportunities for research development, demonstration, and deployment.* EPRI report 1020783. Palo Alto, CA: EPRI.

Category 8: Advanced wind turbine technologies

1. EPRI. 2010. *Advanced wind turbine technology assessment—2010*. EPRI report 1019772. Palo Alto, CA: EPRI.

Category 9: Advanced integration of distributed resources at higher rates

1. EPRI. 2014. *The integrated grid: Realizing the full value of central and distributed energy resources*. EPRI report 3002002733. Palo Alto, CA: EPRI.
2. DOE. 2013. *2013 renewable energy data book*. Oakridge, TN: Oakridge National Laboratory.
3. DOE NREL (National Renewable Energy Laboratory). 2012. *Renewable Energy Futures Study*. Washington, DC: DOE NREL.

Category 10: Carbon capture, transport, and storage

1. MIT (Massachusetts Institute of Technology. 2007. *The future of coal: Options for a carbon-constrained world*. Cambridge, MA: MIT.
2. NACAP (North American Carbon Atlas Partnership), NRCan (Natural Resources Canada), SENER (Mexican Ministry of Energy), and DOE. 2012. *The North American Carbon Storage Atlas*. https://www.netl.doe.gov/File%20Library/Research/Carbon-Storage/NACSA2012.pdf.
3. EPA (Environmental Protection Agency). 2012. *Greenhouse Gas Reporting Program (GHGRP): Subpart PP—suppliers of carbon dioxide*. Based on 2011 data. Washington, DC: EPA. https://www.epa.gov/ghgreporting/subpart-pp-suppliers-carbon-dioxide.
4. EPA (Environmental Protection Agency). 2012. *Greenhouse Gas Reporting Program (GHGRP): Subpart PP—suppliers of carbon dioxide*. Based on 2011 data. Washington, DC: EPA. https://www.epa.gov/ghgreporting/subpart-pp-suppliers-carbon-dioxide. MMT = million metric tons.

Category 11: Advanced natural gas power and combined heat and power (CHP)

1. DOE. 2012. *Combined heat and power: A clean energy solution*. Oakridge, TN: Oakridge National Laboratory.
2. DOE. 2002. *CHP potential at federal sites*. Oakridge, TN: Oakridge National Laboratory.
3. DOE/ICF International Inc. 2016. *U.S. DOE Combined Heat and Power Installation Database*. https://doe.icfwebservices.com/chpdb.
4. EPRI. 2013. *Tracking the demand for electricity from grid services*. EPRI report 3002001497. Palo Alto, CA: EPRI.

Category 12: Water and wastewater treatment

1. Electric Power Research Institute. 2009. *Program on Technology Innovation: Electric Efficiency Through Water Supply Technologies—A Roadmap*. EPRI report 1019360. Palo Alto, CA: EPRI.

Category 13: Advanced nuclear reactors

1. World Nuclear Association. 2013. *Small nuclear power reactors*. http://www.world-nuclear.org/info/Nuclear-Fuel-Cycle/Power-Reactors/Small-Nuclear-Power-Reactors/#.UYPRZKCOs3E.
2. General Atomics. 2013. *EM^2 quick facts*. http://www.ga.com/websites/ga/docs/em2/pdf/FactSheet_QuickFactsEM2.pdf.
3. Choi, H., and R.W. Schleicher. 2011. Design characteristics of the energy multiplier module (EM2). *Transactions of the American Nuclear Society* 104:929-930.
4. Schleicher, R., and C. Back. 2012. Configuring EM^2 to meet the challenges of economics, waste disposition, and nonproliferation confronting nuclear energy in the U.S. *Transactions of Fusion Science and Technology* 61(1T):144-149.
5. Parmentola, J., and J. Rawls. 2012. Energy Multiplier Module (EM^2)—capping the waste problem and using the energy in U-238. *Transactions of Fusion Science and Technology* 61(1T):9-14.
6. Halfinger, J.A., and M.D. Hagherty. 2012. The B&W mPower^{TM} scalable practical nuclear reactor design. *Nuclear Technology* 178(2):164-169.
7. Ingersoll, D. 2011. *An overview of the safety case for the small modular reactors*. Presented at ASME SMR 2011 Conference, Washington, DC, September 29.
8. World Nuclear Association. 2014. *Nuclear power in China*. http://www.world-nuclear.org/info/Country-Profiles/Countries-A-F/China--Nuclear-Power.
9. NucNet. 2013. *China begins construction of first generation IV HTR-PM unit*. http://www.nucnet.org/all-the-news/2013/01/07/china-begins-construction-of-first-generation-iv-htr-pm-unit.
10. IAEA (International Atomic Energy Agency). 2013. *IAEA update on KLT-40S*. http://www.iaea.org/NuclearPower/Downloadable/aris/2013/25.KLT-40S.pdf.
11. Fadeev, Y. 2011. *KLT-40S reactor plant for the floating CNPP FNU*. Presented to the IAEA. http://www.iaea.org/NuclearPower/Downloads/Technology/meetings/2011-Jul-4-8-ANRT-WS/2_%D0%9ALT-40S_VBER_OKBM_Afrikantov_Fadeev.pdf.
12. Colbert, C. 2013. *Overview of NuScale design*. Presented at Technical Meeting on Technology Assessment of SMRs for Near-Term Deployment, Chengdu, China, September 2-4.

http://www.iaea.org/NuclearPower/Downloadable/Meetings/2013/2013-09-02-09-04-TM-NPTD/20_usa_colbert_nuscale.pdf.
13. NuScale Power. 2014. *NuScale Integral System Test Facilities (NIST)*. http://www.nuscalepower.com/testfacilities.aspx.
14. Kim, S.H., K.K. Kim, J.W. Yeo, M.H. Chang, and S.Q. Zee. 2013. *Design verification program of SMART*. Presented at GENES4/ANP2003 Conference, Kyoto, Japan, September 15-19. http://www.uxc.com/smr/Library%5CDesign%20Specific/SMART/Papers/2003%20-%20Design%20Verification%20Program%20of%20SMART.pdf.
15. Seo, J.T. 2013. *Small and modular reactor development, safety and licensing in Korea*. Presented to the IAEA. http://www.uxc.com/smr/Library/Design%20Specific/SMART/Presentations/2013%20-%20SMR%20Development,%20Safety%20and%20Licensing%20in%20Korea.pdf.
16. World Nuclear Association. 2014. *Nuclear power in South Korea*. http://www.world-nuclear.org/info/Country-Profiles/Countries-O-S/South-Korea.
17. Zrodnikov, A.V., G.I. Toshinskya, O.G. Komleva, V.S. Stepanovb, and N.N. Klimovb. 2011. SVBR-100 module-type fast reactor of the IV generation for regional power industry. *Journal of Nuclear Materials* 415(3):237-244.
18. Zrodnikov, A.V., G.I. Toshinskii, O.G. Grigor'ev, Y.G. Dragunov, V.S. Stepanov, N.N. Klimov, I.I. Kopytov, V.N. Krushel'nitskii, and A.A. Grudakov. 2004. SVBR-75/100 multipurpose modular low power fast reactor with lead bismuth coolant. *Atomic Energy* 97(2):528-533.
19. IAEA. 2013. *Super-safe, small and simple reactor (4S, Toshiba design)*. https://aris.iaea.org/sites/..%5CPDF%5C4S.pdf.
20. Ishii, K., H. Matsumiya, and N. Handa. 2011. Activities for 4S USNRC licensing. *Progress in Nuclear Energy* 53(7):831-834.
21. Hirsch, B. 2006. *Review of Toshiba 4S sodium-cooled nuclear power reactor proposed for Galena, Alaska*. Letter from Union of Concerned Scientists. https://www.yumpu.com/en/document/view/29644438/subject-review-of-toshiba-4s-sodium-cooled-nuclear-power-reactor.

Category 14: Small modular nuclear reactors

See references for category 13

Category 15: Long-term operation of existing nuclear power plants

See references for category 13

Category 16: Advanced high-voltage direct current (HVDC) technologies

1. EPRI. 2006. *Advanced HVDC systems for voltages at +/-800kV and above*. EPRI report 1013857. Palo Alto, CA: EPRI.

Category 17: Reducing electricity use in power systems (production and delivery)

1. EPRI. 2010. *The power to reduce CO_2 emissions: Transmission system efficiency*. EPRI report 1020142. Palo Alto, CA: EPRI.
2. EPRI. 2011. *Program on technology innovation electricity use in the electric-sector opportunities to enhancer electric energy efficiency in the production and delivery of electricity*. Palo Alto, CA: EPRI.

Category 18: Smart-grid technologies (grid modernization)

1. EPRI. 2011. *Estimating the costs and benefits of the Smart Grid: A preliminary estimate of the investment requirement and resultant benefits of a fully functioning Smart Grid*. EPRI report 1022519. Palo Alto, CA: EPRI.

Category 19: Increased power flow in transmission systems

1. DOE NREL. 2012. *Renewable Energy Futures Study*. Washington, DC: DOE NREL.

Category 20: Advanced power electronics—smart-grid-ready inverters for distributed power resources

1. http://ww.astrumsolar.com/the-basics/environmental-benefits.
2. EPRI. 2013. *Grid impacts of distributed generation with advanced-inverter functions: Hosting capacity of large-scale solar photovoltaic using smart inverters*. EPRI report 3002001246. Palo Alto, CA: EPRI.
3. EPRI. 2014. *Distribution management systems and advanced inverters: Autonomous versus integrated PV control*. EPRI report 3002003275. Palo Alto, CA: EPRI.

Category 21: Efficient electrical technologies for buildings and industry

1. EPRI. 2012. *Electrotechnology reference guide: Revision 4*. EPRI report 1025038. Palo Alto, CA: EPRI.
2. EPRI. 2012. *Electrotechnology applications in industrial process heating*. EPRI report 1024338. Palo Alto, CA: EPRI.
3. EPRI. 2009. *Assessment of achievable potential from energy efficiency and demand response programs in the U.S. (2010-2030)*. EPRI report 1016987. Palo Alto, CA: EPRI.
4. EPRI. 2009. *The potential to reduce CO_2 emissions by expanding end-use applications of electricity*. EPRI report 1018871. Palo Alto, CA: EPRI.
5. EPRI. 2012. *Plug-in electric vehicle adoption and load forecasting*. EPRI report 1024103. Palo Alto, CA: EPRI.

Appendix E

Glossary of Acronyms and Abbreviations

ACEEE	American Council for an Energy-Efficient Economy
AEE	Advanced Energy Economy
AEIC	American Energy Innovation Council
AEO	*Annual Energy Outlook*
AMI	advanced metering infrastructure
APPA	American Public Power Association
ARPA-E	Advanced Research Projects Agency-Energy
ARRA	American Recovery and Reinvestment Act of 2009
ASCE	American Society of Civil Engineers
ASHRAE	American Society of Heating, Refrigerating and Air-Conditioning Engineers
BEETIT	Building Energy Efficiency Through Innovative Thermodevices
BLS	Bureau of Labor Statistics
BNEF	Bloomberg New Energy Finance
Btu	British thermal unit
CAFE	Corporate Average Fuel Economy
CBO	Congressional Budget Office
CCS	carbon capture and storage
CCSP	Climate Change Science Program
CEER	Council of European Energy Regulators
CEIR	Center for Integrative Environmental Research
CES-21	California Electric Systems for the 21st Century
CESP	customer energy service provider
CHP	combined heat and power
CO_2	carbon dioxide
COP21	21st yearly session of the Conference of the Parties to the 1992 United Nations Framework Convention on Climate Change
C-PACE	commercial property assessed clean energy

CPP	Clean Power Plan
CSP	concentrating solar power
DARPA	Defense Advanced Research Projects Agency
DELTA	Delivering Efficient Local Thermal Amenities
DG	distributed generation
DHS	Department of Homeland Security
DoD	Department of Defense
DOE	Department of Energy
DSM	demand-side management
DSO	distribution system operator
EAC	electricity advisory committee
EDA	Economic Development Agency
EEG	Erneuerbare-Energien-Gesetz (German Renewable Energy Sources Act)
EEI	Edison Electric Institute
EERE	Office of Energy Efficiency and Renewable Energy
EERS	energy-efficiency resource standard
EIA	Energy Information Administration
ELCON	Electric Consumers Resource Council
ELI	Environmental Law Institute
EOR	enhanced oil recovery
EPA	Environmental Protection Agency
EPACT	Energy Policy Act of 2005
EPRI	Electric Power Research Institute
ERCOT	Electric Reliability Council of Texas
ESCO	energy service company
ESPC	energy savings performance contract
ETIS	energy technology innovation system
FDA	Food and Drug Administration
FEMP	Federal Energy Management Program
FERC	Federal Energy Regulatory Commission
FTC	Federal Trade Commission
FY	fiscal year
GAO	Government Accountability Office
GB	Great Britain
GDP	gross domestic product
GENI	Green Electricity Network Integration
GHG	greenhouse gas
GPT	general-purpose technology
GSA	General Services Administration

GW	gigawatt
GWh	gigawatt hours
HECA	Hydrogen Energy California Project
HUD	Department of Housing and Urban Development
HVAC	heating, ventilation, and air conditioning
HVDC	high-voltage, direct current
IAEA	International Atomic Energy Agency
ICC	International Commerce Commission
ICS-CERT	Industrial Control Systems Cyber Emergency Response Team
IDC	intangible drilling cost
IDDRI	Institute for Sustainable Development and International Relations
IEA	International Energy Agency
IECC	Interventional Energy Conservation Code®
IEEE	Institute of Electrical and Electronics Engineers
IGCC	integrated gasification (coal)-combined cycle
IOU	investor-owned electric company
IPCC	Intergovernmental Panel on Climate Change
IRC	Internal Revenue Code
IRM	Innovation Roll-out Mechanism
IRS	Internal Revenue Service
ISO	independent system operator
ITC	investment tax credit
kWh	kilowatt hour
LACE	levelized avoided cost of electricity
LBD	learning by doing
LBS	learning by (re)searching
LBW	learning by waiting
LCOE	levelized cost of electricity
LED	light-emitting diode
LEED	Leadership in Energy and Environmental Design
LLNL	Lawrence Livermore National Laboratory
LTC	load tap changer
LWR	light water reactor
MACRS	Modified Accelerated Cost Recovery System
MEPS	minimum efficiency performance standards
METI	Ministry of Economy, Trade and Industry (Japan)
MHK	marine hydrokinetic

MIT Massachusetts Institute of Technology
MLP master limited partnership
MMBtu 1 million British thermal units
MMcf 1 million cubic feet
Mtpa million tons per annum
MW megawatt
MWh megawatt hour

NAE National Academy of Engineering
NAECA National Appliance Energy Conservation Act
NAM National Academy of Medicine
NAREIT National Association of Real Estate Investment Trusts
NARUC National Association of Regulatory Utility Commissioners
NAS National Academy of Sciences
NASA National Aeronautics and Space Administration
NCRDS National Coal Resources Data System
NEI Nuclear Energy Institute
NEMS National Energy Modeling System
NERC North American Electric Reliability Corporation
NETL National Energy Technology Laboratory
NGO nongovernmental organization
NH_3 ammonia
NIA Network Innovation Allowance
NIC Network Innovation Competition
NIST National Institute of Standards and Technology
NMSS Office of Nuclear Material Safety and Safeguards
NNATCET National Network for Advancing Translational Clean Energy Technologies
NNMI National Network for Manufacturing Innovation
NO_x oxide of nitrogen
NRC National Research Council
NREL National Renewable Energy Laboratory
NSF National Science Foundation

O&M operations and maintenance
OECD Organisation for Economic Co-operation and Development
Ofgem Office of Gas and Electric Markets (United Kingdom)

PAA Price-Anderson Act
PACE property assessed clean energy
PCAST President's Council of Advisors on Science and Technology
PLR private letter ruling
PM particulate matter

PNNL	Pacific Northwest National Laboratory
PPA	power purchase agreement
PSERC	Power Systems Engineering Research Center
PTC	production tax credit
PV	photovoltaic
R&D	research and development
RAP	Regulatory Assistance Project
RD&D	research, development, and demonstration
REC	renewable energy credit
REIDI	regional energy innovation and development institute
REIT	real estate investment trust
REV	"Reforming the Energy Vision"
RFF	Resources for the Future
RFI	request for information
RIDF	Regional Innovation Demonstration Fund
RIIO	Revenue set to deliver strong Incentives, Innovation and Outputs
RNM	reference network model
RPS	renewable portfolio standard
RTO	regional transmission organization
SBA	Small Business Administration
SBIC	Small Business Investment Company (program)
SBIR	Small Business Innovation Research
SDSN	Sustainable Development Solutions Network
SEIA	Solar Energy Industries Association
SF_6	sulfur hexafluoride
SMR	small modular reactor
SO_x	oxide of sulfur
SWITCHES	Strategies for Wide-Bandgap, Inexpensive Transistors for Controlling High-Efficiency Systems
TCEP	Texas Clean Energy Project
TRL	technology readiness level
U.S. NRC	U.S. Nuclear Regulatory Commission
UAE	United Arab Emirates
UESC	utility energy service contract
UN	United Nations
UNFCCC	United Nations Framework Convention on Climate Change
USGS	United States Geological Survey

VAR	volt/volt ampere reactive
VDO	venture development organization
VOCs	volatile organic compounds
VVO	volt/volt ampere reactive optimization
WACC	weighted average cost of capital
WBDG	Whole Building Design Guide